Control of Non-conventional Synchronous Motors

Control of Non-conventional Synchronous Motors

Edited by
Jean-Paul Louis

First published 2012 in Great Britain and the United States by ISTE Ltd and John Wiley & Sons, Inc.

Apart from any fair dealing for the purposes of research or private study, or criticism or review, as permitted under the Copyright, Designs and Patents Act 1988, this publication may only be reproduced, stored or transmitted, in any form or by any means, with the prior permission in writing of the publishers, or in the case of reprographic reproduction in accordance with the terms and licenses issued by the CLA. Enquiries concerning reproduction outside these terms should be sent to the publishers at the undermentioned address:

ISTE Ltd
27-37 St George's Road
London SW19 4EU
UK

www.iste.co.uk

John Wiley & Sons, Inc.
111 River Street
Hoboken, NJ 07030
USA

www.wiley.com

© ISTE Ltd 2012

The rights of Jean-Paul Louis to be identified as the author of this work have been asserted by him in accordance with the Copyright, Designs and Patents Act 1988.

Library of Congress Cataloging-in-Publication Data

Control of non-conventional synchronous motors / edited by Jean-Paul Louis.
 p. cm.
 Includes bibliographical references and index.
 ISBN 978-1-84821-331-9
 1. Electric motors, Synchronous--Automatic control. I. Louis, Jean-Paul, 1945-
 TK2787.C66 2011
 621.46--dc23
                                                                          2011041315

British Library Cataloguing-in-Publication Data
A CIP record for this book is available from the British Library
ISBN: 978-1-84821-331-9

Printed and bound in Great Britain by CPI Group (UK) Ltd., Croydon, Surrey CR0 4YY

Table of Contents

Introduction. xi
Jean-Paul LOUIS

**Chapter 1. Self-controlled Synchronous Motor: Principles
of Function and Simplified Control Model**. 1
Francis LABRIQUE and François BAUDART

    1.1. Introduction. 1
    1.2. Design aspects specific to the self-controlled
    synchronous machine. 2
    1.3. Simplified model for the study of steady state operation 3
    1.4. Study of steady-state operation 6
    1.5. Operation at nominal speed, voltage and current. 12
    1.6. Operation with a torque smaller than the nominal torque. 15
    1.7. Operation with a speed below the nominal speed 15
    1.8. Running as a generator 16
    1.9. Equivalence of a machine with a commutator and brushes. 17
    1.10. Equations inferred from the theory of circuits
    with sliding contacts 22
    1.11. Evaluation of alternating currents circulating in steady state
    in the damper windings 26
    1.12. Transposition of the study to the case of
    a negative rotational speed 28
    1.13. Variant of the base assembly. 28
    1.14. Conclusion 29
    1.15. List of the main symbols used 29
    1.16. Bibliography 30

Chapter 2. Self-controlled Synchronous Motor: Dynamic Model Including the Behavior of Damper Windings and Commutation Overlap. 33
Ernest MATAGNE

    2.1. Introduction. 33
    2.2. Choice of the expression of N^k 35
    2.3. Expression of fluxes. 40
    2.4. General properties of coefficients $<X>$, $<Y>$ and $<Z>$. 46
    2.5. Electrical dynamic equations 48
    2.6. Expression of electromechanical variables 51
    2.7. Expression of torque 53
    2.8. Writing of equations in terms of coenergy 54
    2.9. Application to control. 56
    2.10. Conclusion 60
    2.11. Appendix 1: value of coefficients $<X>$, $<Y>$ and $<Z>$ 60
    2.12. Appendix 2: derivatives of coefficients $<X>$, $<Y>$ and $<Z>$. 61
    2.13. Appendix 3: simplifications for small μ 62
    2.14. Appendix 4: List of the main symbols used in Chapters 1 and 2. . . . 63
    2.15. Bibliography 65

Chapter 3. Synchronous Machines in Degraded Mode 67
Damien FLIELLER, Ngac Ky NGUYEN, Hervé SCHWAB and Guy STURTZER

    3.1. General introduction 67
        3.1.1. Analysis of failures of the set converter-machine: converters with MOSFET transistors. 68
    3.2. Analysis of the main causes of failure 68
        3.2.1. Failure of the inverter 68
        3.2.2. Other failures 72
    3.3. Reliability of a permanent magnet synchronous motors drive 72
        3.3.1. Environmental conditions in the motor industry. 72
        3.3.2. The two reliability reports: MIL-HdbK-217 and RDF2000 73
        3.3.3. Failure rate of permanent magnet synchronous motors actuators 75
    3.4. Conclusion 76
    3.5. Optimal supplies of permanent magnet synchronous machines in the presence of faults 77
        3.5.1. Introduction: the problem of a-b-c controls. 77
    3.6. Supplies of faulty synchronous machines with non-sinusoidal back electromagnetic force 77
        3.6.1. Generalization of the modeling. 77
        3.6.2. A heuristic approach to the solution 80
        3.6.3. First optimization of ohmic losses without constraint on the homopolar current 82

3.6.4. Second optimization of ohmic losses with the sum
of currents of non-faulty phases being zero 91
3.6.5. Third optimization of ohmic losses with a homopolar
current of zero (in all phases). 96
3.6.6. Global formulations 102
3.7. Experimental learning strategy in closed loop to obtain
optimal currents in all cases. 113
3.8. Simulation results 116
3.9. General conclusion 118
3.10. Glossary. 119
3.11. Bibliography 121

**Chapter 4. Control of the Double-star Synchronous Machine
Supplied by PWM Inverters.** 125
Mohamed Fouad BENKHORIS

4.1. Introduction. 125
4.2. Description of the electrical actuator 127
4.3. Basic equations. 128
  4.3.1. Voltage equations. 128
  4.3.2. Equation of the electromagnetic torque 131
4.4. Dynamic models of the double-star synchronous machine 131
  4.4.1. Dynamic model in referential $d_1q_1d_2q_2$ 131
  4.4.2. Dynamic model in referential $dqz_1z_2z_3z_4$. 136
4.5. Control of the double-star synchronous machine. 146
  4.5.1. Control in referential $d_1q_1d_2q_2$ 147
  4.5.2. Control in referential $dqz_1z_2z_3z_4$ 156
4.6. Bibliography 158

**Chapter 5. Vectorial Modeling and Control of Multiphase Machines
with Non-salient Poles Supplied by an Inverter** 161
Xavier KESTELYN and Éric SEMAIL

5.1. Introduction and presentation of the electrical machines 161
5.2. Control model of inverter-fed permanent
magnet synchronous machines 163
  5.2.1. Characteristic spaces and generalization of the notion
  of an equivalent two-phase machine 163
  5.2.2. The inverter seen from the machine 182
5.3. Torque control of multiphase machines 189
  5.3.1. Control of currents in the natural basis 189
  5.3.2. Control of currents in a decoupling basis 193
5.4. Modeling and torque control of multiphase machines
in degraded supply mode 203

5.4.1. Modeling of a machine with a supply defect 203
5.4.2. Torque control of a faulty machine 203
5.5. Bibliography 204

Chapter 6. Hybrid Excitation Synchronous Machines 207
Nicolas PATIN and Lionel VIDO

6.1. Description 207
   6.1.1. Definition 207
   6.1.2. Classification 208
6.2. Modeling with the aim of control 220
   6.2.1. Setting up equations 220
   6.2.2. Formulation in components 224
   6.2.3. Complete model 228
6.3. Control by model inversion 230
   6.3.1. Aims of the torque control 230
   6.3.2. Current control of the machine 231
   6.3.3. Optimization and current inputs 233
6.4. Overspeed and flux weakening of synchronous machines 235
   6.4.1. Generalities 235
   6.4.2. Flux weakening of synchronous machines
   with classical magnets 236
   6.4.3. The unified approach to flux weakening
   using "optimal inputs" 237
6.4. Conclusion 237
6.5. Bibliography 239

Chapter 7. Advanced Control of the Linear Synchronous Motor 241
Ghislain REMY and Pierre-Jean BARRE

7.1. Introduction 241
   7.1.1. Historical review and applications in the field of linear motors . . . 241
   7.1.2. Presentation of linear synchronous motors 243
   7.1.3. Technology of linear synchronous motors 245
   7.1.4. Linear motor models using sinusoidal
   magneto-motive force assumption 246
   7.1.5. Causal ordering graph representation 248
   7.1.6. Advanced modeling of linear synchronous motors 249
7.2. Classical control of linear motors 253
   7.2.1. State-of-the-art in linear motor controls 253
   7.2.2. Control structure design using the COG inversion principles 255
   7.2.3. Closed-loop control 256
7.3. Advanced control of linear motors 265
   7.3.1. Multiple resonant controllers in a two-phase reference frame . . . 265

7.3.2. Feed-forward control for the compensation of detent forces 271
7.3.3. Commands by the nth derivative for sensorless control......... 273
7.4. Conclusion 279
7.5. Nomenclature 280
7.6. Acknowledgment 281
7.7. Bibliography 281
7.8. Appendix: LMD10-050 Datasheet of ETEL 285

Chapter 8. Variable Reluctance Machines: Modeling and Control 287
Mickael HILAIRET, Thierry LUBIN and Abdelmounaïm TOUNZI

8.1. Introduction... 287
8.2. Synchronous reluctance machines 289
   8.2.1. Description and operating principle 289
   8.2.2. Hypotheses and model of a Synchrel machine......... 291
   8.2.3. Control of the Synchrel machine......................... 293
   8.2.4. Applications 303
8.3. Switched reluctance machines................................. 303
   8.3.1. Description and principle of operation 303
   8.3.2. Hypotheses and direct model of the SRM 307
   8.3.3. Control 310
   8.3.4. Applications 321
8.4. Conclusion 323
8.5. Bibliography 323

Chapter 9. Control of the Stepping Motor 329
Bruno ROBERT and Moez FEKI

9.1. Introduction... 329
9.2. Modeling 329
   9.2.1. Main technologies 329
   9.2.2. The modeling hypotheses 330
   9.2.3. The model 332
9.3. Control in open loop 335
   9.3.1. The types of supply...................................... 335
   9.3.2. The supply modes 336
   9.3.3. Case of slow movement................................ 339
   9.3.4. Case of quick movement 344
9.4. Controls in closed loop 350
   9.4.1. Linear models 350
   9.4.2. Servo-control of speed.................................. 357
9.5. Advanced control: the control of chaos 361
   9.5.1. Chaotic behavior 361
   9.5.2. The model 361

9.5.3. Orbit stabilization. 363
9.5.4. Absolute stability 365
9.5.5. Synthesis of the controller. 366
9.5.6. Examples. 368
9.6. Bibliography 371

Chapter 10. Control of Piezoelectric Actuators 375
Frédéric GIRAUD and Betty LEMAIRE-SEMAIL

10.1. Introduction 375
   10.1.1. Traveling wave ultrasonic motors: technology and usage 375
   10.1.2. Functioning features 377
   10.1.3. Models 378
10.2. Causal model in the supplied voltage referential 380
   10.2.1. Hypotheses and notations 380
   10.2.2. Kinematics of the ideal rotor 381
   10.2.3. Generation of the motor torque 385
   10.2.4. Stator's resonance. 386
   10.2.5. Calculation of modal reaction forces. 387
   10.2.6. Complete model 388
10.3. Causal model in the referential of the traveling wave 389
   10.3.1. Park transform applied to the traveling wave motor. 389
   10.3.2. Transformed model 391
   10.3.3. Study of the motor stall 394
   10.3.4. Validation of the model 396
   10.3.5. Torque estimator 398
10.4. Control based on a behavioral model 400
10.5. Controls based on a knowledge model 401
   10.5.1. Inversion principle 402
   10.5.2. Control structure inferred from the causal model:
   emphasis on self-control 402
   10.5.3. Practical carrying out of self-control. 406
10.6. Conclusion 407
10.7. Bibliography 407

List of Authors 411

Index 413

Introduction

For the countless operations of its machine tools, its robots and its "special machines", modern industrial production is in dire need of electrical motors that can be used as "actuators": these machines must impose a torque, speed or position on objects in movement (most often in rotation), all of which are determined by a high level decisional element. Execution speed and accuracy are necessary for high-quality productivity. Electrical motors have thus taken a dominant place in the "axis control" that we find everywhere in modern industrial production. We also find such motors in several "consumer product" applications, even if we will consider mainly professional applications in this book. Electrical motors are also present in the motorization of several modes of transport: terrestrial vehicles, free or guided, and ships. The present work specifically covers the "non-conventional" controls of synchronous actuators, and the control of "non-conventional" synchronous motors.

This book is part of a series published by ISTE-Wiley and Hermes-Lavoisier. The first book, devoted to the *Control of Synchronous Motors* [LOU 11], which studies "conventional" synchronous motors, has already been published in English. Let us add that another volume has been devoted to the identification and to the observation of electrical machines [FOR 10]. A further volume has presented general methods regarding the control of electrical machines [HUS 09], and still another has focused on technological problems [LOR 03]. The control of electrical actuators goes with the control of static converters (here, inverters); it has been the topic of a volume [MON 11] that deals with modulations and current controls. Finally, a work has been published on the design of electrical drives [JUF 10] and a book on the diagnostics of electrical machines [TRI 11].

In the previous book, entitled *Control of Synchronous Motors* [LOU 11], we presented the main advantages of electrical actuators, referred to as "synchronous

---

Introduction written by Jean-Paul LOUIS.

motors", in detail. The essential advantage comes from their high performances in terms of mass torque ratio, their maneuverability and their ease of use – these are all relative, but real, factors of their use. Thus particular uses of the synchronous motor that have been created have received various names, such as: the "self-controlled synchronous motor"; often known in industry as the "brushless DC motor". This first work on synchronous motors dealt specifically with the modeling and control of synchronous actuators *when these are (more or less) conventional synchronous motors*.

Hence it is in this book [LOU 11] that we find the classical models of synchronous machines that either fulfill the hypothesis of the first harmonic: the Magnetomotive Force (MMF) wave is sinusoidal (leading to the usual "Park model") or do not fulfill it ("extension" of Park models when the MMF wave is non-sinusoidal). On the basis of these models, the first volume presented fundamental concepts such as "self-control" and classical torque control, in the natural referential (a-b-c) or the rotor referential ($d - q$). These controls are often referred to as "vector controls" and they are often applied to "axis control" – the speed regulations and/or position regulations of a "mechanical load": a tool, for instance, or a wheel. We have considered the problems with their implementation by digital components. We consider that these controls are now "classical". This volume therefore presents the "advanced controls" that are still applied to conventional synchronous motors: *direct torque control* and *predictive controls*. Finally, the self-control – which lies at the centre of most classical controls – normally requires a mechanical position sensor. As, for many reasons, we often do not want to install one, "sensorless" controls, deterministic or statistic, are also presented.

This present book, the second about the control of synchronous or similar motors, has two parts. The first part deals with *non-conventional uses* of the synchronous motor (motors supplied by thyristor bridges, motors in degraded modes, and multiphase motors). The second part deals with the control of actuators that the community of specialists usually refer to as *non-conventional motors* (hybrid motors, linear motors, variable reluctance motors, stepping motors, and piezoelectric motors). These deserve to be associated with synchronous actuators, because their structures most often link them to synchronous machines, with or without excitation. In the title of this work, the adjective "non-conventional" hence has two meanings.

The first part contains five chapters. It starts with two chapters that complement each other on "self-controlled synchronous machines" supplied with thyristor bridges. These assemblies are, historically, the first to have appeared, since the thyristor was the first component in controllable modern power electronics (after the era of vacuum tubes) that appeared industrially. With the development of GTO (gate turn-off thyristor) and of different types of power transistors (Insulated Gate Bipolar

Introduction   xiii

Transistor (IGBT), etc.), this assembly has been restricted to high and very-high powered machines. Furthermore, it remains present in old industrial assemblies that need to be maintained. This device was ingenious and robust. Two thyristor bridges allowed operation in the four quadrants. It allowed genuine self-control, but was complex because the thyristor bridge is − approximately − a pulsed current source, and we need to take the overlap phenomenon in a thyristor bridge into account when it works as a line commutated inverter, which is conceptually more delicate than the rectifier mode. Besides, it was necessary to deal with synchronous machines with dampers and coiled excitation, supplied with a line commutated inverter. There is therefore a very noticeable scientific culture and know-how in this area, which Chapters 1 and 2 present in essential detail.

Chapter 1, by Francis Labrique and François Baudart, is entitled "Self-controlled synchronous motor: principles of function and simplified control model". It presents this device within the framework of approximations often adopted by engineers: thus "transient inductance" is defined and simple electrical diagrams are given for modeling the synchronous machine/thyristor bridge set. However, the currents generated are pseudo pulses (the importance of overlap), so we also need to take into account current harmonics. Thus, analysis, even in steady state, leads to non-trivial, nonlinear models. Chapter 1 analyzes different modes of operation: at the nominal regime and below the nominal regime, operation as a motor and operation as a generator.

For the analysis of transient regimes, a powerful approach consists of performing equivalence between the synchronous machine supplied with the self-controlled thyristor bridge and a direct current (DC) machine with a collector with three segments. The position of the brushes is imposed by the self-control and the rotor position. We can then apply the theory of sliding contact circuits (due to Manuel da Silva Garrido and Ernest Matagne, see also the first chapter of [LOU 04a]) by "averaging" the signals on each switching interval. We then obtain a global dynamic model of the set synchronous machine/thyristor bridge/self-control. It allows us, for instance, to determine the alternating currents in the dampers.

Chapter 2, by Ernest Matagne, is titled "Self-controlled synchronous motor: dynamic model including the behavior of damper windings and commutation overlap". The author studies the same device as in Chapter 1 without keeping the classic engineering approximations. The first issue to deal with is that of the overlap, which plays a sensitive role in the current form and in the torque harmonics. The second concerns the flux dynamics, including those in the dampers. The author finally presents a dynamic model of electrical variables that are coupled with electromechanical variables imposed by control of the thyristor converter. The author thus obtains an accurate torque equation. This approach using equations leads to block diagrams that are adapted to control problems and to the determination of transients. These models, which are powerful and accurate, are completely original and contribute to knowledge on the tools that enable the synthesis of high-powered self-controlled synchronous machines.

Chapter 3, "Synchronous machines in degraded mode", by Damien Flieller, Ngac Ky Nguyen, Hervé Schwab and Guy Sturtzer, also deals with the non-conventional operation of a synchronous actuator. This chapter is about the study of optimization of operations when there are defects in a converter-synchronous machine set. It is an opportunity to remind ourselves of the importance of safety issues in electrical drives: a variator must be able to carry on working, even in the case of a partial breakdown. Several studies have been devoted to these issues, and the present chapter presents an overview of possible failures, the most frequent happening at the level of power transistors (the accidents mostly generate short-circuits and, less frequently, open circuits). Accidents at the level of the windings are far less numerous. These defects normally generate important perturbating torques, which are added to the cogging torque.

In the spirit of the methods to supply "healthy" synchronous machines presented in Chapters 2 and 3 of *Control of Synchronous Motors* [LOU 11], the authors present a general approach of the following problem: how can we optimally supply the "healthy phases" of a machine to compensate for the defects due to accidents? The authors of Chapter 3 in this current book exploit a general optimization method in the case of a *n*-phase machine, with smooth poles and sinusoidal or non-sinusoidal induced *back emfs*, with at least one faulty phase. In this phase, the current can be: 0 (open circuit) or equal to a saturation value, or be a short-circuit current. The general optimization method allows us to calculate the current to be applied in the healthy phases to impose the desired torque while minimizing Joule losses, and fulfilling the constraints on the zero sequence component of the current (free or 0). Despite the multiplicity of possible cases, the authors show the existence of a global formulation marking the intervening back emfs and a scalar coefficient of proportionality, $k_{opt}(p\theta)$, which varies periodically as a function of position. Coefficient $k_{opt}(p\theta)$ can be determined theoretically when we know the defect conditions. But these are often difficult to determine in real time. Another good strategy involves identifying $k_{opt}(p\theta)$ using a learning device, which is allowed by neural networks (the Adaline network). This self-adaptive control allows the currents to reach their optimal values, which are adapted to the defect after a few periods.

The following chapters show an extension to non-conventional synchronous motors. The classic optimization of electrical machines has led to the design and domination of three-phase AC electrical machines. In fact, the three-phase machines optimize the power to mass ratio: they minimize the number of of conductors (and the copper is expensive) and the mass of magnetic circuits (which are made of iron). The three-phase circuits also optimize the number of components in power electronics converters that supply the machines.

New optimization criteria have, however, appeared, particularly in sensitive domains such as air and naval transport, where propulsion equipment failure has catastrophic consequences. It has become necessary to improve *the reliability and safety of operation* by segmenting power. This can be done using a technical solution using machines with a greater number of phases, supplied with static converters with the appropriate number of legs. If a phase (of the machine or converter) is faulty (for reasons of aging or because of accidents with internal or external origins), the phases that remain in working state allow the system to continue working properly. Several types of technical solutions exist for multiphase synchronous motors; we present them in Chapters 4 and 5.

In Chapter 4, "Control of the synchronous double-star machine supplied by PWM inverters" written by Mohamed Fouad Benkhoris, if a three-phase coil is faulty, the system can still work thanks to the other three-phase coil and thus ensure continuity of service. We can see its advantage, for example, in the propulsion of a ship, for instance. The two coils can be "overlapped" or "shifted". These different solutions exist and distinguish these machines from "conventional multiphase machines" where the phases are all regularly shifted. As it is supplied by inverters controlled in PWM (Pulse Width Modulation), this operating mode allows the generation of sinusoidal currents and the limitation of torque ripples. This property, well known for conventional three-phase machines, presents intrinsic difficulties for these non-conventional machines because of magnetic couplings between their two three-phase coils.

The author describes the two approaches to modeling this machine for control design. The first approach consists of acting as if the machine was made of two classical three-phase machines (each being modeled and controlled with the help of classic Park theory – see the first chapters in *Control of Synchronous Motors*, the first book in this series) and taking into account the couplings between these two machines at a higher level. The second approach introduces the coupling from the initial modeling of the six-phase machine and defining an equivalent machine by an extension of the Park approach, and inferring control by "inverting" this model. The mathematical models that allow us to represent these special machines in a user-friendly way are detailed, the essential element being to diagonalize the stator inductance matrix of the double-star machine in order to define a "good" referential frame and "useable" state variables. The controls are inferred from this modeling by an approach that generalizes the "vector control", which is well known in conventional machines, with the "regulation" terms of currents and the "decoupling" terms. Fouad Benkhoris shows that different solutions are possible. He details and validates the main ones: partial diagonalization and total diagonalization.

Chapter 5, by Xavier Kestelyn and Éric Semail, is called "Vector modeling and control of multiphase machines with non-salient poles supplied by an inverter". The synchronous machines considered have phases that are regularly shifted, like conventional three-phase machines, and show two variants: machines where the phases are independent and single-star machines. The excitation is generated by surface-mounted permanent magnets. The reference example has five phases – but to model the machines, the authors use a vector formalism that can easily be generalized to n phases. It is shown that we can determine a referential frame where the inductance matrix is diagonal, which allows us to implement high-performance control.

Thanks to their approach, the authors generalize the well-known properties for the three-phase conventional machine by defining "fictitious machines" that can be equivalent two-phase machines (that take part in the energy conversion) or single-phase machines of "zero sequence" type. In parallel, the theoretical analysis provides useful information: it shows the advantage of star couplings for machines with an odd number of phases, and shows the use of multi-star couplings for machines with an even number of phases. The analysis also answers the questions that arise with the use of higher-order harmonics and shows the necessity of knowing the leakage inductances well. This is very important at the time of design and supply of the machine, and when elaborating control laws to optimize the torque. The authors present different optimal strategies to control the currents of multiphase machines, in different appropriate frames of reference, in order to supply a given torque by taking into account the problem of the distribution of torques between the different fictitious machines. All the methods are validated by experimental results.

The second part of this book contains five chapters that discuss increasingly non-conventional machines; their proximity to synchronous motors is relative, but they are closer in some aspects.

Chapter 6, by Nicolas Patin and Lionel Vido, is entitled "Hybrid excitation synchronous machines", concerns an immediate extension of the synchronous motor. Even if it remains minor with respect to the conventional synchronous motor with a single magnet excitation, its importance is significant since it extends the performances of the actuator by playing on its excitation. We are in dire need of this in applications linked to transport (terrestrial vehicles and avionics). The problem of the overspeed and flux weakening of synchronous motors is frequently considered, and this chapter presents it as a particular case of the general case considered of the hybrid excitation synchronous machine: a first (fixed) excitation is generated by magnets, and a second (adjustable) excitation is generated by winding coils.

The chapter distinguishes two large families: machines in "series" and machines in "parallel", according to whether the field lines generated by the windings pass through the magnets or not. The machines in series are high-performance for flux weakening, but more sensitive to demagnetization. There is an advantage of machines in parallel, which can be of short-circuit or non short-circuit type. Short-circuit type machines improve operating safety during an accident involving the legs of the supply bridge, but part of the flux is lost. No short-circuit type machines partially correct the previous inconvenience. The excitation winding coils can be placed at the stator or rotor, and the structures can be with "flux concentration" or "flux-switching". Whatever structure is adopted, the modeling for control design is a simple extension of conventional modeling of the synchronous machine where the flux embraced by the stator coils is the sum of two fluxes – one generated by the magnets and the other by the windings. The Park representation is well adapted for simplicity and efficiency, even if it (naturally) leads to more complex equations than that which have been given for conventional machines in Chapter 1 of *Control of Synchronous Motors*. The authors infer a simulation functional diagram in the form of block diagrams. The model being obtained, the control law is obtained by its inversion. Since there are more unknown variables (three currents) than equations (one torque), the solution is given by optimization (minimization of losses). The authors give the control diagrams with different regulation loops and the "optimizers". This approach is also applied to the easier case of flux weakening in the classic single excitation actuator.

Chapter 7, by Ghislain Remy and Pierre-Jean Barre, presents the "Advanced control of the linear synchronous motor". Initially restricted to induction motors, the linear structure has largely been developed for synchronous motors thanks to permanent magnets and the associated electronics (power and control). We thus have motors with direct linear drive (without reducer or transmission systems – these are costly and introduce defects) at our disposal that are adapted to specific solutions. These include, for example, actuating ultra-fast machine tools (with very high accelerations and speeds) that we want to be extremely rigid and very precise. The authors describe the building structures specific to this class of motors, the "permanent magnet synchronous linear motor", (PMSLM) that is monolateral with short primary. Then they give models. The first model, in the first harmonic sense (as for conventional rotating motors), is presented and the authors infer a dynamic model formalized by a causal ordering graph (COG). Advanced models allow us to describe phenomena specific to linear motors: the rank 5 and 7 harmonics, the nonlinearities of inductances during the quick current transients, the cogging forces due to the end-effect forces – all these phenomena introduce ripples on the thrust (cogging).

The authors present the conventional controls of the synchronous linear motor: scalar controls (with V/f = constant) and vector controls with diverse variants. The authors develop controls by "model inversion" (obtained from COG), which imposes structures with cascade loops. The current loop is inferred from the first harmonic model and dimensioned according to the capabilities of the inverter. To compensate for the cogging forces, the authors present advanced controls such as the use of multiple resonant controllers or feed-forward controls. Finally, rank n derivative controls (or open loop precontrol, in the sensorless case) are presented, which require an analysis and very precise models. The authors show that the generation of the reference value is the major problem for control of the motor when we take into account nonlinear phenomena.

Chapter 8, by Mickael Hilairet, Thierry Lubin and Abdelmounaïm Tounzi, is entitled "Variable reluctance machines: Modeling and control". Here, we move even further from conventional synchronous motors, since only the reluctance effects (which exist on "salient pole" synchronous machines) intervene. We move further into the category of "non-conventional machines" with constructive variants that are far more numerous than those of conventional machine variants. This is why this chapter focuses on the two families of variable reluctance machines that are the most promising and the most frequently used in industry: reluctant synchronous machines (or synchro reluctance, Synchrel) that have coils *distributed* at the stator; and switched reluctance motors (SRMs), which have coils that are *concentrated* at the stator. There is no excitation at the rotor, which makes these machines very robust, particularly at high speed (when the rotor is often massive). It is this quality that is looked for, because these machines are susceptible to disadvantages (involving torque pulsation and noise for SRMs), also we need to optimize the structure of the machines and their supply.

This chapter first describes the main structures of the Synchrel that must maximize the saliency effect. The authors propose their modeling, which is largely analytical, for control design. The torque is proportional to the product of currents (stator) i_d and i_q, which defines only one equation, so we need to optimize a criterion. Thus, if we optimize the dynamics, we can work at constant i_d. If we maximize the torque for a given current, we need $i_d = i_q$. The authors also present the maximization of the power factor. We often find "vector control" structures that compensate for perturbations terms between the d and q axes. SRMs have a larger field of application than Synchrel machines, their performances being better. SRM operation is situated largely in zones where the machine is saturated. The optimization (of its design and supply) and the modeling must be done within nonlinear frameworks, even if a linear model can be considered for a first synthesis. The nonlinear modelings require the use of coenergy. The implantation makes use of precalculated tables. The SRM only works if it is self-controlled with specific self-control rules. Control strategies – "instantaneous controls" and "average value

Introduction   xix

controls" – are presented whose performances (including in terms of complexity of implantation) are analyzed and compared.

Chapter 9, written by Bruno Robert and Moez Feki, concerns the control of stepping motors. Stepping actuators are particularly well appreciated industrially because we can use them in open loop with a very easy and inexpensive voltage supply. Such a control will not be high-performance so very precise studies must sometimes be done to properly set the limits of use (for a given cost), and in cases where we want to introduce (a little) closed loop. The authors start by defining a generic stepping motor model that is actually a variant of the salient pole two-phase synchronous motor and the torque is described by three terms. The three most common types of stepping motors are:

– the permanent magnet motor (where the dominant torque term is the electromagnetic torque);

– the variable reluctance motor (where the dominant term is the reluctance torque); and

– the hybrid motor, where the cogging torque is taken into account.

The absence of a position sensor (for economical reasons), which is common, means that the control model remains in the "natural referential" (here: α and β, see Chapter 1 of *Control of Synchronous Motors*) and it is *strongly nonlinear*. The authors indicate the approximations that must be carried out on each type of motor.

In the generic case, the authors study which types of control can be executed: the open loop controls with a voltage supply (the most economical) or with a current supply (more costly), and the closed loop controls. For the first case (voltage supply) the modes allowing different operating types – the electrical quarter turn modes and the mixed modes – are presented. The second case (current supply, with a motor that is specially built for) allows microsteps. The authors detail the case of slow movements, the methods for starting, the oscillating responses, microsteps and optimal control ("bang-bang"). The study of quick movements allows us to define the different operating zones of the stepping motors: the stop-start zone and the drive zone. The authors show us how to obtain different speed and acceleration profiles that can be integrated into low-cost processors. Closed-loop controls, which require knowledge of the position, either by a coder or by estimation, are then discussed (see the last two chapters of *Control of Synchronous Motors*). They are therefore more costly and their economical advantage must be justified. The author presents an angle control, that is high-performance but costly. We can also infer frequency control, which is more economical because it can be implanted in open loop. The latter can be completed using a speed control (we therefore need to measure or estimate the speed); the authors provide methods of controller synthesis.

Controlled in open loop, the stepping motor has a strongly nonlinear behavior that restricts its performances at high speeds, where instabilities and "chaotic" phenomena appear. An advanced control method, referred to as "chaos control", is presented that allows us to widen the operating zone towards higher speeds.

Chapter 10 concerns an actuator that is the furthest from classic synchronous motors. It is not electromagnetic, but electrostatic. Its control calls for analog concepts, such as self-control and vector control. This chapter is entitled "Control of piezoelectric actuators", and has been written by Frédéric Giraud and Betty Lemaire-Semail. It concerns traveling-wave piezoelectric motors. These motors are an alternative to electromagnetic motors for small dimensions where they can bring a gain of factor 10 for the mass torque ratio. The motors considered are two-phase and are associated with a position sensor. The rotor and stator are in contact, and the mechanical friction phenomena play an important role that considerably complicates the modeling. The dynamic modelings are complex because there is a double energy conversion: electromechanical (indirect piezoelectric effect) and mechano-mechanical (by contact). The authors present the modeling using equivalent electrical diagrams that are restricted to steady states, and hybrid models that associate electrical and mechanical equations – the latter are more precise but too complex. The authors also develop a model that is specifically adapted to control in real time. In the frame of reference of equations of stator supply voltages, a "direct model" makes four domains appear:

– the electrical domain;

– the stator domain;

– the "ideal rotor" domain; and

– the "real rotor" domain.

The energy conversion is situated at the border of the stator domain with the ideal rotor domain, and we can observe an analogy (with duality) with the equations of the synchronous motor when we write its equations in referential α and β (see Chapter 1 of *Control of Synchronous Motors*). This analogy is taken advantage of: the authors infer a model in the referential frame of the travelling wave to obtain equations of $d - q$ type.

This model is formally easier and is well adapted to the determination of a control that will have to have self-control (as for controls of AC machines). It is also useful for defining a torque estimator. The authors then present the large families of control methods of the piezoelectric motor. The first, and most frequent, is based on a behavioral (black box) model, the control variables being able to be the frequency, the amplitude of two-phase voltages and dephasing between these two voltages. The relationships between these variables and the speed are not linear and can present dead zones that need to be dealt with specifically. This situation legitimizes the (less frequent) use of methods based on a knowledge model. This second model family is detailed by the authors, who rely on the "inversion" of direct models. This approach shows the advantage of control on the tangential axis and on the normal axis. It leads the authors to distinguish three strategies according to the variable considered: the normal speed of the ideal rotor, the pulsation of the supply voltages, and the voltage on path d in open loop. The authors show us how to create the self-control with the help of a phase locking loop. These controls allow us to avoid stalling phenomena and compensate for effects due to thermal drifts.

Bibliography: works in the EGEM-Hermes and ISTE–John Wiley treatise on electrical motors and actuators

[FOR 10] DE FORNEL B., LOUIS J.-P., *Electrical Actuators: Identification and Observation*, ISTE Ltd., London and John Wiley & Sons, New York, 2010.

[HUS 09] HUSSON R., *Control Methods for Electrical Machines*, ISTE Ltd., London and John Wiley & Sons, New York, 2009.

[JUF 10] JUFER M., *Les Entraînements Électriques: Méthodologie de Conception*, Hermès-Lavoisier, Paris, France, 2010.

[LOR 03] LORON L., *Commande des Systèmes Électriques: Perspectives Technologiques* (Traité EGEM, Série Génie Électrique), Hermès-Lavoisier, Paris, 2003.

[LOU 04a] LOUIS J.-P., *Modélisation des Machines Électriques en vue de Leur Commande, Concepts Généraux* (Traité EGEM, Série Génie Électrique), Hermès-Lavoisier, Paris, 2004.

[LOU 04b] LOUIS J.-P., *Modèles pour la Commande des Actionneurs Électriques* (Traité EGEM, Série Génie Électrique), Hermès-Lavoisier, Traité EGEM, Paris, 2004.

[LOU 11] LOUIS J.-P., *Control of Synchronous Motors*, ISTE Ltd., London and John Wiley & Sons, New York, 2011.

[MON 11] MONMASSON E., *Power Electronic Converters: PWM Strategies and Current Control Techniques*, ISTE Ltd., London and John Wiley & Sons, New York, 2011.

[REZ 11] REZZOUG A., ZAÏM, M.E.H., *Non-conventional Electrical Machines*, ISTE Ltd., London and John Wiley and Sons, New York, 2011.

[TRI 11] TRIGEASSOU J.-C., *Electrical Machines Diagnosis*, ISTE Ltd., London and John Wiley & Sons, New York, 2011.

Chapter 1

Self-controlled Synchronous Motor: Principles of Function and Simplified Control Model

1.1. Introduction

Every synchronous machine supplied by an electronic power converter and functioning with variable speed needs to ensure control of the converter that supplies it, in the sense that the rotational speed of its rotor imposes the frequency of the voltages and of the currents the converter provides to the stator windings. However the name "self-synchronous machine" usually refers to the case where the electronic power converter that supplies it is a thyristor bridge whose commutations are ensured by the voltages the machine develops at its terminals [BON 97, BUH 77, CHA 88, KLE 80, PAL 99]. This bridge is supplied with direct current (DC) via a smoothing inductance by an input converter that is itself usually a thyristor bridge connected to the alternating current (AC) mains. We thus obtain the principle diagram in Figure 1.1.

The synchronous machine can run as a motor or as a generator:

− to run as a motor, the input bridge works as a rectifier and supplies the intermediate DC circuit with electric power that is transferred to the synchronous machine by the bridge that supplies it, which works as an inverter.

− to run as a generator (braking operation) the bridge connected to the synchronous machine works as a rectifier and transfers the electric power generated

---

Chapter written by Francis LABRIQUE and François BAUDART.

by the synchronous machine to the network via the input bridge, which works as an inverter.

Operation with a positive or negative speed is possible since a change in rotation sense at the level of the synchronous machine corresponds to a permutation of the order of phase succession and to an inversion of the polarity of their voltages. This means that we only have to adapt the control sequence of the thyristors of the bridge that supplies it accordingly.

The assembly in Figure 1.1 therefore corresponds to a drive system that can work in the four quadrants of the torque-speed plane: running as a motor or a brake is possible in both rotational directions [BON 97, PAL 99].

Taking into account the remark made on the effect of the change in direction of rotation, for the sake of simplicity, all of the examples presented in this chapter will be done for a positive rotational speed and we will point out at the end of the study how to transpose the results in the case of a negative rotational speed.

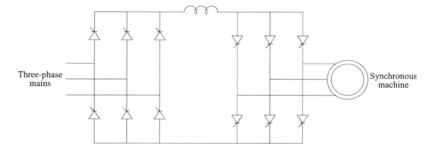

Figure 1.1. *Principle diagram of the self-controlled synchronous machine*

1.2. Design aspects specific to the self-controlled synchronous machine

The self-controlled synchronous machine is used for strong to very strong power applications and systematically uses machines with field winding [BON 01]. This type of machine is characterized by the high value of cyclical inductance L_{CS} from the stator windings, whose impedance at the nominal working frequency ranges from 1 to 1.5 per unit[1] [CHA 83].

---

[1] In the 'per unit' system, the values of statoric impedances at the nominal frequency are divided by the value of a base impedance (equal to the phase nominal voltage divided by the nominal current) in order to obtain dimensionless values.

Such a cyclical inductance value is not compatible with supply by a thyristor bridge that is functioning as a current commutator, as it will attempt to impose pulsed currents in the windings. Using appropriate constructive measures, it is therefore advisable to decrease the apparent inductance of the windings towards the abrupt variations of the currents flowing into them. We obtain this result by providing the rotor with large dimensioned damper windings [BON 00, KLE 80].

1.3. Simplified model for the study of steady state operation

To establish this model we are going to consider that the machine has a cylindrical rotor, that there is no saturation effect, that the stator is three-phase with a Y connection of its windings and that the machine rotor circuits consist of a field winding and two damper windings in short-circuit: one (d-axis damper winding) aligned with the field winding; and the other (q-axis damper winding) in quadrature with the former. The field winding is assumed to be supplied by a current source I_f (see Figure 1.2). We also assume that all the mutual inductances between the windings of the stator and rotor sinusoidally vary according to the electric angular position of the rotor, θ_{em}, which is equal to its mechanical angular position θ_m multiplied by the number of pole pairs p of the machine. The modeling principles have been presented in [LOU 04, LOU 10].

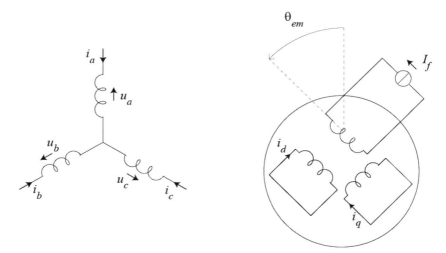

Figure 1.2. *Simplified representation of the machine windings*

In steady state, the electromotive forces (emfs) generated by the field winding in the phases of the armature are sinusoidal functions whose period T is set by the electric speed of the machine rotor $p\omega_m = p\dot{\theta}_m$:

$$T = \frac{2\pi}{p\omega_m}$$

If M_f represents the maximal value of the mutual inductance between the field winding and a phase of the stator, the magnitude E_0 of these emfs is equal to:

$$E_0 = p\omega_m M_f I_f \qquad [1.1]$$

In steady state, the currents in the stator windings are periodic time functions with the same period T as the emfs.

We can represent these currents as a development in Fourier series, which leads us to distinguish the following in each:

– a fundamental component of angular frequency $p\omega_m$;

– harmonic components of angular frequency, $kp\omega_m$, with k being an integer >1.

The fundamental components of the currents, with angular frequency, $p\omega_m$, generate an armature reaction rotating field whose rotational speed is synchronous with that of the rotor. For these components, the inductance of the stator windings is equal to the cyclical inductance, L_{CS}.

The harmonic components with angular frequency, $kp\omega_m$, generate armature reaction rotating fields whose rotational speed with respect to the rotor is equal to $(k\pm 1)\omega_m$ according to whether these harmonics form inverse or direct systems. For these components, if we neglect the resistance of the damper windings, the apparent inductance of the stator windings is equal to:

$$L_C = L_{CS}\left(1 - \frac{M_0^2}{L_{CS}L_R}\right) \qquad [1.2]$$

where:

– L_R is the inductance of a damper winding; and

– M_0 is equal to $\sqrt{3/2}$ times the maximal value M of the mutual inductance between a damper winding and a stator winding.

Inductance L_C has a value of around 15% of the cyclical inductance and constitutes what we call the transient inductance of the machine [BON 00, CHA 83].

The equivalent circuit of one phase of the machine, towards the fundamental component of the current flowing into it, is made up of the cyclical inductance and the resistance of the winding in series with the emf, which is generated therein by direct current I_f that flows in the field winding (see Figure 1.3).

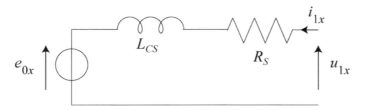

Figure 1.3. *Equivalent circuit of a phase towards the fundamental component of the current flowing into it*

In this diagram, e_{0x} is the electromotive force induced by the field winding in phase x, $x \in [a,b,c]$; i_{1x} is the fundamental component of the current that flows into it; and u_{1x} is the fundamental component of the voltage at its terminals.

If $\overline{E}_0$ is the phasor representative of emfs e_{0x}, and $\overline{I}_1$ and $\overline{U}_1$ are the phasors representative of currents and voltages i_{1x} and u_{1x}, we have:

$$\overline{U}_1 = \overline{E}_0 + jp\omega_m L_{CS}\overline{I}_1 + R_S\overline{I}_1 \qquad [1.3]$$

The equivalent circuit of a phase towards the harmonic of rank k of the current flowing into it is reduced to resistance R_S in series with an inductance equal to L_C, (see Figure 1.4).

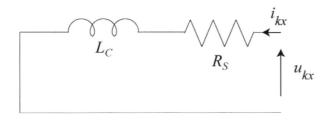

Figure 1.4. *Equivalent circuit of the machine towards the harmonic components of rank k of the current in phase x and the voltage at its terminals*

We can combine the equivalent circuit, which is valid for the fundamental components of phase currents and voltages, with those valid for the harmonic components of these variables by rewriting equation [1.3] as follows:

$$\bar{U}_1 = \bar{E}_0 + jp\omega_m(L_{CS} - L_C)\bar{I}_1 + jp\omega_m L_C \bar{I}_1 + R_S \bar{I}_1 \quad [1.4]$$

From this equation, we can in fact substitute a new equivalent circuit for the circuit in Figure 1.3, where an inductance L_C and a resistance R_S are put in series with an emf e_{cx} with an angular frequency $p\omega_m$, whose representative phasor $\bar{E}_C$ corresponds to the first two terms on the right-hand side of equation [1.4] (see Figure 1.5):

$$\bar{E}_C = \bar{E}_0 + jp\omega_m(L_{CS} - L_C)\bar{I}_1 \quad [1.5]$$

With the voltage and current components of angular frequency $kp\omega_m$, the voltage source e_{cx} of angular frequency $p\omega_m$ appears to be a short-circuit due to the superposition principle, which is applicable since we assume that the machine is unsaturated. For current harmonics, the circuit in Figure 1.5 thus amounts to that in Figure 1.4. We can therefore apply the whole current i_x (fundamental + harmonics) to it to find the total voltage u_x at the phase terminals.

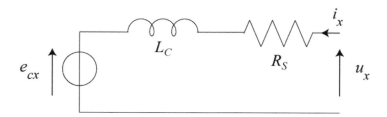

Figure 1.5. *Modified equivalent circuit*

It is from this equivalent circuit – known as "a commutation electromotive force behind the commutation inductance" [BON 00, CHA 88], where we will disregard resistance R_S – that we are going to study the functioning of the rectifier bridge that supplies the machine and from it how the machine functions.

1.4. Study of steady-state operation

To perform this study (see Figure 1.6), we are going to suppose, as indicated in the introduction, that the rotational speed ω_m is positive, that the current I at the input of the bridge that supplies the machine is perfectly smooth and that the thyristor bridge works in binary commutation. We can refer to the classical study of rectifiers [SEG 92] but consider that the reference directions of the voltage U and phase currents i_x have been inverted with respect to those normally used for the study of rectifier circuits. This is done in order to take into account the fact that, in the case of a self-controlled synchronous motor, normal operation corresponds to electric power supplied to the machine, i.e. going from the DC to the AC circuit, and hence to the operation as an inverter of the bridge that supplies the machine.

In these conditions, with the adopted reference directions, to ensure commutations we need to turn on each thyristor with a sufficient advance firing angle with respect to the point where:

– the commutation emf of the phase to which it is linked stops being smaller than that of the preceding phase if it belongs to thyristors with common anodes (+ input terminal);

– the commutation emf of the phase to which it is linked stops being greater than that of the preceding phase if it belongs to one of the thyristors with common cathodes (- input terminal).

The advance firing angle α is sufficient if the current in the phase, whose thyristor is turned on, reaches the value I before its commutation emf is crossed with that of the preceding phase. The angle ξ, between the point where the current in the phase takes the value I and the point of crossing of the emf, is called the extinction angle. We generally give this angle a value between $\pi/9$ and $\pi/6$.

According to the rotor electrical position angle $\theta_{em} = p\omega_m t$, the commutation emfs can be written as:

$$e_{ca} = E_C \cos(p\omega_m t + \psi_C)$$

$$e_{cb} = E_C \cos(p\omega_m t + \psi_C - 2\pi/3) \qquad [1.6]$$

$$e_{cc} = E_C \cos(p\omega_m t + \psi_C - 4\pi/3)$$

where E_C and ψ_C are two parameters linked to the field winding current I_f, to the rotational speed ω_m of the rotor, to the extinction angle ξ and to the current I imposed at the input of the bridge supplying the machine. We will later see how E_C and ψ_C are linked to current I_f, ω_m, ξ and I.

8 Control of Non-conventional Synchronous Motors

If we consider the commutation of current I from phase c to phase a at the + input terminal as an example, this commutation starts at:

$$p\omega_m t = -\pi/3 - \psi_C - \alpha$$

when thyristor T_{a1} is turned on (see Figure 1.6).

From this point until:

$$p\omega_m t = -\pi/3 - \psi_C - \xi$$

where it ends, we can write:

$$e_{ca} + L_C \frac{di_a}{dt} = e_{cc} + L_C \frac{di_c}{dt} \qquad [1.7]$$

with:

$$i_a + i_c = I$$

From this we can infer the equation that rules the evolution of i_a during the commutation; as $dI/dt = 0$ (since we assume that current I is constant), we obtain:

$$\frac{di_a}{dt} = \frac{e_{cc} - e_{ca}}{2L_C} \qquad [1.8]$$

By calculating the integral of this equation, we get the evolution of i_a during the commutation.

Taking into account that:

$$i_a(-\pi/3 - \psi_C - \alpha) = 0$$

we have:

$$i_a = \frac{\sqrt{3}E_C}{2L_C p\omega_m}\left[\cos(p\omega_m t + \pi/3 + \psi_C) - \cos\alpha\right] \qquad [1.9]$$

When:

$$p\omega_m t = -\pi/3 - \psi_C - \xi$$

current i_a becomes equal to I and current i_c, which is equal to $I - i_a$, cancels out leading to the blocking of T_{c1} and to the end of the transfer of current I from phase c to phase a.

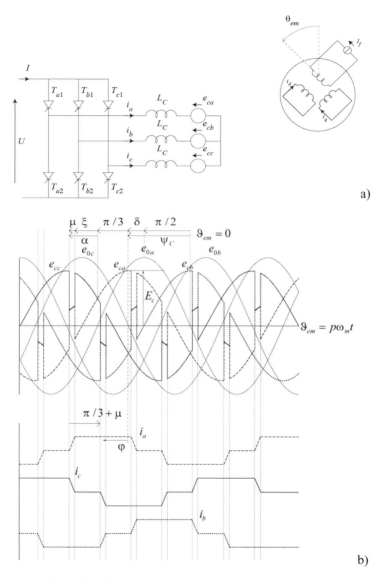

Figure 1.6. *Evolution of voltages and currents in steady state*

We have:

$$I = \frac{\sqrt{3}E_C}{2L_C p\omega_m}[\cos\xi - \cos\alpha] \qquad [1.10]$$

or, by introducing an overlap angle μ equal to $\alpha - \xi$, which corresponds to the interval of simultaneous conduction of T_{a1} and T_{c1}:

$$I = \frac{\sqrt{3}E_C}{2L_C p\omega_m}[\cos\xi - \cos(\xi + \mu)] \qquad [1.11]$$

Relationship [1.10] according to the values of E_c, ω_m and I, allows us to fix the value we need to give to the advance firing angle α to have a given extinction angle ξ:

$$\alpha = \xi + \mu = \arccos\left[\cos\xi - \frac{2L_C p\omega_m I}{\sqrt{3}E_C}\right] \qquad [1.12]$$

Voltage U, at the input of the thyristor bridge that supplies the machine, consists of, at each output terminal, three sinusoidal arches with a notch due to commutation at the beginning of each arch. The voltage at the − terminal is symmetrical to the voltage at the + terminal, with a shift of $\pi/3$. The mean value of the voltage at the + terminal with respect to the neutral point of the stator windings (taken as a reference point for drawing the voltages in Figure 1.6) is half the mean value of voltage U. We obtain this mean value by considering the third of the period, for instance, which begins with the firing of thyristor T_{a1}:

$$U_{av} = \frac{1}{\pi/3}\left[\int_{-(\alpha+\psi_C)-\pi/3}^{-(\alpha+\psi_C)+\pi/3} E_C \cos(p\omega_m t + \psi_C) dp\omega_m t \right.$$

$$\left. + \int_{-(\alpha+\psi_C)-\pi/3}^{-(\alpha+\psi_C)-\pi/3+\mu} L_C p\omega_m \frac{di_a}{dp\omega_m t} dp\omega_m t\right] \qquad [1.13]$$

where the second term reflects the notch effect due to commutation and corresponds to the drop in voltage on L_c due to current i_a during its growth from zero to I.

We obtain:

$$U_{av} = \frac{3}{\pi}\sqrt{3}E_C \cos\alpha + \frac{3}{\pi}L_C p\omega_m I \qquad [1.14]$$

since:

$$i_a(-\alpha - \psi_C - \pi/3) = 0$$

and:

$$i_a(-\alpha - \psi_C - \pi/3 + \mu) = I$$

By using relationship [1.10], equation [1.14] can be written as:

$$U_{av} = \frac{3\sqrt{3}}{\pi} E_C \cos\alpha + \frac{3\sqrt{3}}{2\pi} E_C [\cos\xi - \cos\alpha] \qquad [1.15]$$

or:

$$U_{av} = \frac{3\sqrt{3}}{\pi} E_C \left[\frac{\cos\alpha + \cos\xi}{2} \right] \qquad [1.16]$$

By introducing angle μ, equation [1.16] becomes:

$$U_{av} = \frac{3\sqrt{3}}{\pi} E_C \left[\frac{\cos\alpha + \cos(\alpha - \mu)}{2} \right] = \frac{3\sqrt{3}}{\pi} E_C \cos(\alpha - \mu/2)\cos\mu/2 \qquad [1.17]$$

If, as is the case in Figure 1.6, the conditions of functioning correspond to a low μ value, we can admit that:

– we will not make a significant error by assuming that $\cos\mu/2 = 1$, which leads to:

$$U_{av} = \frac{3\sqrt{3}}{\pi} E_C \cos(\alpha - \mu/2) \qquad [1.18]$$

– the amplitude of the peak value I_1 of the fundamental components of currents i_a, i_b and i_c is almost equal to that of pulsed currents of amplitude I and width $2\pi/3$:

$$I_1 = \frac{2\sqrt{3}}{\pi} I \qquad [1.19]$$

– phase shift φ (leading) to the fundamental components of currents i_a, i_b and i_c with respect to voltages e_{ca}, e_{cb} and e_{cc} is almost equal to:

$$\varphi = \frac{\alpha + \xi}{2} = \alpha - \frac{\mu}{2} = \xi + \frac{\mu}{2} \qquad [1.20]$$

12 Control of Non-conventional Synchronous Motors

– the fundamental components of voltages u_a, u_b and u_c at the terminals of the phases of the machine are almost equal to the commutation emfs e_{ca}, e_{cb} and e_{cc} since the voltages at the terminals of the machine only depart from these emfs by the commutation notches whose width, equal to μ, is small.

1.5. Operation at nominal speed, voltage and current

Admitting that, for nominal conditions of operation, the overlap angle is small (we will check this *a posteriori* by taking into account the order of magnitude of the commutation inductance), on the basis of the approximations introduced at the end of the previous section, we can consider that the nominal operating point will be reached when the machine is running at its nominal speed ω_{mN} if:

– we set direct current I to a value I_N equal to $\dfrac{\pi}{2\sqrt{3}} I_{1N}$, where I_{1N} is the nominal amplitude (i.e. the peak value) of the fundamental component of phase currents (see equation [1.19]). This occurs when we supply the input rectifier with a control circuit to control the current it delivers;

– we set the commutation emfs to a value E_{CN} equal to U_{1N}, where U_{1N} is the nominal amplitude (i.e. the peak value) of phase voltages. This occurs via the effect of the field current I_f on the value E_0 of emfs induced by the field winding.

By substituting $\dfrac{\pi}{2\sqrt{3}} I_{1N}$ in relationship [1.12] for I and U_{1N} for E_c, we have:

$$\alpha_N = \xi_N + \mu_N = \arccos\left[\cos\xi - \frac{\pi}{3} L_C p \omega_{mN} \frac{I_{1N}}{U_{1N}}\right] \qquad [1.21]$$

As $L_C p \omega_{mN}$ is roughly equal to 0.15 p.u. (see section 1.3), the second term in square brackets is about equal to 0.16. For an extinction angle ξ of $\pi/9$ (20°), relationship [1.21] gives a value of roughly $\pi/9 + \pi/10$ (38°) for the advance firing angle of the thyristors, $\alpha_N = \xi_N + \mu_N$, and hence a μ_N angle of roughly $\pi/10$ (18°). This value is small, as initially assumed.

To find value E_{0N} of E_0, which will lead to the desired amplitude of E_c, we use relationship [1.5] and the vectorial diagram that is associated with it (see Figure 1.7) drawn by using this approximation:

$$\varphi_N = (\alpha_N + \xi_N)/2 = \alpha_N - \mu_N/2$$

We have:

$$E_{0N} = p\omega_{mN}M_f I_{fN} = \sqrt{(E_{CN} + p\omega_{mN}(L_{CS}-L_C)I_{1N}\sin\varphi_N)^2 + (p\omega_{mN}(L_{CS}-L_C)I_{1N}\cos\varphi_N)^2}$$
[1.22]

$$\delta_N = atg\frac{p\omega_{mN}(L_{CS}-L_C)I_{1N}\cos\varphi_N}{E_{CN} + p\omega_{mN}(L_{CS}-L_C)I_{1N}\sin\varphi_N}$$
[1.23]

Equation [1.22] sets amplitude E_{0N} to give to emfs e_{0x} and hence that to be given to current I_f.

Equation [1.23] gives the leading phase shift angle δ_N of the commutation emfs on emfs e_{0x} induced by the field winding. Since the emf e_{0a} of phase a is in advance of $\pi/2$ on the electric position θ_{em} of the rotor, if we select a position for zero for which the flux induced by the field winding in phase a is maximum, the value of angle ψ_C that appears in equations [1.6] is equal to:

$$\psi_{CN} = \pi/2 + \delta_N$$
[1.24]

at the nominal operating point.

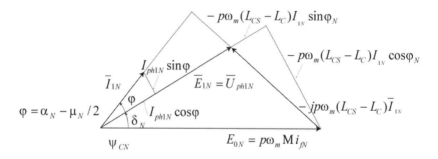

Figure 1.7. *Vectorial diagram linking the fundamental components of voltages and currents*

We can then link the firing instants of the thyristors to the electrical position of the rotor. Thus, thyristor T_{a1} must be fired for the following positions of the rotor:

$$-(\pi/3 + \alpha_N + \delta_N + \pi/2) + 2k\pi \text{ with } k \in [1,2,...].$$

14   Control of Non-conventional Synchronous Motors

The positions of the rotor corresponding to the firing of the other thyristors can easily be inferred from that of T_{a1}, considering that:

– the firing of T_{b1} and T_{c1} are respectively shifted by $2\pi/3$ and $4\pi/3$ with respect to those of T_{a1};

– the firing of T_{a2} is shifted by π with respect to those of T_{a1};

– the firing of T_{b2} and T_{c2} are respectively shifted by $2\pi/3$ and $4\pi/3$ with respect to those of T_{a2}.

The mean value of the electromagnetic torque produced by the machine is obtained by expressing that the mechanical power produced $C_{em}\omega_m$ is equal to the active power received by sources e_{0x}. Thus (see Figure 1.7), we have:

$$C_{emN}\omega_{mN} = \frac{3}{2}E_{0N}I_{1N}\cos(\delta_N + \alpha_N - \mu_N/2) \qquad [1.25]$$

Taking into account that:

$$E_{0N} = p\omega_m M i_{fN}$$

$$I_{1N} = \frac{2\sqrt{3}}{\pi}I_N$$

equation [1.25] leads to:

$$C_{emN} = \frac{3\sqrt{3}}{\pi}pMi_{fN}I_N\cos(\delta_N + \alpha_N - \mu_N/2) \qquad [1.26]$$

The mean value of DC voltage at the input of the bridge that supplies the motor U_{av} (see equation [1.18]) is equal to:

$$U_{av} = \frac{3\sqrt{3}}{\pi}E_{CN}\cos(\alpha_N - \mu_N/2)$$

It can be written:

$$U_{av} = \frac{3\sqrt{3}}{\pi}E_{0N}\cos(\alpha_N + \delta_N - \mu_N/2) \qquad [1.27]$$

as shown in Figure 1.7, or:

$$U_{av} = \frac{3\sqrt{3}}{\pi} p\omega_m Mi_{fN} \cos(\alpha_N + \delta_N - \mu_N/2) \qquad [1.28]$$

By multiplying this voltage by the current I to find the power at the input of the bridge that supplies the machine, we again find the value of the power received by sources e_{0x}. This is normal, since we have neglected all of the sources of loss.

The instantaneous torque, like the instantaneous voltage at the input of the bridge that supplies the machine, shows a ripple at six times the frequency of the phase voltage and currents.

1.6. Operation with a torque smaller than the nominal torque

If, at nominal speed ω_{mN} and nominal field current current I_{fN}, we decrease the value of current I without modifying the positions of the firing angles of the thyristors with respect to the position of the rotor, the amplitude of fundamental components of phase currents is decreased proportionally to the decrease in I. The phase shift (equal to $\delta_N + \alpha_N - \mu/2$) increases slightly with respect to the emfs produced by the field winding, insofar as the overlap angle μ slightly decreases. If we ignore this variation, which is small, we obtain a decrease in torque that is proportional to the decrease in current I, just like in a DC machine.

1.7. Operation with a speed below the nominal speed

Insofar as we can disregard the voltage drops on the winding resistances with respect to those on the winding inductances, a decrease in the speed of rotation at constant currents I and I_f and at constant advance firing angle α leads to a decrease in voltages at the machine terminals proportional to the decrease in speed. Indeed, both the emfs induced by the field winding and the voltage drops on the stator winding inductances are proportional to the electric speed of the rotor; whereas the value of angle μ is unaffected by the variation in frequency.

However, when the rotational speed drops below a small proportion of the nominal speed (in practice from 5 to 10%, depending on the machine parameters) we can no longer disregard the effects on resistances. The emfs at the machine terminals become insufficient to properly ensure the commutations of current I from one machine phase to the other.

We then resort to the following artifact: each time we need to commute current I from one phase to the next (at the + or − terminal), we cancel out this current by altering the control of the input bridge. Canceling out current I leads to all the thyristors of the bridge that supplies the motor being blocked. We then control the firing of the thyristors through which current I has to circulate after commutation and, thanks to the input bridge current, we can restore I. To speed up this process, we generally provide smoothing inductance with a free wheeling thyristor that is triggered each time we want to cancel out the circulation of current I in the stator windings of the motor, see Figure 1.8 [CHA 88].

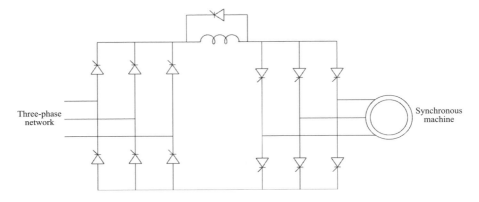

Figure 1.8. *Addition of a free-wheeling thyristor on the smoothing inductance*

1.8. Running as a generator

When the synchronous machine is run as a generator there is an inversion of the direction of energy transfer between the machine and the DC circuit.

The power received by the machine becomes negative and corresponds to an energy transfer from the machine to the DC circuit. Since current I has to be positive due to the irreversibility of current in thyristors, the transition of the synchronous machine from being used as a motor to being used as a generator involves the mean value of voltage U at the input of the bridge that supplies the machine being inverted by action on the thyristors' value of advance firing angle α. At set values of I, I_f and ω_m, by giving α a value equal to $\pi - \xi$, we invert the value of the DC voltage. At the level of the windings, the transition of α from a value below $\pi/2$ to a value close to π modifies accordingly the phase shift of the fundamental components of the currents with respect to the emfs induced by the field winding. The phase shift goes from a value below $\pi/2$ to a value above $\pi/2$, and thus changes the sign of $\cos(\alpha_N + \delta_N - \mu_N/2)$.

1.9. Equivalence of a machine with a commutator and brushes

The connection that the thyristor bridge – which supplies the machine – establishes between the DC circuit and the statoric windings is equivalent to that established by a commutator with three segments $2\pi/3$ wide equipped with a pair of brushes of μ wide whose position ρ_s follows electric position θ_{em} of the rotor (see Figure 1.9). We will notice the presence of a diode in series with the pair of brushes on this figure. This takes the irreversibility of the current of the thyristor bridge into account.

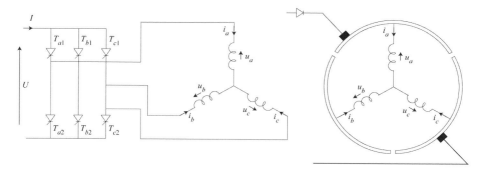

Figure 1.9. *Commutator-brush system equivalent to the thyristor bridge supplying the machine stator*

If ρ_S is the position of the center of the brush linked to the + terminal and we match $\rho_S = 0$ with the center of the commutator segment linked to phase *a*, this brush links:

– phase *a* to the + terminal from $\rho_S = -\pi/3 - \mu/2$ to $\rho_S = +\pi/3 + \mu/2$;

– phase *b* to the + terminal from $\rho_S = +\pi/3 - \mu/2$ to $\rho_S = \pi + \mu/2$;

– phase *c* to the + terminal from $\rho_S = \pi - \mu/2$ to $\rho_S = \pi + 2\pi/3 + \mu/2$.

From then on, it is enough for ρ_S to be linked to the electric position of the rotor by the following relationship:

$$\rho_S = \theta_{em} + (\alpha + \delta + \pi/2) - \mu/2 \qquad [1.29]$$

So the intervals during which each stator winding is linked to the + terminal by the commutator-brush system correspond to intervals during which the common anode thyristor connected to the winding is ON.

18   Control of Non-conventional Synchronous Motors

Similarly, since the brush linked to the − input terminal is shifted by π with respect to the brush linked to the + terminal, it connects each stator winding to the − terminal during the intervals corresponding to conduction of the common cathode thyristor to which the winding is connected.

If we neglect width μ of the overlap intervals, the brushes become punctual and the circuit seen from the access terminals of the brushes (armature circuit) contains at every instant two stator windings in series; on one turn of the brushes, we see six successive configurations, each corresponding to a commutation interval (the interval separating two successive changes in configuration, see Figure 1.10):

− from $\rho_S = -\pi/3$ to $\rho_S = 0$, the armature is made of windings a and b in series;

− from $\rho_S = 0$ to $\rho_S = \pi/3$, the armature is made of windings a and c in series;

− from $\rho_S = \pi/3$ to $\rho_S = 2\pi/3$, the armature is made of windings b and c in series;

− from $\rho_S = 2\pi/3$ to π, the armature is made of windings b and a in series;

− from $\rho_S = \pi$ to $\rho_S = \pi + \pi/3$, the armature is made of windings c and a in series;

− from $\rho_S = \pi + \pi/3$ to $\rho_S = \pi + 2\pi/3$, the armature is made of windings c and b in series.

Since the position of the brushes is linked to the electric position of the rotor by relationship [1.29], and since we assume that the overlap angle has a negligible value, the armature circuit is made of the series connection:

− of phases a and b for $-\pi/3 - (\alpha + \delta + \pi/2) < \theta_{em} < -(\alpha + \delta + \pi/2)$;

− of phases a and c for $-(\alpha + \delta + \pi/2) < \theta_{em} < \pi/3 - (\alpha + \delta + \pi/2)$;

− of phases b and c for $\pi/3 - (\alpha + \delta + \pi/2) < \theta_{em} < 2\pi/3 - (\alpha + \delta + \pi/2)$;

− of phases b and a for $2\pi/3 - (\alpha + \delta + \pi/2) < \theta_{em} < \pi - (\alpha + \delta + \pi/2)$;

− of phases c and a for $\pi - (\alpha + \delta + \pi/2) < \theta_{em} < \pi + \pi/3 - (\alpha + \delta + \pi/2)$;

− of phases c and b for $\pi + \pi/3 - (\alpha + \delta + \pi/2) < \theta_{em} < \pi + 2\pi/3 - (\alpha + \delta + \pi/2)$.

This circuit possesses a resistance R_A equal to $2R_S$ and a self-inductance L_A equal to $2L_{CS}$.

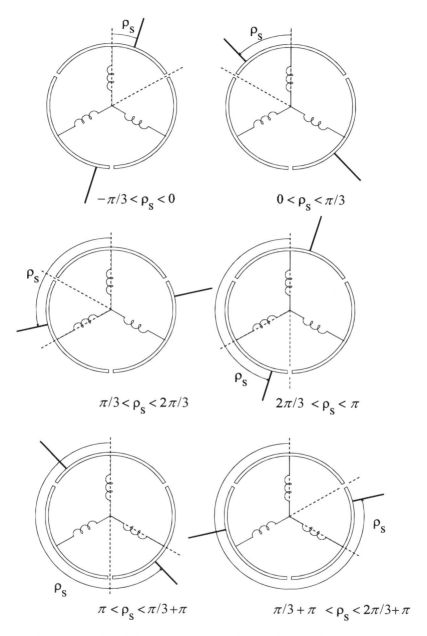

Figure 1.10. *The different armature circuits seen from the input terminals*

On the commutation interval corresponding to:

$$-\pi/3-(\alpha+\delta+\pi/2) < \theta_{em} < -(\alpha+\delta+\pi/2)$$

where the armature circuit is made of phases a and b in series, the mutual inductance M_{Af} between this circuit and the field winding is equal to:

$$M_{Af} = M_{af} - M_{bf} = M_f \cos\theta_{em} - M_f \cos(\theta_{em} - 2\pi/3) \quad [1.30]$$

namely:

$$M_{Af} = -\sqrt{3}M_f \sin(\theta_{em} - \pi/3) = \sqrt{3}M_f \cos(\theta_{em} + \pi/6) \quad [1.31]$$

This mutual inductance varies between:

$$\sqrt{3}M_f \cos(-(\alpha+\delta+\pi/2) - \pi/6)$$

and:

$$\sqrt{3}M_f \cos(-(\alpha+\delta+\pi/2) + \pi/6)$$

Its mean value $< M_{Af} >$ on the given interval is equal to:

$$< M_{Af} > = \frac{3\sqrt{3}M_f}{\pi} \cos(-(\alpha+\delta+\pi/2)) \quad [1.32]$$

On interval $-(\alpha+\delta+\pi/2) < \theta_{em} < -((\alpha+\delta+\pi/2)+\pi/3)$, the mutual inductance M_{Af} is equal to:

$$M_{Af} = M_{af} - M_{cf} = M_f \cos\theta_{em} - M_f \cos(\theta_{em} - 4\pi/3) \quad [1.33]$$

namely:

$$M_{Af} = -\sqrt{3}M_f \sin(\theta_{em} - 2\pi/3) = \sqrt{3}M_f \cos(\theta_{em} - \pi/6) \quad [1.34]$$

Mutual inductance M_{Af} varies again between:

$$\sqrt{3}M_f \cos(-(\alpha+\delta+\pi/2) - \pi/6)$$

and:

$$\sqrt{3}M_f \cos(-(\alpha+\delta\pi/2)+\pi/6)$$

with a mean value equal to:

$$<M_{Af}> = \frac{3\sqrt{3}}{\pi} M_f \cos(-(\alpha+\delta+\pi/2))$$

and so on.

The calculation of the inductance value M_{Af} shows that on a commutation interval, position ρ_A of the armature with respect to the field winding varies with speed $\dot{\theta}_{em}$ from a position equal to $-(\alpha+\delta+\pi/2)-\pi/6$ at the beginning of the interval to $-(\alpha+\delta+\pi/2)+\pi/6$ at the end of the interval (see Figure 1.11).

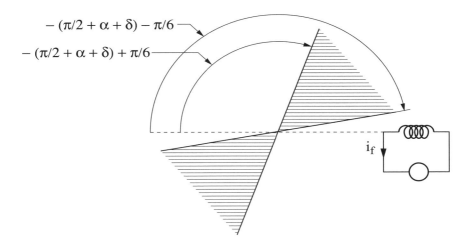

Figure 1.11. *Armature position with respect to the field winding*

At this moment, a change in configuration and hence armature circuit is imposed by the transition of one of the brushes from a commutator segment to the following (see Figure 1.10) and leads to the armature circuit returning to a position with respect to the field winding that is equal to $-(\alpha+\delta+\pi/2)-\pi/6$. This configuration change causes a transfer of current I from the armature circuit, which stops being in

22 Control of Non-conventional Synchronous Motors

service, to the armature circuit that is activated. Insofar as we neglect duration μ of commutations, this transfer is assumed to take place instantaneously.

Similarly to what has been said on its position towards the field winding, during each commutation interval the position of the armature towards the *d*-axis damper winding varies with speed $\dot{\theta}_{em}$ from $-(\alpha+\delta+\pi/2)-\pi/6$ to $-(\alpha+\delta+\pi/2)+\pi/6$ and its position towards *q*-axis damper winding changes from $-(\alpha+\delta)-\pi/6$ to $-(\alpha+\delta)+\pi/6$, since axis *q* is shifted by $\pi/2$ with respect to axis *d*.

In a referential linked to the electric position of the rotor, we can consider the synchronous machine and the diode bridge that supplies it as a machine with a commutator and brushes. The circuit seen from the brushes (armature circuit) possesses a resistance R_A equal to $2R_S$. It has a self-inductance L_A equal to $2L_{CS}$ and mutual inductances M_{Af}, M_{Ad} and M_{Aq}, with the field and damper windings whose values depend on position ρ_A of the armature with respect to the field winding (axis *d* of the rotor). This position varies repeatedly, with speed $\dot{\theta}_{em}$, from $-(\alpha+\delta+\pi/2)-\pi/6$ to $-(\alpha+\delta+\pi/2)+\pi/6$.

1.10. Equations inferred from the theory of circuits with sliding contacts

By referring to the works of M. Garrido on the general equations of electrical machines [GAR 68, GAR 71, GAR 72, MAT 04], the description self-controlled synchronous machine function (see section 1.9) involves considering the behaviour of this machine to be that of a non-lineic circuit with sliding contacts and discontinuous commutation.

Without proving them, we are going to use the results of the studies presented in [GAR 72] to establish a macroscopic model of the behaviour of the self-controlled synchronous machine. This model, as indicated in the reference, is based on the transition to mean values of voltages and currents, taken on a commutation interval, but also the mean values of the machine parameters, which vary during the interval.

Thus, at the level of machine parameters we replace:

– the instantaneous position ρ_A of the armature, which varies from $-(\alpha+\delta+\pi/2)-\pi/6$ to $-(\alpha+\delta+\pi/2)+\pi/6$ during a commutation interval, with its mean position during this interval:

$$<\rho_A> = -(\alpha+\delta+\pi/2) \qquad [1.35]$$

– the value of mutual inductance M_{Af} between the armature and the field winding, which varies from $\sqrt{3}M_f\cos(-(\alpha+\delta+\pi/2)-\pi/6)$ to $\sqrt{3}M_f\cos(-(\alpha+\delta+\pi/2)+\pi/6)$, with its mean value during this interval:

$$<M_{Af}> = \frac{3\sqrt{3}}{\pi}M_f\cos(-(\alpha+\delta+\pi/2)) = \frac{3\sqrt{3}}{\pi}M_f\cos(<\rho_A>) \qquad [1.36]$$

– the value of mutual inductance M_{Ad} between the armature and the d-axis damper winding, which varies from $\sqrt{3}M\cos(-(\alpha+\delta+\pi/2)-\pi/6)$ to $\sqrt{3}M\cos(-(\alpha+\delta-\pi/2)+\pi/6)$, with its mean value:

$$<M_{Ad}> = \frac{3\sqrt{3}}{\pi}M\cos(-(\alpha+\delta+\pi/2)) = \frac{3\sqrt{3}}{\pi}M\cos(<\rho_A>) \qquad [1.37]$$

– the value of mutual inductance M_{Aq} between the armature and the q-axis damper winding, which varies from $\sqrt{3}M\cos(-(\alpha+\delta)-\pi/6)$ to $\sqrt{3}M\cos(-(\alpha+\delta)+\pi/6)$, with its mean value:

$$<M_{Aq}> = \frac{3\sqrt{3}}{\pi}M\cos(-(\alpha+\delta)) = \frac{3\sqrt{3}}{\pi}M\cos(<\rho_A> -\pi/2) \qquad [1.38]$$

At the level of the different voltages and currents, we replace their instantaneous values with a sliding mean taken during a time interval ΔT_C corresponding to the duration of a commutation interval:

$$\Delta T_C = \frac{\pi/3}{\dot\theta_{em}} = \frac{\pi/3}{p\omega_m}$$

We thus obtain the following values:

$$<I> = \frac{1}{\Delta T_C}\int_{t-\Delta T_C/2}^{t+\Delta T_C/2} I\,dt \qquad [1.39]$$

$$<U> = \frac{1}{\Delta T_C}\int_{t-\Delta T_C/2}^{t+\Delta T_C/2} U\,dt \qquad [1.40]$$

$$<I_f> = \frac{1}{\Delta T_C}\int_{t-\Delta T_C/2}^{t+\Delta T_C/2} I_f\,dt \qquad [1.41]$$

$$<i_d> = \frac{1}{\Delta T_C} \int_{t-\Delta T_C/2}^{t+\Delta T_C/2} i_d \, dt \qquad [1.42]$$

$$<i_q> = \frac{1}{\Delta T_C} \int_{t-\Delta T_C/2}^{t+\Delta T_C/2} i_q \, dt \qquad [1.43]$$

According to the average parameters and the macroscopic variables that have just been defined, by neglecting the terms of second order we obtain the following macroscopic electrical equations:

$$\begin{aligned}<U> = {} & R_A <I> + L_A \frac{d<I>}{dt} + (\dot{\theta}_{em} - <\dot{p}_A>) \frac{\partial <M_{Af}>}{\partial <p_A>} <I_f> \\ & + (\dot{\theta}_{em} - <\dot{p}_A>) \frac{\partial <M_{Ad}>}{\partial <p_A>} <i_d> \\ & + (\dot{\theta}_{em} - <\dot{p}_A>) \frac{\partial <M_{Aq}>}{\partial <p_A>} <i_q> \\ & + \frac{d<M_{Af}><i_f>}{dt} + \frac{d<M_{Ad}><i_d>}{dt} + \frac{d<M_{Aq}><i_q>}{dt}\end{aligned} \qquad [1.44]$$

$$<u_f> = R_f <i_f> + L_f \frac{d<i_f>}{dt} + \frac{d<M_{Af}><I>}{dt} + <M_{df}> \frac{d<i_d>}{dt} \qquad [1.45]$$

$$0 = R_R <i_d> + L_R \frac{d<i_d>}{dt} + \frac{d<M_{Ad}><I>}{dt} + M_{df} \frac{d<i_f>}{dt} \qquad [1.46]$$

$$0 = R_R <i_q> + L_R \frac{d<i_q>}{dt} + \frac{d<M_{Aq}><I>}{dt} \qquad [1.47]$$

In these equations:

– R_f and L_f are the resistance and self-inductance of the field winding;

– R_R and L_R are the resistance and self-inductance of a damper winding;

– M_{df} is the mutual inductance between the d-axis damper winding and the field winding.

The equivalent diagram in Figure 1.12 corresponds to equations [1.44] to [1.47].

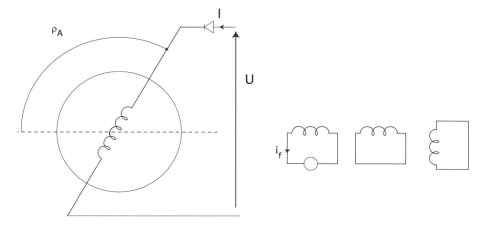

Figure 1.12. *Direct current machine corresponding to the macroscopic model*

In steady state $<\rho_A> = -(\alpha + \delta + \pi/2)$, and $\dot{\theta}_{em} = p\omega_m$. The equations become:

$$<U> = R_A <I> + p\omega_m \frac{3\sqrt{3}}{\pi} M_f I_f \cos(-(\alpha+\delta)) <I_f>$$ [1.48]

$$<u_f> = R_f <I_f>$$ [1.49]

$$<i_d> = 0$$ [1.50]

$$<i_q> = 0$$ [1.51]

The power converted from electrical into mechanical energy $C_{em}\omega_m$ is equal to the product of current $<I>$ by the voltage produced by sliding (the terms in $(\dot{\theta}_{em} - <\dot{\rho}_A>)$) in the equation of the armature circuit.

By comparing equation [1.48], after having neglected the voltage drop $R_A <I>$, with equation [1.28] of the "classic" study, where $\mu/2 = 0$ to take into account the hypothesis of instantaneous commutation, we can notice the equivalence between the results given by the macroscopic model inferred from the theory of circuits with sliding contacts and the "classic" study based on the theory of rectification.

1.11. Evaluation of alternating currents circulating in steady state in the damper windings

The macroscopic model shows that, in steady state, the mean values of currents in the damper windings are zero. These currents therefore do not have a DC component. The variation in the position of the armature with respect to the field winding, and hence with respect to the damper windings within each commutation interval, is however going to induce AC currents in the damper windings.

If during each commutation interval we consider a time t' whose origin matches the beginning of the interval, we can write:

$$M_{Ad} = \sqrt{3}M\cos(<\rho_A> - \pi/6 + p\omega_m t') \quad ; \quad 0 < p\omega_m t' < \pi/3 \quad [1.52]$$

$$M_{Aq} = \sqrt{3}M\cos(<\rho_A> - \pi/2 - \pi/6 + p\omega_m t') \quad ; \quad 0 < p\omega_m t' < \pi/3 \quad [1.53]$$

Taking into account these variations in mutual inductances and the fact that in steady state I and I_f are constant (see the hypotheses made on these currents in sections 1.3 and 1.4), the equations of the damper windings become [BUY 82]:

$$0 = R_R i_d + L_R \frac{di_d}{dt} + \frac{d}{dt}(\sqrt{3}M\cos(<\rho_A> - \pi/6 + p\omega_m t')I) \quad [1.54]$$

$$0 = R_R i_q + L_R \frac{di_q}{dt} + \frac{d}{dt}(\sqrt{3}M\cos(<\rho_A> - \pi/2 - \pi/6 + p\omega_m t')I) \quad [1.55]$$

If we disregard the resistance R_R of damper windings, we obtain:

$$i_d = -\frac{\sqrt{3}M}{L_R}I\cos(<\rho_A> - \pi/6 + p\omega_m t') + K \quad ; \quad 0 < p\omega_m t' < \pi/3 \quad [1.56]$$

$$i_q = -\frac{\sqrt{3}M}{L_R}I\sin(<\rho_A> - \pi/6 + p\omega_m t') + K' \quad ; \quad 0 < p\omega_m t' < \pi/3 \quad [1.57]$$

The values of constants K and K' are set by the fact that the mean values of i_d and i_q are zero:

$$K = -\frac{3}{\pi}\frac{\sqrt{3}MI}{L_R}[\sin(<\rho_A> - \pi/6) - \sin(<\rho_A> + \pi/6)] \quad [1.58]$$

$$K' = -\frac{3}{\pi}\frac{\sqrt{3}MI}{L_R}[\cos(<\rho_A>+\pi/6)-\cos(<\rho_A>-\pi/6)] \qquad [1.59]$$

Figure 1.13 gives the evolution as a function of $\omega t'$ of $\dfrac{i_d L_R}{\sqrt{3}MI}$ and $\dfrac{i_q L_R}{\sqrt{3}MI}$ on a commutation interval. The abrupt variations these currents undergo at the changes in commutation intervals come from the transfers of current I from one armature circuit to the following, which take place at these times.

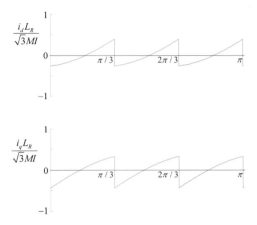

Figure 1.13. *Currents in the damper windings (in normalised value)*

The AC currents that circulate in the damper windings are induced by the harmonic components of the currents circulating in the stator phases of the machine[2].

The calculation of these currents, which as we have just seen is easy when we use the commutator machine model, is mainly used to ensure dimensioning of the damper windings that takes into account the conditions of functioning peculiar to the supply of the machine with a thyristor bridge.

---

2 If we disregard the duration of the commutations, the currents in the machine phases are almost rectangular and have harmonics of rank $6k\pm1$, k being an integer, [SEG 92] and the resulting currents in the damper windings have angular frequencies equal to $6kp\omega_m$. The currents in the damper windings thus have a frequency equal to six times that of statoric currents, as shown in Figure 1.13.

1.12. Transposition of the study to the case of a negative rotational speed

In the case where the rotational speed is negative, we need only invert the direction of reference adopted for position θ_m of the rotor, so that the rotational speed becomes positive once again.

We only need to take into account that this change in direction of reference reverses the polarity of the emfs induced by the field winding (which corresponds to a phase shift of π) and swaps the role of phases b and c (see Figure 1.14). All of the results of the analysis we have performed in the case of a positive rotational speed are applicable.

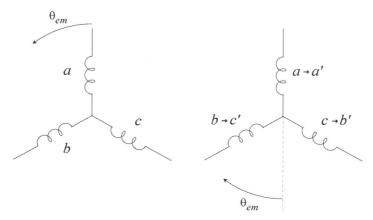

Figure 1.14. *Effect of an inversion of the direction of reference adopted to define the position of the rotor*

1.13. Variant of the base assembly

For very high power applications, the base assembly that has just been studied is adapted using a machine equipped with two three-phase stator windings, mechanically shifted by 30 electrical degrees, one in relation to the other. Each stator winding system is supplied with an input rectifier via a DC smoothing inductance (see Figure 1.15) [BON 01].

With this solution, we decrease the torque ripple during normal operation but reduce the problem of canceling out torque during low-speed commutations (see section 1.7). Indeed, due to the 30° shift between the two winding systems, the commutations of the two systems are shifted by $1/12^{th}$ of an electrical period. When one cancels out the current in a winding system, the current carries on circulating in the other and we retain 50% of the torque.

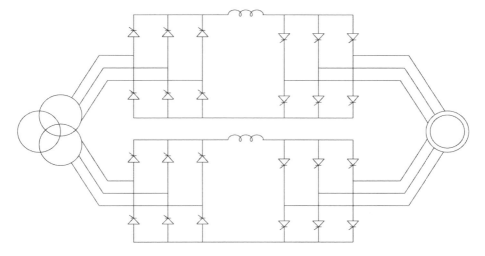

Figure 1.15. *Double stator machine*

1.14. Conclusion

In this chapter, we have studied how the self-controlled synchronous motor functions. We started with a classical study based on a "transient emf in series with a commutation inductance"-type stator model. We have then shown the similarities existing between the functioning of the rectifier supplying the machine and that of a three-segment commutator equipped with a pair of brushes. We have ended up with a model of "non-lineic circuit with sliding contacts and discontinuous commutation"-type for stator windings and the bridge that supplies them. Using Garrido's work, we have then built an equivalent model of the self-controlled synchronous machine that is similar to that of a DC machine.

The results of the classical study, as much as the model of the equivalent DC machine, can be used to design the control system and regulation of the self-controlled synchronous machine.

1.15. List of the main symbols used

A list of the main symbols used in this chapter and in Chapter 2 is given at the end of Chapter 2.

1.16. Bibliography

[BON 97] BONAL J., *Entraînements Électriques à Vitesse Variable*, Technique et Documentation, Paris, France, 1997.

[BON 00] BONAL J., *Entraînements Électriques à Vitesse Variable: Interactions Convertisseur-réseau et Convertisseur-moteur-charge*, Technique et Documentation, Paris, France, 2000.

[BON 01] BONAL J., *Utilisation Inductrielle des Moteurs à Courant Alternatif*, Technique et Documentation, Paris, France, 2001.

[BUH 77] BUHLER H., *Einführung in die Theorie Geregelten Drehstromantriebe*, Birkhauser, Bale, Switzerland, 1977.

[BUY 81] BUYSE H., THIRY J.M., GARRIDO M., "Une nouvelle méthode de commutation du courant d'induit d'une machine synchrone autopilotée", *Actes du Colloque Européen du Carrefour de la Force Motrice*, pp. 127-133, Paris, France, December 10-11, 1981.

[BUY 82] BUYSE H., THIRY J.M., "Contribution to the transient analysis of the self controlled synchronous machine", *Proceedings of the International Conference on Electrical Machines ICEM'82*, pp. 546-549, Budapest, Hungary, September 5-9, 1982.

[CHA 83] CHATELAIN J., *Machine Électriques*, Presses Polytechniques Romandes, Lausanne, Switzerland, 1983.

[CHA 88] CHAUPRADE R., MILSANT F., *Electronique de Puissance. 2. Commande Moteurs à Courant Alternatif*, Eyrolles, Paris, France, 1988.

[GAR 68] GARRIDO M.S., Contribution à la théorie dynamique des systèmes électromécaniques, Thesis, Institut National Polytechnique de Lorraine, Nancy, France, 1968.

[GAR 71] GARRIDO M.S., "Les équations générales des machines électriques déduites de l'électromagnétisme", *Revue E*, vol. VI, no. 9, pp. 269-273, 1971.

[GAR 72] GARRIDO M.S., GUDEFIN E., "Equations des circuits à commutation non linéiques", *Revue E*, vol. VII, no. 3, pp. 53-60, 1972.

[KLE 80] KLEINRATH H., *Stromrichtegespeiste Drehfeldmaschine*, Springer, Vienna, Austria, 1980.

[LOU 04] LOUIS J.-P., *Modélisation des Machines Électriques en Vue de Leur Commande: Concepts Généraux*, Hermès-Lavoisier (Traité EGEM, série Génie électrique), Paris, France, 2004.

[LOU 10] LOUIS J.-P., *Commandes Classiques et Avancées des Actionneurs Synchrones*, Hermès-Lavoisier (Traité EGEM, série Génie électrique), Paris, France, 2010.

[MAT 04] MATAGNE E., GARRIDO M., "Conversion électronique d'énergie: du phénomène physique à la modélisation dynamique" in : *Modélisation des Machines Électriques en Vue de Leur Commande*, LOUIS J-P (ed.), Hermès-Lavoisier (Traité EGEM), Paris, France, 2004.

[PAL 99] PALMA J., *Accionamentos Electricos de Velocedade Variavel*, Fundacao Colousta Gulbenkian, Lisbon, Portugal, 1999.

[SEG 92] SEGUIER G., *Les Convertisseurs de l'Électronique de Puissance. 1. La Conversion Alternatif Continu*, Technique et Documentation, Paris, France, 1992.

Chapter 2

Self-controlled Synchronous Motor: Dynamic Model Including the Behavior of Damper Windings and Commutation Overlap

2.1. Introduction

This chapter studies the same system examined in Chapter 1 – a synchronous motor supplied by a thyristor bridge operating as a current inverter and a control system that links the starting angle of the thyristors to the rotor position, as can be seen in Figure 2.1.

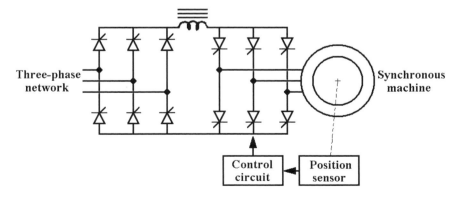

Figure 2.1. *Principle diagram of the self-controlled synchronous machine (identical to the diagram in Figure 1.1)*

Chapter written by Ernest MATAGNE.

34 Control of Non-conventional Synchronous Motors

The aim is to develop a macroscopic model similar to that presented in section 1.10. This will take into account the transient regime due to the damper windings, without making the hypothesis that the overlap angle (the electric angle during which the armature current is transferred from one phase to the other during the commutation) is zero. As in Chapter 1, we will assume that the synchronous machine has a cylindrical rotor, the couplings between the windings are sinusoidal, the field winding is supplied with a current source and that there is no magnetic saturation.

Since the overlap angle is not assumed to be zero, the model obtained will have at least one more degree of freedom, and thus need an additional equation to set this. It will have to remain simple enough to be used in a control system. To define the armature circuit, we will use a variable ρ that, in association with overlap angle μ, sets the beginning and end times of commutations, as shown in Figure 2.2, which uses some elements included in Figure 1.6.

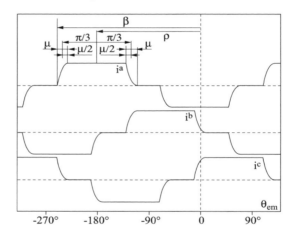

Figure 2.2. *Definition of angles ρ and μ. Drawing done for $\rho = 182°$ and $\mu = 18°$*

Angle ρ can be considered the position of the center of the fictitious brushes introduced in section 1.9, but defined with respect to the magnetic axis of the field winding. It therefore plays a role comparable to that of angle ρ_S considered in section 1.9, that was defined with respect to the magnetic axis of phase a. The link between these two variables is:

$$\rho = <\rho_S> - \theta_{em} \qquad [2.1]$$

where θ_{em} is the electric position of the rotor.

In section 1.10, neglecting overlap angle μ enables us to well-define the armature circuit at each moment: it consists of the anti-series connection of the two stator windings in which current flows. This circuit is completely defined for each value of rotor electric position θ_{em}, by position ρ_S of the fictitious brushes. This is no longer the case if we consider an overlap angle μ other than zero. To extend the notion of the armature circuit, we will follow a formalism inspired by a method often used to calculate circuit type quantities at the end of a field calculation (see, for example, [DUF 97, REM 05 or MAT 02]). The armature circuit will be defined by giving three sharing coefficients N^a, N^b and N^c. The currents in the three stator phases will then be defined by:

$$i^k = N^k I \qquad k \in \{a,b,c\} \qquad [2.2]$$

In this formula, the upper indices have been attributed to currents i^k [KRO 59, p. 101, MYL 08] and to parameters N^k. This is to remind us that they are contravariant variables [SYN 49].

We will try to express coefficients N^k as a function of θ_{em} and a small number of variables considered constant at the scale of a commutation, as is the case in particular for ρ and μ. In practice, for this purpose we will choose simplified expressions of functions N^k, so that formula [2.2] will be approximate. The choice of these functions is arbitrary in a way. It results from a compromise between simplicity of the expressions used and the wish to obtain, in conditions of normal operation, evolutions of the currents close to their exact evolutions using equations [2.2].

2.2. Choice of the expression of N^k

In this chapter, we will choose functions N^k so that the currents i^k, when computed with the approximations introduced in Chapter 1 to study commutations, can be exactly represented with the help of functions N^k. These approximations are that current I at the input of the bridge that supplies the motor (armature current) is constant (infinite inductance in series with the supply), that the resistances of the damper windings are negligible and that steady-state operation is established. These assumptions have allowed us to define commutation electromotive forces (emfs) e_a, e_b and e_c. Using these emfs, we can define angles α, ξ, φ, ν, ψ_C and δ, as shown in Figure 2.3, which uses some elements from Figure 1.6.

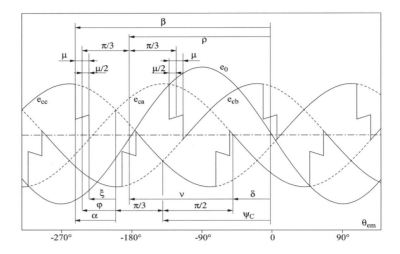

Figure 2.3. *Definition of angles α, φ, ξ, v and δ. Drawing done for $\rho = 182°$, $\mu = 18°$ and $\delta = 50°$*

Let us recall that, in section 1.4, the overlap angle is defined as the difference between angle α at the start of a commutation and angle ξ at the end of a commutation. These are referred to in Chapter 1 with respect to the crossing time of commutation emfs of the phases involved in the commutation process:

$$\mu = \alpha - \xi \qquad [2.3]$$

All the angles in Figure 2.3 can be defined as a function of ρ, μ and δ. In particular, we have:

$$\alpha = \rho - \delta - \frac{\pi}{2} + \frac{\mu}{2} \qquad [2.4]$$

With equation [2.3], from equation [2.4] we get:

$$\xi = \rho - \delta - \frac{\pi}{2} - \frac{\mu}{2} \qquad [2.5]$$

Furthermore, similarly to [1.24], we can write:

$$\psi_C = \frac{\pi}{2} + \delta \qquad [2.6]$$

Dynamic Model of a Self-controlled Synchronous Motor 37

In accordance with what is indicated in section 1.4, the commutation of current I from phase c to a at the + input terminal takes place during the interval defined by:

$$-\frac{\pi}{3} - \rho - \frac{\mu}{2} < \theta_{em} < -\frac{\pi}{3} - \rho + \frac{\mu}{2} \qquad [2.7]$$

where θ_{em} is the rotor electric position angle, which is equal to $p\,\omega_m\,t$ for steady-state operation.

During this commutation, current i^a is given by formula [1.9], which becomes:

$$i^a = \frac{\sqrt{3}\,E_C}{2\,L_C\,p\,\omega_m}[\cos(p\,\omega_m\,t + \frac{\pi}{3} + \frac{\pi}{2} + \delta) - \cos(\rho - \delta - \frac{\pi}{2} + \frac{\mu}{2})] \qquad [2.8]$$

whereas equation [1.10] is written as:

$$I = \frac{\sqrt{3}\,E_C}{2\,L_C\,p\,\omega_m}[\cos(\rho - \delta - \frac{\pi}{2} - \frac{\mu}{2}) - \cos(\rho - \delta - \frac{\pi}{2} + \frac{\mu}{2})] \qquad [2.9]$$

By eliminating E_C between these two equations, after a few simplifications we get:

$$i^a = \frac{\sin(\rho - \delta + \frac{\mu}{2}) + \sin(p\,\omega_m\,t + \frac{\pi}{3} + \delta)}{\sin(\rho - \delta + \frac{\mu}{2}) - \sin(\rho - \delta - \frac{\mu}{2})}\,I \qquad [2.10]$$

The form of this equation leads us to choose:

$$N^a = \frac{\sin(\rho - \delta + \frac{\mu}{2}) + \sin(\theta_{em} + \frac{\pi}{3} + \delta)}{\sin(\rho - \delta + \frac{\mu}{2}) - \sin(\rho - \delta - \frac{\mu}{2})} \qquad [2.11]$$

Expression [2.11] is function of ρ in a complicated way and of θ_{em} in a very different way. We can obtain a simpler expression by replacing variable δ with variable v:

$$v = \rho - \delta \qquad [2.12]$$

Equation [2.11] is then written:

$$N^a = \frac{\sin(\nu + \frac{\mu}{2}) + \sin(\theta_{em} + \rho + \frac{\pi}{3} - \nu)}{\sin(\nu + \frac{\mu}{2}) - \sin(\nu - \frac{\mu}{2})}$$ [2.13a]

During the interval defined by [2.7], all of the current I returns via phase b, so we will take:

$$N^b = -1$$ [2.13b]

and since current I is shared between phases c and a, for N^c we will take the complement to 1 of equation [2.13a], i.e.:

$$N^c = \frac{-\sin(\theta_{em} + \rho + \frac{\pi}{3} - \nu) - \sin(\nu - \frac{\mu}{2})}{\sin(\nu + \frac{\mu}{2}) - \sin(\nu - \frac{\mu}{2})}$$ [2.13c]

We will notice that ν intervenes in expressions [2.13] only by its value module 180°. Indeed, subtracting or adding a value of 180° to ν leaves the value of these expressions unchanged.

Within the framework of the new model, the signification of angle δ, or ν which replaces it, must be redefined. This will be done later on. At the moment, these angles represent an additional degree of freedom that we can use.

Figure 2.4 shows the variation of N^a on the interval defined by [2.7] for different values of ν.

We see that ν degree of freedom, which replaces δ via equation [2.12], does not have any influence on the limits [2.7] of the overlap interval. Via equation [2.13] it gives the evolution of way the current I is shared between the phases during the overlap interval. We will notice that the current evolution is almost linear on the interval defined by equation [2.7] when $\nu = 180°$. In this case, in the middle of the overlap interval, we have $N^a = N^c = 0.5$.

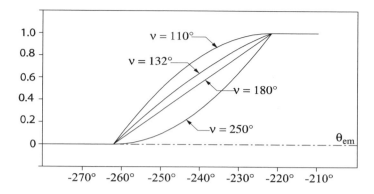

Figure 2.4. *Drawing of N^a for different values of v, for $\rho = 182°$ and $\mu = 40°$*

The current sharing defined by equation [2.13] is not the only option. Another interesting choice would be to consider a linear commutation, or a linear variation of N^a between 0 and 1 on the interval defined by equation [2.7] and N^c from 1 to 0 on the same interval. This has the advantage of not involving variable v, but is a little less accurate. A quasi-linear commutation law can also be obtained by setting v to 180° (or 0°) in formulae [2.13].

The interval that follows the one set by equation [2.7] is defined by:

$$-\frac{\pi}{3} - \rho + \frac{\mu}{3} < \theta_{em} < -\rho - \frac{\mu}{2} \qquad [2.14]$$

During this interval, we have:

$$N^a = 1 \qquad [2.15a]$$

$$N^b = -1 \qquad [2.15b]$$

$$N^c = 0 \qquad [2.15c]$$

The union of intervals given by equations [2.7] and [2.14] forms the interval defined by:

$$-\frac{\pi}{3} - \rho - \frac{\mu}{2} < \theta_{em} < -\rho - \frac{\mu}{2} \qquad [2.16]$$

40 Control of Non-conventional Synchronous Motors

This interval has a length of $\pi/3$ radians. The sharing of current I between the phases during the following intervals of $\pi/3$ is similar to that described by equations [2.13] and [2.15] to within about a permutation of indices abc and possibly a statoric electrical variables sign change. We can hence restrict the study to the interval defined by equation [2.16].

The current sharing obtained from equations [2.13] and [2.15] possesses several properties that will ease the analysis:

N^k are continuous functions of θ_{em} [2.17a]

and $\dfrac{\partial \, N^k}{\partial \, \rho} = \dfrac{\partial \, N^k}{\partial \, \theta_{em}}$ [2.17b]

Property [2.17b] is due to the choice of variables ρ, μ and ν which depend on N^k. We will notice that properties [2.17] would have remained valid if we had used the hypothesis of a linear commutation.

Another noticeable property of sharing using equations [2.13] and [2.15], which is a general property of collector machines, is that:

$$\sum_{k} N^k = 0 \qquad [2.18]$$

From now on, we will no longer assume that the forms of sharing coefficients N^k are known, but we will only use general properties [2.17] and [2.18]. We will refer to the analysis made in Chapter 1 to set the equations that govern the electromechanical variables ρ, μ and ν in section 2.6. The particular case in [2.13] and [2.15] will be developed in the appendices and used for the numerical simulations in section 2.9.

2.3. Expression of fluxes

When the inductances rapidly vary over time, it is interesting to take fluxes ψ as state variables. Indeed, their time derivative can be obtained directly by using Faraday's law:

$$\dfrac{d\psi}{dt} = u - e \qquad [2.19]$$

where ψ is the flux of a given circuit branch, u its voltage and e its emf.

Dynamic Model of a Self-controlled Synchronous Motor 41

The fluxes are linked to the currents by a relationship that does not involve time derivatives, so at each step we can calculate the currents as a function of fluxes.

Assuming, as indicated before, that the machine studied has sinusoidal coupling and linear materials, the fluxes of the five circuits of this machine are linked to the currents by:

$$\psi_a = L_S\, i^a + M_S\, i^b + M_S\, i^c + M\, \cos(p\, \theta_m)\, i^D \\ - M\, \sin(p\, \theta_m)\, i^Q + M_f\, \cos(p\, \theta_m)\, I^f \quad [2.20a]$$

$$\psi_b = M_S\, i^a + L_S\, i^b + M_S\, i^c + M\, \cos\left(p\, \theta_m - \frac{2\pi}{3}\right) i^D \\ - M\, \sin\left(p\, \theta_m - \frac{2\pi}{3}\right) i^Q + M_f\, \cos\left(p\, \theta_m - \frac{2\pi}{3}\right) I^f \quad [2.20b]$$

$$\psi_c = M_S\, i^a + M_S\, i^b + L_S\, i^c + M\, \cos\left(p\, \theta_m + \frac{2\pi}{3}\right) i^D \\ - M\, \sin\left(p\, \theta_m + \frac{2\pi}{3}\right) i^Q + M_f\, \cos\left(p\, \theta_m + \frac{2\pi}{3}\right) I^f \quad [2.20c]$$

$$\psi_D = M\, \cos(p\, \theta_m)\, i^a + M\, \cos\left(p\, \theta_m - \frac{2\pi}{3}\right) i^b \\ + M\, \cos\left(p\, \theta_m + \frac{2\pi}{3}\right) i^c + L_R\, i^D + M_{Rf}\, I^f \quad [2.21a]$$

$$\psi_Q = -M\, \sin(p\, \theta_m)\, i^a - M\, \sin\left(p\, \theta_m - \frac{2\pi}{3}\right) i^b \\ - M\, \sin\left(p\, \theta_m + \frac{2\pi}{3}\right) i^c + L_R\, i^Q \quad [2.21b]$$

where:

- L_S is the self-inductance of one phase of the stator;
- M_S is the mutual inductance between two phases of the stator; and
- M_{Rf} is the mutual inductance between the damper winding of the main axis and the field winding.

The other parameters, which are defined in Chapter 1, are:

- M, the maximum value of the inductance between a phase of the stator and a damper winding;

– M_f, the maximum value of the inductance between a phase of the stator and the field winding;

– L_R, the self-inductance of a damper winding.

Figure 2.5 shows the model considered.

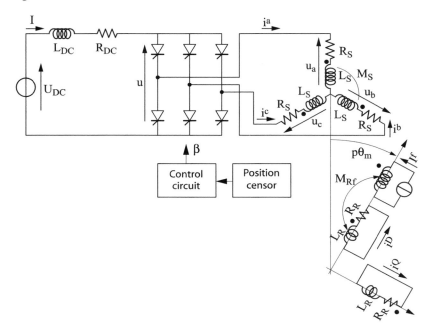

Figure 2.5. *Detailed model considered (the mutual inductances between the rotor and the stator have not been shown)*

Since the resistances of the rotoric circuits are small, fluxes ψ_D and ψ_Q vary slowly. Currents i^D and i^Q, in contrast, vary quickly. It is therefore interesting to eliminate these currents in function of the corresponding fluxes in equations [2.20]. Knowing that the sum of the phase currents is zero, we thus obtain the three following equations:

$$\psi_a = L_C\, i^a + \psi_{DC} \cos(p\, \theta_m) - \psi_{QC} \sin(p\, \theta_m) \qquad [2.22a]$$

$$\psi_b = L_C\, i^b + \psi_{DC} \cos(p\, \theta_m - \frac{2\pi}{3}) - \psi_{QC} \sin(p\, \theta_m - \frac{2\pi}{3}) \qquad [2.22b]$$

Dynamic Model of a Self-controlled Synchronous Motor 43

$$\psi_c = L_C \ i^c + \psi_{DC} \cos(p \ \theta_m + \frac{2\pi}{3}) - \psi_{QC} \sin(p \ \theta_m + \frac{2\pi}{3}) \quad [2.22c]$$

where:

$$L_C = L_S - M_S - \frac{3}{2} \frac{M^2}{L_R} \quad [2.23]$$

is the parameter already defined in equation [1.2], taking into account that [GRE 01]:

$$L_{CS} = L_S - M_S \quad [2.24]$$

and where:

$$\psi_{DC} = (M_f - \frac{M \ M_{Rf}}{L_R}) \ I^f + \frac{M}{L_R} \psi_D \quad [2.25a]$$

$$\psi_{QC} = \frac{M}{L_R} \psi_Q \quad [2.25b]$$

Assuming that the fluxes given by equations [2.25] are constant, and hence equal to their mean value (from now on, we will refer to mean values with capital letters), by taking the time derivative of [2.22] and comparing the result obtained from equations [1.6] and Figure 1.5, E_C, ψ_C and δ may be redefined as follows:

$$E_C = p \ \omega_m \sqrt{\Psi_{DC}^2 + \Psi_{QC}^2} \quad [2.26]$$

$$\psi_C = \pi - arc \ tg \frac{\Psi_{DC}}{\Psi_{QC}} \quad [2.27]$$

and, taking into account equation [2.6]:

$$\delta = arc \ tg \frac{\Psi_{QC}}{\Psi_{DC}} \quad [2.28]$$

The calculation of fluxes Ψ_D and Ψ_Q is important since they intervene, via fluxes [2.25], in expression [2.22] of the armature fluxes and, if we choose functions [2.13], in the calculations of these functions via angle δ (in equation [2.28]).

44   Control of Non-conventional Synchronous Motors

By substituting equation [2.2] in expression [2.21] of fluxes ψ_D and ψ_Q, we can write:

$$\psi_D = M \ X \ I + L_R \ i^D + M_{Rf} \ I^f \quad [2.29a]$$

$$\psi_Q = -M \ Y \ I + L_R \ i^Q \quad [2.29b]$$

where we have set:

$$X = N^a \cos(p \ \theta_m) + N^b \cos(p \ \theta_m - \frac{2\pi}{3}) + N^c \cos(p \ \theta_m + \frac{2\pi}{3}) \quad [2.30a]$$

$$Y = N^a \sin(p \ \theta_m) + N^b \sin(p \ \theta_m - \frac{2\pi}{3}) + N^c \sin(p \ \theta_m + \frac{2\pi}{3}) \quad [2.30b]$$

By substituting equations [2.29] in [2.25], we obtain:

$$\psi_{DC} = \frac{M^2}{L_R} \ X \ I + M_f \ I^f + M \ i^D \quad [2.31a]$$

$$\psi_{QC} = -\frac{M^2}{L_R} \ Y \ I + M \ i^Q \quad [2.31b]$$

In equations [2.29] and [2.31], X, Y, i^D and i^Q are the only variables that significantly vary during the interval defined by equation [2.16]. By making the approximation that the other variables are constant during this interval, we can switch to mean values, which leads:

$$\Psi_D = M \ <X> \ I + L_R \ I^D + M_{Rf} \ I^f \quad [2.32a]$$

$$\Psi_Q = -M \ <Y> \ I + L_R \ I^Q \quad [2.32b]$$

$$\Psi_{DC} = \frac{M^2}{L_R} <X> I + M_f \ I^f + M \ I^D \quad [2.33a]$$

$$\Psi_{QC} = -\frac{M^2}{L_R} <Y> I + M \ I^Q \quad [2.33b]$$

Dynamic Model of a Self-controlled Synchronous Motor   45

To obtain an equation of the form [2.19] for the armature circuit, we need to define the flux of this circuit. We obtain this flux by taking the sum of phase fluxes [2.22] weighted by coefficients N^k. The choice of this definition of the armature flux will be justified later on by the forms of macroscopic equations that follow on from it. By again introducing equation [2.2], we thus obtain:

$$\psi = L_C \; Z \; I + X \; \psi_{DC} - Y \; \psi_{QC} \qquad [2.34]$$

where we have defined:

$$Z = (N^a)^2 + (N^b)^2 + (N^c)^2 \qquad [2.35]$$

The switch to mean values made in equation [2.34] leads to the macroscopic equation of the armature flux. We will refer to $<X>$, $<Y>$ and $<Z>$ as the mean values of coefficients X, Y and Z. By adding the smoothing inductance L_{DC}, shown in Figure 2.5, in the armature circuit, we get:

$$\Psi = [L_{DC} + L_C <Z>] \; I + <X> \Psi_{DC} - <Y> \Psi_{QC} \qquad [2.36]$$

Introducing L_{DC} in equation [2.36] allows us to take into account the fact that this inductance is not infinite in the dynamic equations.

We can write the armature flux as a function of the currents by introducing [2.33] therein. We get:

$$\Psi = [L_{DC} + L_C <Z> + \frac{M^2}{L_R}(<X>^2 + <Y>^2)] \; I \\ + M_f <X> I^f + M <X> I^D - M <Y> I^Q \qquad [2.37]$$

The expression of the field winding flux can be found without difficulty:

$$\Psi_f = M_f <X> I + L_f \; I^f + M_{Rf} \; I^D \qquad [2.38]$$

Coefficients $<X>$, $<Y>$ and $<Z>$, present in equations [2.37], [2.38] and [2.32], depend on ρ, μ and ν alone. These coefficients are calculated assuming that ρ, μ and ν are constant during the averaging interval defined by equation [2.16]. Their analytical expression is given in the appendix for the choice of functions N^k given by equation [2.13]. Let us notice here and now that $<Z>$ normally takes a value close to 2. Indeed, taking into account the definition [2.35] of $<Z>$, equation [2.18] and assuming that $|N^k| \leq 1$, we know that, during the interval defined by equation

46   Control of Non-conventional Synchronous Motors

[2.16], Z varies between 2 and a minimum value of $(1/2)^2 + 1 + (1/2)^2 = 1.5$. Moreover, during this interval, except during the overlap interval which is defined by [2.14], Z keeps the constant value 2. Its mean value will therefore normally be slightly below 2.

2.4. General properties of coefficients $<X>$, $<Y>$ and $<Z>$

Before developing the model further, it is useful to establish a few properties of the coefficients that can be found therein. We are going to establish these properties by assuming that equations [2.17] and [2.18] are fulfilled. This is more general than assuming that coefficients N^k are given by equation [2.13].

Returning to definition [2.30b], we have:

$$\begin{aligned}
Y &= N^a \sin(p\theta_m) + N^b \sin(p\theta_m - \frac{2\pi}{3}) + N^c \sin(p\theta_m + \frac{2\pi}{3}) \\
&= -[N^a \frac{\partial \cos(p\theta_m)}{\partial(p\theta_m)} + N^b \frac{\partial \cos(p\theta_m - \frac{2\pi}{3})}{\partial(p\theta_m)} + N^c \frac{\partial \cos(p\theta_m + \frac{2\pi}{3})}{\partial(p\theta_m)}] \\
&= -\frac{\partial}{\partial(p\theta_m)}[N^a \cos(p\theta_m) + N^b \cos(p\theta_m - \frac{2\pi}{3}) + N^c \cos(p\theta_m + \frac{2\pi}{3})] \\
&\quad + \frac{\partial N^a}{\partial(p\theta_m)} \cos(p\theta_m) + \frac{\partial N^b}{\partial(p\theta_m)} \cos(p\theta_m - \frac{2\pi}{3}) + \frac{\partial N^c}{\partial(p\theta_m)} \cos(p\theta_m + \frac{2\pi}{3})
\end{aligned}$$
[2.39]

The first term of equation [2.39], being the derivative of a periodic function, has a mean of 0. Hence, using equation [2.17b], we obtain:

$$<Y> = <\frac{\partial X}{\partial \rho}> \quad [2.40]$$

namely:

$$<Y> = \frac{\partial <X>}{\partial \rho} \quad [2.41a]$$

A similar development leads to:

$$<X> = -\frac{\partial <Y>}{\partial \rho} \quad [2.41b]$$

A consequence of equations [2.41] is that:

$$\frac{\partial}{\partial \rho}(<X>^2 + <Y>^2)$$
$$= 2<X>\frac{\partial <X>}{\partial \rho} + 2<Y>\frac{\partial <Y>}{\partial \rho} = 0 \qquad [2.42]$$

From equations [2.41], we can infer that $<X>$ and $<Y>$ are sinusoidal functions of variable ρ, which are identical to within a phase shift of about 90°.

Moreover, by using definition [2.35] we have:

$$\frac{\partial <Z>}{\partial \rho} = <\frac{\partial Z}{\partial \rho}> = <2\sum_k N^k \frac{\partial N^k}{\partial \rho}> \qquad [2.43]$$

namely, taking into account equation [2.17b]:

$$\frac{\partial <Z>}{\partial \rho} = <2\sum_k N^k \frac{\partial N^k}{\partial \theta_{em}}> = <\frac{\partial Z}{\partial \theta_{em}}> \qquad [2.44]$$

Since Z is a periodic function of θ_{em}, the mean value of its derivative is 0. We thus obtain:

$$\frac{\partial <Z>}{\partial \rho} = 0 \qquad [2.45]$$

When the overlap interval μ tends to 0, we can calculate:

$$<X> = \frac{3\sqrt{3}}{\pi}\cos\rho \qquad [2.46a]$$

$$<Y> = -\frac{3\sqrt{3}}{\pi}\sin\rho \qquad [2.46b]$$

$$<Z> = 2 \qquad [2.46c]$$

Equations [2.46a-c] fulfill the properties demonstrated in this section.

2.5. Electrical dynamic equations

Equations [2.32] and [2.37] can be used to calculate I, I^D and I^Q at each time. To describe the evolution of fluxes Ψ_D and Ψ_Q, we obtain equations of the form [2.19]. Since the damper windings are in short-circuit, their voltage is 0. Their emf is purely ohmic, since they are filiform circuits. We hence obtain:

$$\frac{d\Psi_D}{dt} = -R_R\,I^D \qquad [2.47a]$$

$$\frac{d\Psi_Q}{dt} = -R_R\,I^Q \qquad [2.47b]$$

To retain the formalism of equation [2.19], the evolution of Ψ must be determined by an equation of the form:

$$\frac{d\Psi}{dt} = U_{DC} - E \qquad [2.48]$$

where U_{DC} is the supply voltage. E is a mean emf whose expression will be determined. The expression of the left-hand side of [2.48] is easily inferred from [2.37]. It remains to obtain an analytical expression of the voltage U_{DC} to be able to infer that of the emf from equation [2.48].

The different armature phases are filiform circuits (according to the meaning given in [MAT 04]). We therefore obtain their voltage by taking the derivative of their flux (equation [2.22]) and adding the ohmic voltage drops. We thus have:

$$u_a = L_C\frac{d\,i^a}{dt} - p\,\omega_m\,\psi_{DC}\sin(p\,\theta_m) - p\,\omega_m\,\psi_{QC}\cos(p\,\theta_m)$$
$$+\dot{\psi}_{DC}\cos(p\,\theta_m) - \dot{\psi}_{QC}\sin(p\,\theta_m) + R_S\,i^a \qquad [2.49a]$$

$$u_b = L_C\frac{d\,i^b}{dt} - p\,\omega_m\,\psi_{DC}\sin(p\,\theta_m-\frac{2\pi}{3}) - p\,\omega_m\,\psi_{QC}\cos(p\,\theta_m-\frac{2\pi}{3})$$
$$+\dot{\psi}_{DC}\cos(p\,\theta_m-\frac{2\pi}{3}) - \dot{\psi}_{QC}\sin(p\,\theta_m-\frac{2\pi}{3}) + R_S\,i^b \qquad [2.49b]$$

$$u_c = L_C\frac{d\,i^c}{dt} - p\,\omega_m\,\psi_{DC}\sin(p\,\theta_m+\frac{2\pi}{3}) - p\,\omega_m\,\psi_{QC}\cos(p\,\theta_m+\frac{2\pi}{3})$$
$$+\dot{\psi}_{DC}\cos(p\,\theta_m+\frac{2\pi}{3}) - \dot{\psi}_{QC}\sin(p\,\theta_m+\frac{2\pi}{3}) + R_S\,i^c \qquad [2.49c]$$

Dynamic Model of a Self-controlled Synchronous Motor 49

The armature voltage is given by the sum of these three voltages weighted by coefficients N^k. It should be noted that the value of the voltages given by equations [2.49] for the phases under commutation are not necessarily equal, as they would be if we used the exact expressions of coefficients N^k. This is not a problem with the above-mentioned way of computing the armature voltage. By using equations [2.30] and [2.35], we thus obtain the armature voltage:

$$u = p\,\omega_m\, L_C\, \frac{1}{2}\frac{\partial Z}{\partial \theta_{em}} I + \dot{\rho}\frac{1}{2} L_C \frac{\partial Z}{\partial \rho} I$$
$$+ \dot{\mu}\frac{1}{2} L_C \frac{\partial Z}{\partial \mu} I + \dot{v}\frac{1}{2} L_C \frac{\partial Z}{\partial v} I$$
$$+ L_C Z \frac{dI}{dt} - p\,\omega_m\, \psi_{DC}\, Y - p\,\omega_m\, \psi_{QC}\, X$$
$$+ \dot{\psi}_{DC}\, X - \dot{\psi}_{QC}\, Y + R_S\, Z\, I$$

[2.50]

By switching to mean values, the first term in equation [2.50] will disappear since, Z being a periodic function of θ_{em}, the mean of its derivative is 0. The second term will be canceled out by equation [2.45]. The following five terms, as well as the last one, have only one rapidly variable factor, so switching to mean values does not represent any difficulty. Even if the fluxes vary slowly, the case is not the same for their derivatives. These are linked to ohmic voltage drops $R_R i^D$ and $R_R i^Q$, which have a rapidly variable part.

To find the expression of currents i^D and i^Q, we can calculate the difference between equations [2.29] and [2.32], then disregard differences $\psi_D - \Psi_D$ and $\psi_Q - \Psi_Q$. We thus obtain:

$$i^D - I^D = -\frac{1}{L_R} M\,(X - <X>)\, I$$

[2.51a]

$$i^Q - I^Q = \frac{1}{L_R} M\,(Y - <Y>)\, I$$

[2.51b]

By switching to the time derivative of the difference between equations [2.29] and [2.32], using equations [2.51] we obtain:

$$\dot{\psi}_D = \dot{\Psi}_D + \frac{R_R}{L_R} M (X - <X>) I \qquad [2.52a]$$

$$\dot{\psi}_Q = \dot{\Psi}_Q - \frac{R_R}{L_R} M (Y - <Y>) I \qquad [2.52b]$$

and, by [2.25],

$$\dot{\psi}_{DC} = \dot{\Psi}_{DC} + \frac{R_R M^2}{L_R^2} (X - <X>) I \qquad [2.53a]$$

$$\dot{\psi}_{QC} = \dot{\Psi}_{QC} - \frac{R_R M^2}{L_R^2} (Y - <Y>) I \qquad [2.53b]$$

By substituting equation [2.53] in [2.50], then switching to mean values, by adding the armature circuit inductance L_{DC} and resistance R_{DC} of the smoothing inductor, we get:

$$\begin{aligned} U_{DC} = & [L_{DC} + L_C <Z>]\frac{dI}{dt} + <X> \dot{\Psi}_{DC} - <Y> \dot{\Psi}_{QC} \\ & - p\omega_m <Y> \Psi_{DC} - p\omega_m <X> \Psi_{QC} \\ & + \dot{\mu}\frac{1}{2} L_C \frac{\partial <Z>}{\partial \mu} I + \dot{v}\frac{1}{2} L_C \frac{\partial <Z>}{\partial v} I \\ & + [R_{DC} + R_S <Z> + R_{add}] I \end{aligned} \qquad [2.54]$$

where:

$$R_{add} = R_R \frac{M^2}{L_R^2} (<X^2> + <Y^2> - <X>^2 - <Y>^2) \qquad [2.55]$$

which can also be written by using equation [2.18] [MAT 11]:

$$R_{add} = R_R \frac{M^2}{L_R^2}(\frac{3}{2}<Z>-<X>^2-<Y>^2) \qquad [2.56]$$

Expression [2.54] can be broken down into one term of the form $d\Psi/dt$ and one emf term. By subtracting the time derivative of flux Ψ computed using [2.36] from equation [2.54], and by using general properties [2.41] and [2.43], we obtain the expression of the emf:

$$\begin{aligned} E = &- (p\,\omega_m + \dot{\rho})\,[<Y>\Psi_{DC} + <X>\Psi_{QC}] \\ &- \dot{\mu}\,[\frac{1}{2}L_C\frac{\partial <Z>}{\partial \mu}I + \frac{\partial <X>}{\partial \mu}\Psi_{DC} - \frac{\partial <Y>}{\partial \mu}\Psi_{QC}] \\ &- \dot{v}\,[\frac{1}{2}L_C\frac{\partial <Z>}{\partial v}I + \frac{\partial <X>}{\partial v}\Psi_{DC} - \frac{\partial <Y>}{\partial v}\Psi_{QC}] \\ &+ [R_{DC} + R_S<Z>+R_{add}]\,I \end{aligned} \qquad [2.57]$$

Expression [2.57] does not involve any time derivative of the electrical variables, which justifies *ex post facto* the choice of macroscopic expressions of flux and voltage. At each step we can thus evaluate the derivatives of electromechanical variables ρ, μ and v, calculate the value of E by using [2.57] and introduce it in equation [2.48] to obtain the evolution of the armature flux.

We will notice that emf (equation [2.57]) is not restricted to the ohmic term. The armature circuit is not a filiform circuit (according to the meaning given in [MAT 04]).

The first term of [2.57] is a sliding term: it cancels out if $d\rho/dt = -p\,\omega_m$, i.e. when the armature circuit (the equivalent circuit in which current I is assumed to flow), does not move with respect to the stator windings. The second and third terms are non-lineic terms, which take into account variations in the form of the armature circuit. The fourth term is an ohmic term.

2.6. Expression of electromechanical variables

The electrical equations developed in the previous sections are only useful if we add a calculation procedure of electromechanical variables to them, i.e. of angle ρ and, with the expression of N^* introduced in section 2.2, angles μ and v. The macroscopic model alone does not allow us to obtain the expressions of the electromechanical variables: it is necessary to refer to a detailed model. Therefore in this section we will sometimes refer to the approximate analysis in section 2.3.

In order to determine angles ρ, μ and ν we need to have three equation functions of these angles (or the angle defined in Figure 2.3, which amounts to the same thing taking into account the relationships that exist between the two sets of angles) at our disposal. Relation [2.28] is a first constraint on the electromechanical variables. Equation [1.10] gives an approximate value of a second constraint, which can be written as:

$$\cos\xi - \cos\alpha = \frac{2L_C}{\sqrt{3}\sqrt{\Psi_{DC}^2 + \Psi_{QC}^2}} I \qquad [2.58]$$

There is therefore only one degree of freedom left on angles ρ, ν and μ, which can only be set by the control system of the thyristors. This control system can only act on the turn on times of the thyristors, i.e. on angle β (the firing angle of thyristor T_{a1} measured with respect to the axis of the field winding). For the purpose of analysis, we will infer the value of angle α from the value of β, by using the value of δ given by equation [2.28]:

$$\alpha = \beta - \frac{5\pi}{6} - \delta \qquad [2.59]$$

The value of ξ is then easy to obtain by using equation [2.59], then that of μ by using equation [2.3]. We can then calculate ν using an expression inferred from equations [2.4] and [2.12]:

$$\nu = \alpha + \frac{\pi}{2} - \frac{\mu}{2} \qquad [2.60]$$

and eventually ρ by expression [2.12].

In addition to the fact that angles ρ, ν and μ must be computed to determine the coefficients defined in section 2.3, they also serve to verify that the conditions of good commutation of the thyristors are fulfilled, i.e.:

$$\xi > 0 \qquad [2.61a]$$

$$\alpha > \xi \qquad [2.61b]$$

and:

$$\mu < 60° \tag{2.61c}$$

constraints that must be verified with a safety margin in order to take into account the imperfections of the modeling and analysis.

We can also set the value of ξ and calculate all the angles, including the control angle β of thyristors, by following a procedure that is the inverse of the one described above. However, this type of control can lead to instabilities because, when we increase the current I in order to increase the power converted, angle μ and hence ρ are also going to increase. This decreases the value of the sliding emf and hence the effect of the current increase.

Other control strategies can also be considered. For instance, we can set the value of ρ and infer the value of ν from it using equation [2.12]. The value of μ is then obtained by expression [2.58], which can be re-written in the form:

$$I = \sqrt{3} \frac{\sqrt{\Psi_{DC}^2 + \Psi_{QC}^2}}{L_C} \cos(\rho - \delta) \sin\frac{\mu}{2} \tag{2.62}$$

namely:

$$\mu = 2 \arcsin \frac{L_C I}{\sqrt{3}\sqrt{\Psi_{DC}^2 + \Psi_{QC}^2} \cos(\nu)} \tag{2.63}$$

The other angles are easily obtained, in particular ξ.

2.7. Expression of torque

In the case of filiform circuit systems (according to the meaning given in [MAT 04]) without magnetic saturation, the electromagnetic torque is given by:

$$c_{em} = \sum_{k,m} \frac{1}{2} \frac{\partial L_{k\,m}}{\partial \theta_m} i^k i^m \tag{2.64}$$

which in our case is:

$$c_{em} = -p \begin{bmatrix} \sin(p\theta_m) \ i^a \\ +\sin(p\theta_m - \frac{2\pi}{3}) \ i^b \\ +\sin(p\theta_m + \frac{2\pi}{3}) \ i^c \end{bmatrix} (M_f I^f + M \ i^D)$$

$$-p \begin{bmatrix} \cos(p\theta_m) \ i^a \\ +\cos(p\theta_m - \frac{2\pi}{3}) \ i^b \\ +\cos(p\theta_m + \frac{2\pi}{3}) \ i^c \end{bmatrix} M \ i^Q \qquad [2.65]$$

By using equation [2.2] and definition [2.30], this expression becomes:

$$c_{em} = -pYI(M_f I^f + M i^D) - pXIM i^Q \qquad [2.66]$$

By expressing i^D and i^Q as functions of ψ_{DC} and ψ_{QC}, by using [2.31], we obtain:

$$c_{em} = [-pY \ \Psi_{DC} \ -p \ X \ \Psi_{QC}] \ I \qquad [2.67]$$

or, by switching to the mean value of the torque:

$$C_{em} = [-p<Y> \ \Psi_{DC} \ -p \ <X> \ \Psi_{QC}] \ I \qquad [2.68]$$

The comparison of equations [2.54] and [2.68] shows that, in steady state, the mechanical power is equal to the electrical power to roughly within the ohmic losses.

2.8. Writing of equations in terms of coenergy

By integrating equation system [2.37], [2.38] and [2.32], we obtain the expression of macroscopic coenergy:

$$w_{cm} = \frac{1}{2}[L_{DC} + L_C <Z> + \frac{M^2}{L_R}(<X>^2 + <Y>^2)] \, I^2$$
$$+ \frac{1}{2} L_f \, (I^f)^2 + \frac{1}{2} L_R \, (I^D)^2 + \frac{1}{2} L_R \, (I^Q)^2$$
$$+ M_f <X> I \, I^f + M \, I \, (<X> I^D - <Y> I^Q) + M_{Rf} \, I^f \, I^D$$
[2.69]

The partial derivative of [2.69] with respect to current I gives the flux (equation [2.38]). We can therefore write expression [2.40] in the form:

$$U_{DC} = \frac{d}{dt} \frac{\partial w_{cm}}{\partial I} + E \qquad [2.70]$$

Let us now consider the partial derivatives of [2.69] with respect to electromechanical variables ρ, μ and ν. By using equation [2.33], we have:

$$\frac{\partial w_{cm}}{\partial \rho} = \frac{1}{2} L_C \frac{\partial <Z>}{\partial \rho} I^2 + \frac{M^2}{L_R}(<X> \frac{\partial <X>}{\partial \rho} + <Y> \frac{\partial <Y>}{\partial \rho}) I^2$$
$$+ M_f \frac{\partial <X>}{\partial \rho} I I^f + M I (\frac{\partial <X>}{\partial \rho} I^D - \frac{\partial <Y>}{\partial \rho} I^Q)$$
$$= \frac{1}{2} L_C \frac{\partial <Z>}{\partial \rho} I^2$$
$$+ \frac{\partial <X>}{\partial \rho}[\frac{M^2}{L_R}<X> I + M_f I^f + M I^D] I \qquad [2.71]$$
$$- \frac{\partial <Y>}{\partial \rho}[-\frac{M^2}{L_R}<Y> I + M I^D] I$$
$$= \frac{1}{2} L_C \frac{\partial <Z>}{\partial \rho} I^2 + \frac{\partial <X>}{\partial \rho} \Psi_{DC} I - \frac{\partial <Y>}{\partial \rho} \Psi_{QC} I$$

By using equations [2.41] and [2.45], we can infer that:

$$\frac{\partial w_{cm}}{\partial \rho} = <Y> \Psi_{DC} I + <X> \Psi_{QC} I \qquad [2.72a]$$

In a similar way to equation [2.71], we also obtain:

$$\frac{\partial w_{cm}}{\partial \mu} = \frac{1}{2} L_C \frac{\partial <Z>}{\partial \mu} I^2 + \frac{\partial <X>}{\partial \mu} \Psi_{DC} I - \frac{\partial <Y>}{\partial \mu} \Psi_{QC} I \qquad [2.72b]$$

and:

$$\frac{\partial w_{cm}}{\partial \nu} = \frac{1}{2} L_C \frac{\partial <Z>}{\partial \nu} I^2 + \frac{\partial <X>}{\partial \nu} \Psi_{DC} I - \frac{\partial <Y>}{\partial \nu} \Psi_{QC} I \qquad [2.72c]$$

We can hence write equation [2.57] in the form:

$$E = -(p\omega_m + \dot{\rho}) \frac{1}{I} \frac{\partial w_{cm}}{\partial \rho} - \dot{\mu} \frac{1}{I} \frac{\partial w_{cm}}{\partial \mu} - \dot{\nu} \frac{1}{I} \frac{\partial w_{cm}}{\partial \nu}$$
$$+ [R_{DC} + R_S <Z> + R_{add}] I \qquad [2.73]$$

Similarly, torque expression [2.68] can be written as:

$$C_{em} = -p \frac{\partial w_{cm}}{\partial \rho} \qquad [2.74]$$

The derivatives of w_{cm} with respect to variables μ and ν do not appear in equation [2.74], and in equation [2.73] they are not multiplied by rotational speed ω_m. According to the terminology introduced by Garrido [MAT 04], variables μ and ν are non-lineic.

2.9. Application to control

For a determination of angles ρ, μ and ν in real time, some variables involved in their computation are not measurable, in particular those relating to the damper windings. It is therefore necessary to estimate these variables, which can be done using the equations in this chapter. Figure 2.6 shows the main steps of this calculation in the case where, further to a measure of the rotor position, we have a measure of the armature current. Considering Ψ_D and Ψ_Q as state variables, knowing their value at one time (the preceding time) we can calculate their derivatives and hence estimate their value at the following time (the present time). Figure 2.6 also shows how to calculate the electromechanical variables and update the values of coefficients <X>, <Y>…

Dynamic Model of a Self-controlled Synchronous Motor   57

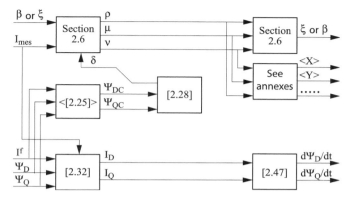

Figure 2.6. *Analysis of variables and coefficients at a given time, where we have a measure of the armature current*

We can improve the precision of the algorithm by using a predictor-corrector method. For this purpose, the calculation of $d\Psi_D/dt$ and $d\Psi_Q/dt$ is repeated using as values of I^D and I^Q the mean of their intial and final values at the step considered.

In the absence of a current sensor in the armature circuit, we can modify the Figure 2.6 to include an estimation of this current, which leads to the diagram given in Figure 2.7. The time derivatives of ρ, μ and ν must be estimated in order to use equation [2.57]. This is done by calculating the difference between the final and initial values of these variables at the step considered. The supply voltage is imposed and the rotational speed is measured.

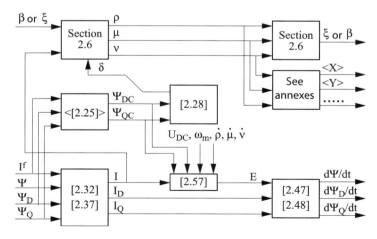

Figure 2.7. *Analysis of the variables and coefficients at a given time where we do not use a measure of armature current*

58   Control of Non-conventional Synchronous Motors

We have applied the diagram in Figure 2.7 to the simulation of the transient operation following an abrupt variation of the supply voltage, by calculating the rotational speed from torque [2.68]. The program used is available on http://perso.uclouvain.be/ernest.matagne/glissant/index.htm [MAT 11]. The data relating to the system are inspired from the literature [KAT 81] relating to a motor of 2.2 kW, four poles, 200 V, 8.0 A, 50 Hz but assuming a symmetry of the direct and quadrature axes (the initial reference is about a salient pole motor). We have decreased the resistance of the damper windings to take account of damping due to the field winding, which is not considered here or in Chapter 1 since the field winding is assumed therein to be supplied by a current source. The electrical parameters chosen for the simulation are:

$L_{DC} = 0.217$ H $\qquad R_{DC} = 0.88\ \Omega \qquad R_S = 0.5\ \Omega$

$L_C = 6.11$ mH $\qquad M = L_R = 0.0109$ H $\qquad R_R = 0.02\ \Omega$

$M_f I^f = 0.4922$ Wb

The mechanical parameters, inertia, viscous friction coefficient and constant antagonist torque, have a respective values of:

$J = 0.0936$ kg m² $\quad R_\omega = 0.018$ Nm/(rad/s) $\quad C_{mec.} = 2.03$ Nm

The voltage jump is calculated in order to switch from a speed of 1,146 rpm to a speed of 1,590 rpm.

A first simulation is done for an angle ξ initially corresponding to 49.5°. The values of the other variables for the initial working conditions are: $\mu = 6.10°$, $\beta = 212°$ and I = 5 A. The supply voltage goes from $U_{dc} = 110$ V to $U_{dc} = 153.5$ V at the initial time.

Figure 2.8 shows the evolution of the main variables during the transient interval.

The evolution of variables described in Figure 2.8 is close to the experimental reading carried out in the same conditions [KAT 81].

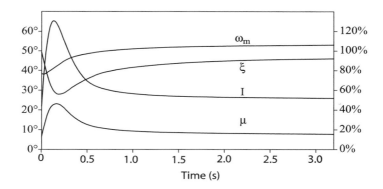

Figure 2.8. *Evolution of main variables after an abrupt variation in voltage U at $\beta_{ref} = 212°$. We have considered that 1 p.u. of speed = 1,500 rpm, and 1 p.u. of voltage = 163.3 V*

By imposing a smaller value of β, we can improve the efficiency of the system: the initial value of the armature current is smaller and hence the Joule losses in the armature and the smoothing inductor are also smaller. Unfortunately, we quickly reach a situation where angle ξ cancels out during the transient phase, which makes it impossible to turn off the thyristors. We can, however, decrease angle β provided that we change the calculation mode of the angles when ξ takes a value that is too small. The simulation represented in Figure 2.9 uses this possibility, by setting the value of ξ to 10° when it tends to go below this value. In this simulation, the initial and final speeds are the same as previously. The reference value of β is decreased to 201.3°, which corresponds to an initial angle ξ of 40°. The supply voltage undergoes an abrupt jump from 136.2 V to 189.7 V. The initial current is decreased to 3.91 A.

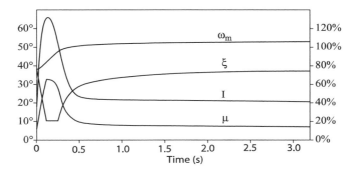

Figure 2.9. *Evolution during an abrupt variation of voltage U at $\beta_{ref} = 201.3°$. We have considered that 1 p.u. of speed = 1,500 rpm, and 1 p.u. of voltage = 163.3 V*

60   Control of Non-conventional Synchronous Motors

As we can see in Figure 2.9, the armature current tends towards a smaller value, which decreases the Joule heating losses.

2.10. Conclusion

In this chapter, we have extended the study of the functioning of the self-controlled synchronous machine from Chapter 1 to take into account the influence of the overlap angle. For this purpose we have applied a general method of definition of circuits to the definition of the armature circuit. This method has led to a dynamic model that takes into account the behavior of the damper windings.

The model obtained has three electromechanical variables. The first, ρ, is a sliding variable defining the position of the armature circuit (which can be represented by the position of the fictitious brushes). The following two, μ and ν, are non-lineic variables according to the meaning of the sliding circuit theory developed by Garrido. Variable μ can be represented as the width of the fictitious brushes, whereas variable ν describes the current form during overlap intervals.

We have presented the theory in the form where the flux of the armature circuit and the fluxes of the damper windings are electrical state variables. This model is sufficiently simple to be used within the framework of control and regulation systems. Finally, we have shown that this model can be put in a Lagrangian form, which opens the way to taking account of magnetic saturation.

2.11. Appendix 1: value of coefficients <X>, <Y> and <Z>

In this and the following appendices, we give exact expressions of coefficients <X>, <Y> and <Z> and their derivatives with respect to μ and ν for the choice of N^k (in equations [2.13] and [2.15]). If the available means of calculation are not sufficient, it is possible to use simplified versions of these expressions, for instance by setting the value of variable ν to 180°.

After a few tedious but easy calculations [MAT 11], for the coefficients corresponding to the choices of functions N^k given in equations [2.13] and [2.15], we obtain the following expressions:

$$<X> = \frac{3\sqrt{3}}{2\pi \cos(\nu)} \left\{ \cos(\rho-\nu)\frac{\mu/2}{\sin(\mu/2)} + \cos(\rho+\nu)\cos(\frac{\mu}{2}) \right\} \quad [2.75]$$

$$<Y> = -\frac{3\sqrt{3}}{2\pi \cos(v)} \{\sin(\rho-v)\frac{\mu/2}{\sin(\mu/2)} + \sin(\rho+v)\cos(\frac{\mu}{2})\} \qquad [2.76]$$

and:

$$<Z> = 2 - \frac{3}{4\pi \sin^2\frac{\mu}{2}} \begin{cases} tg^2(v)[3\sin(\mu) - 2\mu\cos^2(\frac{\mu}{2}) - \mu] \\ + [\sin\mu + 2\mu\sin^2(\frac{\mu}{2}) - \mu] \end{cases} \qquad [2.77]$$

From [2.75] and [2.76], we can calculate:

$$<X>^2 + <Y>^2 = \frac{27}{4\pi^2} \begin{cases} [(\frac{\mu/2}{\sin(\mu/2)})^2 + \cos(\mu)]^2 \\ + tg^2 v[(\frac{\mu/2}{\sin(\mu/2)})^2 - \cos(\mu)]^2 \end{cases} \qquad [2.78]$$

2.12. Appendix 2: derivatives of coefficients <X>, <Y> and <Z>

In addition to their derivatives with respect to ρ given by equations [2.41], [2.42] and [2.46], from the expressions in Appendix 1 we can obtain:

$$\frac{\partial <X>}{\partial \mu} = \frac{3\sqrt{3}}{4\pi \cos v} \begin{cases} \cos(\rho-v)\ [\sin(\frac{\mu}{2}) - \frac{\mu}{2}\cos(\frac{\mu}{2})]/\sin^2(\frac{\mu}{2}) \\ \qquad - \cos(\rho+v)\ \sin(\frac{\mu}{2}) \end{cases} \qquad [2.79]$$

$$\frac{\partial <X>}{\partial v} = \frac{3\sqrt{3}}{2\pi} \frac{\sin \rho}{\cos^2 v} [\frac{\mu/2}{\sin(\mu/2)} - \cos(\frac{\mu}{2})] \qquad [2.80]$$

$$\frac{\partial <Y>}{\partial \mu} = -\frac{3\sqrt{3}}{4\pi \cos v} \begin{cases} \sin(\rho-v)\ [\sin(\frac{\mu}{2}) - \frac{\mu}{2}\cos(\frac{\mu}{2})]/\sin^2(\frac{\mu}{2}) \\ \qquad - \sin(\rho+v)\ \sin(\frac{\mu}{2}) \end{cases}$$

$$[2.81]$$

$$\frac{\partial <Y>}{\partial v} = \frac{3\sqrt{3}}{2\pi} \frac{\cos \rho}{\cos^2 v} [\frac{\mu/2}{\sin(\mu/2)} - \cos(\frac{\mu}{2})] \qquad [2.82]$$

$$\frac{\partial <Z>}{\partial \mu} = -\frac{3}{4\pi} \left\{ \begin{array}{l} tg^2 v \ [-4\sin(\frac{\mu}{2}) - 2 \ \cos^2(\frac{\mu}{2}) \ \sin(\frac{\mu}{2}) + 3 \ \mu \ \cos(\frac{\mu}{2})]/\sin^3(\frac{\mu}{2}) \\ +[-2 \ \sin(\frac{\mu}{2}) + 2\sin^3(\frac{\mu}{2}) + \mu \ \cos(\frac{\mu}{2}) \ ]/\sin^3(\frac{\mu}{2}) \end{array} \right\}$$

[2.83]

$$\frac{\partial <Z>}{\partial v} = -\frac{3}{2\pi} \frac{\sin v}{\cos^3 v} \ [3 \ \sin \ \mu \ - \ \mu \ - \ 2\mu \ \cos^2 \frac{\mu}{2}] \ /\sin^2 \frac{\mu}{2} \quad [2.84]$$

2.13. Appendix 3: simplifications for small μ

When the overlap interval μ is small, the expressions in Appendix 1, limited to the first order, are exactly [2.46a] and [2.46b], as well as:

$$<Z> = 2 - \frac{\mu}{8\pi} \quad [2.85]$$

The mutual inductance between the armature and the field winding, as it comes out of equations [2.37] and [2.38], amounts to:

$$M_f <X> = M_f \frac{3\sqrt{3}}{\pi} \cos \rho \quad [2.86]$$

which is in accordance with the value of $<M_{Af}>$ obtained in section 1.9 if we consider the correspondence between the various angles given by equation [2.4] when $\mu = 0$.

For a small μ, expression [2.78] at the first order becomes:

$$<X>^2 + <Y>^2 \approx \frac{27}{\pi^2} \quad [2.87]$$

Using equations [2.85] and [2.87], expression [2.56] of R_{add} becomes:

$$\begin{aligned} R_{add} &= R_r \frac{M^2}{L_R^2} (3 - \frac{27}{\pi^2} - \frac{3\mu}{2\pi}) \\ &= R_r \frac{M^2}{L_R^2} (0.2643_3 - 0.4774_6 \ \mu) \end{aligned} \quad [2.88]$$

Dynamic Model of a Self-controlled Synchronous Motor

2.14. Appendix 4 – List of the main symbols used in Chapters 1 and 2

Symbol	Description
$<g>$	Mean value of variable g
G	Variable of the macroscopic model corresponding to variable g of the detailed model
C_{em}	Electromagnetic torque developed by the synchronous machine
E	Total electromotive force of the armature circuit, gathering the sliding terms, the non-lineic terms and the ohmic losses (damper windings included)
emf	Electromotive force
$\overline{E}_0$	Phasor representative of the emf induced by the field winding in phase a of the stator
E_0	Amplitude of emf induced by field winding in the stator phases
e_{cx}	Commutation emf of stator phases ($x \in \{a, b, c\}$)
f	Frequency of voltages and currents
I	DC current at the input of the bridge supplying the synchronous machine
i_d or i^D	Current of the D-axis damper winding
i_q or i^Q	Current of the Q-axis damper winding
$\overline{I}_1$	Phasor representative of the fundamental component of phase currents of the stator
i_{1x}	Fundamental components of the currents in the stator phases ($x \in \{a, b, c\}$)
i_{kx}	Harmonic components of rank k in the stator currents ($x \in \{a, b, c\}$)
I_{ph1N}	Nominal amplitude of the fundamental component of phase current
i^x or i_x	Currents in stator phases ($x \in \{a, b, c\}$)
L_A	Armature inductance of the equivalent DC machine
L_C	Transient or commutation inductance
L_{CS}	Cyclical inductance of stator windings
L_{DC}	Smoothing inductance of the supply
L_f	Self-inductance of the field winding
L_R	Self-inductance of a damper winding
L_S	Self-inductance of a stator phase
M	Maximum value of the mutual inductance between a damper winding and a stator winding
M_o	$\sqrt{3}/2$ times M

M_{Ad} or M_{Aq}	Mutual inductances between the armature and the damper windings of axes d and q of the equivalent DC machine
M_{Af}	Mutual inductance between the armature and the field winding of the equivalent DC machine
M_{df} or M_{Rf}	Mutual inductance between the field winding and the damper winding of axis d
M_f	Maximum value of the mutual inductance between the field winding and a stator winding
M_S	Mutual inductance between two stator phases
N	Index indicating that it is about the nominal value of a variable
N^x	Sharing factors of current I between stator phases ($x \in \{a, b, c\}$)
p	Number of pole pairs
R_A	Armature resistance of the equivalent DC machine
R_{add}	Additional resistance to be included in the armature circuit to take into account the effect of damper windings
R_{DC}	Internal resistance of the supply, smoothing inductor included
R_f	Resistance of the field winding
R_R	Resistance of a damper winding
R_S	Resistance of a stator winding
T	Period of voltages and currents
T_C	Length of a commutation interval
T_{ij}	Thyristors of the bridge supplying the machine ($i, j \in \{a, b, c\}$)
u	Voltage at the input of the bridge supplying the synchronous machine
U	Mean value of u
U_{av}	Average value of the voltage at the input of the bridge feeding the synchronous machine
$\overline{U}_1$	Phasor representative of the fundamental component of voltages at the terminals of the stator phases
u_{1x}	Fundamental components of voltages at the terminals of the stator phases ($x \in \{a, b, c\}$)
U_{DC}	Voltage of the ideal voltage source modeling the supply
u_x	Voltage at the terminals of the stator phases ($x \in \{a, b, c\}$)
w_{cm}	Macroscopic magnetic coenergy
X	Factor allowing us to take into account the influence of the overlap interval on the features of the armature circuit along the direct axis
Y	Factor allowing us to take into account the influence of the overlap interval on the features of the armature circuit along the quadrature axis

Dynamic Model of a Self-controlled Synchronous Motor

z	Factor allowing us to take into account the influence of the overlap interval on the self-features of the armature circuit
α	Advance firing angle of thyristors with respect to field winding
β	Advance firing angle of thyristors with respect to the field winding
δ	Leading angle of commutation emf e_{cx} on emf e_{0x} induced by the field winding
θ_{em}	Electric position of the rotor of the synchronous machine
μ	Overlap angle
ν	Position angle of the brushes with respect to a zero of commutation emf
ξ	Extinction angle
ρ_S	Position of the brush linked to the + input terminal with respect to the center of the commutator segment linked to phase a
ρ	Position with respect to the field winding of the brushes of the equivalent collector-brush system
φ	Leading angle of the fundamental components of phase currents on commutation emf
ψ_x	Fluxes of stator phases (x $\in \{a, b, c\}$)
ψ_D	Flux of the D-axis damper winding
ψ_{DC}	Main flux along axis D
ψ_Q	Flux of the Q-axis damper winding
ψ_{QC}	Main flux along axis Q
ψ_C	Leading angle of emf e_{cx} with respect the zero position of the rotor
ω_m	Mechanical speed of the rotor in radians/sec

2.15. Bibliography

The short bibliography below complements the one in Chapter 1, to which this chapter often refers.

[DUF 97] DUFOUR S., Calcul numérique des paramètres externes de la machine à reluctance variable, PhD thesis of INPL, Nancy, 9 January 1997.

[GRE 01] GRENIER D., LABRIQUE F., BUYSE H., MATAGNE E., *Electromécanique, Convertisseurs d'Énergie et Actionneurs*, Dunod, Paris, France, 2001.

[KAT 81] KATAOKA T., NISHIKATA S., "Transient performance analysis of self-controlled synchronous motors", *IEEE Transactions on Industry Applications*, vol. 1A, no. 2, pp. 152-159, 1981.

[KRO 59] KRON G., *Tensor for Circuits* (Second Edition), Dover, New York, USA, 1959.

[MAT 02] MATAGNE E., "Physique interne des convertisseur électromécaniques", http://perso.uclouvain.be/ernest.matagne/ELEC2311/INDEX.HTM

[MAT 04] MATAGNE E., GARRIDO M., "Conversion électronique d'énergie: du phénomène physique à la modélisation dynamique en modélisation des machines électriques en vue de leur commande", LOUIS J-P (ed.), *Modélisation des machines électriques en vue de leur commande : concepts généraux*, Traité EGEM, Editions Hermès-Lavoisier, Paris, France, 2004.

[MAT 11] MATAGNE E., Supplement to this chapter, in: "Compléments à ce chapitre", http://perso.uclouvain.be/ernest.matagne/GLISSANT/INDEX.HTM.

[MYL 08] MYLNIKOV A., "Tensor-geometric methods for problems of the circuit theory", *Journal of Mathematical Sciences*, vol. 148, no. 2, pp. 192-258, 2008.

[REM 05] REMY GH., TOUNZI A., BARRE P.J., PIRIOU F., HAUTHIER J.P., "Finite-element analysis of non-sinusoidal electromotive force in a permanent magnet linear synchronous motor", *The 5th International Symposium on Linear Drives for Industry Application, LDIA 2005*, Awaji Yumebutai, Hyogo, Japan, September 25-28, 2005.

[SYN 49] SYNGE J.L., SCHILD A., *Tensor Calculus*, University of Toronto Press, Toronto, Canada, 1949.

Chapter 3

Synchronous Machines in Degraded Mode

3.1. General introduction

In this chapter, we are first going to study the failures in a set of converter-synchronous machines that are used in a particular industrial application. The numerical data comes from [SCH 03] and [SCH 04]. We will see that the most important failures involve power transistors; other failures concern the stator coils of the synchronous machine. Hence, the main faults involve short-circuit or open-circuit problems. We will see that these problems cause perturbator torques, which are added to the cogging torque.

Second, we will study the supply strategies of healthy phases, aiming to compensate for both the perturbator torque and the cogging torque. We will show that in certain conditions it is possible to find optimal currents ensuring the desired torque while minimizing ohmic losses. We will give four main solutions and state six practical rules about these optimal currents.

Finally, we will see an adaptive control strategy in a closed loop based on neural learning to obtain optimal currents, whatever the type of fault. To illustrate the main situations that occur, we will give several examples of applications, which will be the object of simulations.

Chapter written by Damien FLIELLER, Ngac Ky NGUYEN, Hervé SCHWAB, Guy STURTZER.

3.1.1. *Analysis of failures of the set converter-machine: converters with MOSFET transistors*

3.1.1.1. *Introduction*

The goal of this section is to study the operating safety of the permanent magnet three-phase synchronous actuator associated with the static converter. First, we will list all of the failures that can occur and we will study their consequences on the system. Second, we will tackle the problem of faults forecast from data coming from reliability reports and practical data coming from an application in the motor industry. The purpose is to estimate the distribution of faults for different inverter topologies.

3.2. Analysis of the main causes of failure

The analysis of failure modes, their effects and their criticalities is an inescapable operation during the design and manufacturing of a new product. The failure tree (see Figure 3.1) presents a physical breakdown of the system. From here, we can list the types of failure that prevent each element from fulfilling its function. In this section, these faults are described as failure modes, with their effects and their criticality for the set converter-machine being discussed [HAY 99, KRU 03, SCH 03]. The diverse topologies studied are presented in Figure 3.2.

3.2.1. *Failure of the inverter*

3.2.1.1 *Failure of the transistors*

Faults in MOSFETs are known to be particularly frequent. Generally, the four modes by which transistors fail are:

– the short-circuit;

– the open circuit;

– the resistance increase between the drain and the source;

– resistant behavior.

Synchronous Machines in Degraded Mode 69

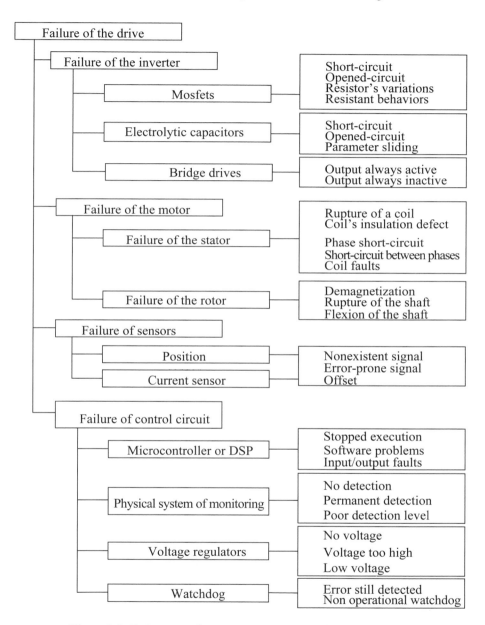

Figure 3.1. *Failure tree of a permanent magnet synchronous actuator*

70 Control of Non-conventional Synchronous Motors

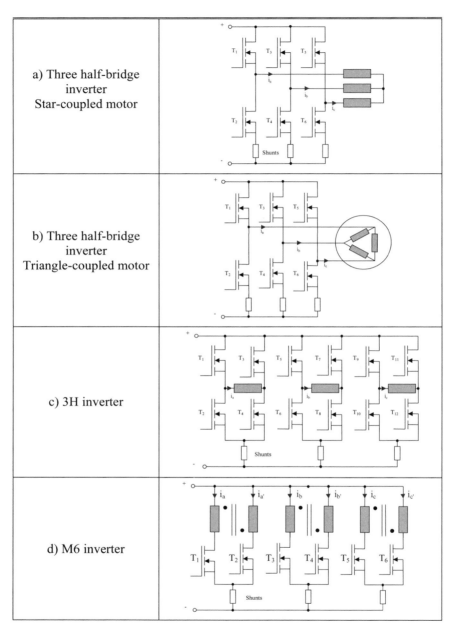

Figure 3.2. *Different topologies for the permanent magnet synchronous motor converter-machine set*

15% of failures are open circuits and 85% short circuits [BIR 96, WAL 88, SCH 04]. The variation of the drain-source resistance of the transistor often leads to a loss of balance between phases. This imbalance can be compensated for through vector control and so does not have any particular influence, except that a higher current appears in the failing phase implying a decrease in the maximum torque. The resistant behavior of the transistor while working in a blocked mode leads us to model it using resistance. The most unfavorable case is the frank short-circuit case, when the value of resistance is 0. This failure mode is accompanied by an increase in temperature and generally quickly leads to an additional short circuit or an open circuit. Here, we only consider the failing modes cited by the RDF 2000 report [UTE 00], namely the short circuit and the open circuit.

3.2.1.2. *Failure of the bridge drivers*

Failure of the bridge drivers has a similar effect to that due to transistor failure. In fact, when the output of the bridge driver is permanently active, the transistor is always closed – hence it is equivalent to a short circuit. In the case where the driver output is always inactive, the transistor is permanently open. We cannot consider the latest case as an open circuit; however, since the free wheel diode remains functional. This fault is comparable to a fault on the grid side (no grid signal). Figure 3.3 shows the diagram of a three half-bridge actuator with a bridge driver supplying control for the six transistors.

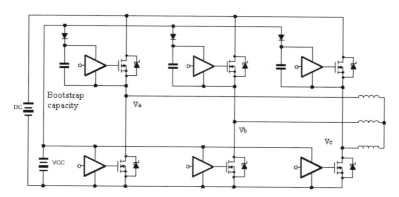

Figure 3.3. *Diagram of a three half-bridge actuator with bridge driver and bootstrap capacities*

Three bootstrap capacities generate the grid-source voltages, which allow us to control the transistors at the top. It must be noted that if an output of the bridge driver of one of the transistors from the top is permanently active, the transistor of the bottom is kept open and the bootstrap capacity can no longer be loaded. The grid-source voltage is then insufficient to close the MOSFET. The two transistors situated in the same bridge arm are therefore permanently open.

3.2.1.3. *Failure of electrolytic capacitors*

In the applications implementing voltages below 350 V, faults in the electrolytic capacitors are distributed in the following way: 30% short-circuit, 30% open circuit and 40% drifting of the capacity value. In fact, over time the capacitor loses part of its electrolyte, which decreases its capacity. This component hence has a limited lifespan that depends in particular on the current that crosses it. The variation of the capacitor's parameters leads to a modification of the filter's features, which can lead to harmonics problems. Issues with electromagnetic compatibility can then occur. The capacitor always allows quick demagnetization of the coils. In order to decrease the currents, however, it is often necessary to have several capacitors in parallel so that the lifespan of the component is sufficient. If a capacitor is in open circuit, this is not particularly dangerous; the filter features are only modified. In the case of a short circuit, however, we have an increase in temperature which can lead to the vaporization of the electrolyte and so to the component exploding. A solution to this problem consists of placing two capacitors in series [HAY 98].

3.2.2. *Other failures*

The most common faults are those linked to the inverter (see section 3.2.1). However, other faults linked to the motor can appear. These are less frequent and the details are presented in [SCH 04]. Stator faults correspond to failures linked to the coils (a short circuit between two turns, short circuit with the frame, short circuit between two phases, etc.). We can also include the demagnetization of the rotor's permanent magnets, failures of the shaft, failures of a sensor and failure of the control electronics.

3.3. Reliability of a permanent magnet synchronous motors drive

In order to determine the occurrence of each fault, we can use the reliability reports and the observed breakdown statistics. We still need to check the extent of which the results obtained are valid.

3.3.1. *Environmental conditions in the motor industry*

The environment of the system and the conditions of use are included in the calculation of the failure rate. The drive studied as an example belongs to the ground mobile systems category [DOD 95]. For this scenario, the most important failures are caused by thermomechanical stress. The use profile of the system is important here, thus we can distinguish three phases: alternative operating phase (on/off); continuous operating phase; and a storing or immobilization phase.

For each phase, the range of variation in temperature is considered. Thus, it is necessary to have weather data at our disposal. The mean global temperature is 14°C and the mean difference in temperatures between day and night is 10°C. A vehicle is used for an average of 500 hours per year, which corresponds to 5.7% of the duration of one year. Moreover, during the operative phase we can consider three different internal temperatures: 34% of the operative duration corresponds to a temperature of 32°C; 26% to 60°C; and finally 40% to 85°C. Thus the mean temperature is situated around 60°C. At the heart of the motor, however, the electronics are often subjected to temperatures close to 90°C. It is therefore necessary to adjust the temperatures by increasing the operative temperature by 30°C. Three different thermal cycles are then considered: a day operative cycle; a night one; and the weather variations during the vehicle's immobilization period. Table 3.1 below summarizes the thermal cycles.

Phase	Number of days per year	Number of uses (thermal cycles) per day	Number of cycles per year	Mean external temperature
Day	330	4 (return trip home to work, morning, noon, evening)	1,340	15°C
Night	330	2 (evening)	660	5°C
Immobilization	30	0 (car not used)	30	5°C and 15°C

Table 3.1. *Annual thermal cycles*

3.3.2. *The two reliability reports: MIL-HdbK-217 and RDF2000*

Reliability reports MIL-HdbK-217 [DOD 95] and RDF2000 [UTE 00] are used with predictive statistical methods. Data about a large number of systems have been listed and their use requires us to know the system environment, the stresses to which it is submitted and the operative conditions.

3.3.2.1. *Reliability report: MIL-HdbK-217*

The MIL-HdbK-217 reliability report [DOD 95] is the oldest and most well-known. Even if it is now considered obsolete, it remains a reference. The failure rate

74   Control of Non-conventional Synchronous Motors

of the system considered is given by formula [3.1]. An initial failure rate of basis λ_b is associated with the considered component by the report. The failure rate is then weighted by different coefficients allowing us to take the operative conditions into account.

$$\lambda = \lambda_b \cdot \pi_T \cdot \pi_S \cdot \pi_Q \cdot \pi_E \cdot \pi_X \qquad [3.1]$$

Coefficient π_T is the temperature factor. It obeys an Arrhenius law [3.2] that is generally used in the field of chemistry to describe the speed of a chemical reaction as a function of temperature. Coefficients π_{T_1} and π_{T_2} are two temperature factors for temperatures T_1 and T_2.

$$\pi_{T_2} = \pi_{T_1} \cdot \exp\left(\frac{E_a}{k_b}\left(\frac{1}{T_1} - \frac{1}{T_2}\right)\right) \qquad [3.2]$$

Coefficient π_S is an electrical load factor. Its value depends on currents and voltage levels in the component with respect to the maximum values specified by the manufacturer. π_Q is a quality factor that depends on the manufacturing quality. π_E is a factor depending on the environment: for the motor industry, we consider the system as belonging to the ground mobile category. Finally, π_X corresponds to other factors that can specifically interact with certain components.

3.3.2.2. *Reliability report: RDF 2000*

Reliability report RDF 2000 [UTE 00] was established by *Union Technique de l'Électricité* in 2000. It uses more complex models than the other reliability reports and takes into account different phases of the mission profile.

Contrary to report MIL-HdbK-217, the quality of components is not taken into account in the predictive calculation of reliability. In fact, we consider that the components are subject to strict controls and tests to ensure a uniform quality. The specifications of components must also be fulfilled. Finally, this norm does not take into account factors such as mechanical shocks or ambient humidity. However, it is proven that in the motor industry these factors are not insignificant.

The failure rate of a component is the sum of three failure rates: λ_{die}, $\lambda_{package}$ and $\lambda_{overstress}$. The first term, λ_{die}, involves the temperature factor calculated by Arrhenius law as well as the operative rate of the transistor. In the motor industry, this operative rate is 0.058 (operative rate of the vehicle). It is therefore insignificant, which is not the case for machines that are permanently working. λ_{die} is the term that takes into account the type of case used, as well as the temperature cycles at each operation. Finally $\lambda_{overstress}$ characterizes the conditions of the electrical environment.

In the case of the inverter, the transistors constitute an interface with the external electrical environment, i.e. the voltage source (car battery). This additional stress has an influence on the reliability.

3.3.3. Failure rate of permanent magnet synchronous motors actuators

Reliability report RDF 2000 (section 0.2) is used to determine the relative failure rates of the diverse components that make up a PMSM actuator in the motor industry. Here, we take into account the envirnonmental conditions defined in section 3.3.1. The system studied is made of a permanent magnet synchronous motor (brushless direct current or BLDC motor), a 3H inverter and its control. The constitutive elements of this system are the following (the number of components is not specified, but is taken into account in the calculation of the failure rate): MOSFET transistors, tantalium capacitors, ceramic capacitors, resistances, microcontroller (Digital Signal Processor (DSP)), Application-Specific Integrateed Circuit (Asic), position sensor, bridge driver, quartz, buffers, operational amplifiers, electrolytic capacitors and bipolar transistors [SCH 04].

From all the data about the system, it is possible to calculate the different failure rates with the help of formulae given by the reliability reports [SCH 03]. These results are then compared to field data charting the failure of electronic components noticed during the analysis of breakdowns encompassing a large number of faulty sytems [SCH 04].

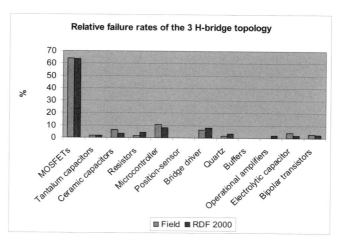

Figure 3.4. *Distribution of failure rates of the electronics estimated with RDF2000 and field data for a 3H topology*

Figure 3.4 represents the fault distribution of the electronic part of the 3H topology, estimated at first from databases and then from the RDF 2000 report. The distribution is similar for both methods, but the absolute values differ.

3.4. Conclusion

The failure rates of electronic components can be determined from reliability reports, which has not been the case for motor coils. Several publications [JOK 98, PEN 94] have dealt with the topic and considered empirically that the main failures were about transistors and coils. This is true within the framework of typical industrial applications, but it is different in the motor industry. In fact, experience shows that the constraints suffered by the insulators are much weaker for a battery voltage of 12 V. The voltage dynamic, which constitutes an important cause of stress, is also decreased [LEB 99]. Moreover, the faults in the coils of BLDC motors are much weaker than for coiled rotor motors (direct or DC motors, winded synchronous or induction motors with coiled rotors); the stator's coils are not submitted to as much mechanical stress as the rotor's coils. The failures of the coils are linked to the winding mode as well as to the thickness of the conductor. This is why the failure rates can differ from one project to another. However as high-voltage tests on accelerating aging can be carried out at the end of the production line, the coils with a winding fault break down quickly and are immediately withdrawn from the production. All of these reasons allow us to explain why failures of the coils in the motor industry are not particularly frequent.

By considering data coming from experience with permanent magnet synchronous motors (PMSM) in the motor industry, it is possible to estimate the proportion of faults in the coils for a typical PMSM actuator, as well as those linked to the mechanics. They cause 9% of all faults a PMSM actuator with three half-bridge inverters is susceptible to. This 9% can be broken down according to the different faults. This distribution is estimated from experience and physical considerations:

– short circuit of several turns in a coil: 3%;

– short circuit between turns belonging to two different phases: 1%;

– short circuit between a turn and the frame: 1%;

– rupture of a coil: 2%;

– mechanical failures (bearing, rupture or flexion of the shaft, etc.): 2%.

3.5. Optimal supplies of permanent magnet synchronous machines in the presence of faults

3.5.1. *Introduction: the problem of a-b-c controls*

The analysis of all failures, and particularly those of the power transistors, was carried out in section 3.2. For several applications, and power steering in particular, it is essential that the torque ripple is as weak as possible. Yet it has been proven in the previous sections that the fault of a transistor leads to a torque ripple that can be significant enough to harm the behavior of the system. The goal of this section is therefore to determine control strategies in failure mode, allowing the minimization of the cogging torque. To enlarge the framework of this study, we will tackle the case of a *n*-phase machine and apply the results to significant cases.

First we will show that we can put the supply problem of fault machines (with one or more failing phases) in terms of compensation of perturbating torque generated by the failing phase, the latter being added to the cogging torque. We feel that it is useful to include current saturation in one phase of this problem. We propose several strategies to compensate for the perturbator torque, according to whether or not there is a homopolar current.

3.6. Supplies of faulty synchronous machines with non-sinusoidal back electromagnetic force

3.6.1. *Generalization of the modeling*

We consider the general case of a smooth pole machine with p pole pairs, non-sinusoidal distribution, a cogging torque C_d and n phases (the phases are numbered from $j=1$ to n, and the rotor is spotted by index f). Without affecting the generality, we will consider that phase k is faulty so that there are only $n-1$ phases left on which it is possible to act (it is possible to consider several faulty phases, and we will give the corresponding formulae for n_d faulty phases provided that certain conditions are fulfilled in Table 3.2). The faults which we will take into account are the following:

– the current in phase k is saturated so that: $i_k = \pm I_{sat}$;

– the current in phase k is 0, which corresponds to a phase k in open circuit (often a fault of the inverter);

– the current in phase k is a short-circuit current. It can be due to an inverter or winding fault of phase k. In the first case, the corresponding current can be one- or two-directional, according to the fault type (blocking of a diode). In the second case, the current is always two-directional.

The model leans on the expression of the total electromagnetic torque C_{em} and on the expressions of the *back electromagnetic force emfs (or counter emf)*. The total electromagnetic torque is the sum of torque generated by the stator currents, which contain the effect of the saliency of the rotor (reluctant torque and excitation), and the cogging torque. We call the first C_{em_stat} (it is a function of stator currents and of position) and the second $C_d = C_d(\theta)$ (it is only a function of the position):

$$C_{em}((i),\theta) = C_{em\_stat}((i),\theta) + C_d(\theta) \qquad [3.3]$$

We can write C_{em_stat} in the form:

$$C_{em\_stat}((i),\theta) = \sum_{j=1}^{n} i_j \cdot \frac{\partial \psi_{jf}}{\partial \theta} + \frac{1}{2} \cdot \sum_{j=1}^{n} \sum_{k=1}^{n} \frac{\partial L_{jk}}{\partial \theta} \cdot i_j \cdot i_k \qquad [3.4]$$

In this chapter, we will restrict our study to the case of smooth pole machines for which the second term of C_{em_stat} is 0 (reluctant torque is disregarded); the relationship between torque C_{em_stat} and *back emfs* induced in the n phases (with $\Omega = \Omega(t)$) is written:

$$C_{em\_stat} = \sum_{j=1}^{n} i_j \cdot \frac{\partial \psi_{jf}}{\partial \theta} = \sum_{j=1}^{n} i_j \cdot \frac{e_j}{\Omega} = \frac{1}{\Omega} \cdot (e)^T \cdot (i) \qquad [3.5]$$

where (i) is the vector of stator currents, (e) the vector of *nback emf* and (e^1) the vector of the fundamental components of (e).

$$(e) = \begin{pmatrix} e_1 & e_2 & \cdots & e_k & \cdots & e_n \end{pmatrix}^T \qquad [3.6]$$

$$(e^1) = \begin{pmatrix} e_1^1 & e_2^1 & \cdots & e_k^1 & \cdots & e_n^1 \end{pmatrix}^T \qquad [3.7]$$

Each $e_j = \dfrac{\partial \psi_{jf}}{\partial \theta}$ component of (e) can be represented by its development in Fourier series, taken here in its most general case:

$$e_j = \sum_{k=1}^{\infty} a_k \cdot \cos\left(k \cdot p \cdot (\theta - (j-1) \cdot \gamma)\right) + b_k \cdot \sin\left(k \cdot p \cdot (\theta - (j-1) \cdot \gamma)\right) \quad [3.8]$$

or:

$$\gamma = 2\pi/(p \cdot n) \text{ and } j = 1, \ldots, n$$

Thus:

$$e_j^1 = a_1 \cdot \cos\left(p \cdot (\theta - (j-1) \cdot \gamma)\right) + b_1 \cdot \sin\left(p \cdot (\theta - (j-1) \cdot \gamma)\right) \quad [3.9]$$

When several phases are failing, we have to distinguish two parts in the current vector:

$$(i) = (i)_{fault} + (i)_{without\_fault} \quad [3.10]$$

such as:

$$\begin{aligned}(i)_{fault} &= (Q') \cdot (i) \\ (i)_{without\_fault} &= (Q) \cdot (i)\end{aligned} \quad [3.11]$$

and

$$(Q') + (Q) = (Q)^{-1} \cdot (Q) = (Id_n) \quad [3.12]$$

The first term gathers the currents of faulty phases and the second gathers the currents of phases without faults.

For instance, for a five-phase machine, $(i) = \begin{pmatrix} i_1 & i_2 & i_3 & i_4 & i_5 \end{pmatrix}^T$, if phases two and five are faulty then:

$$(i)_{fault} = \begin{pmatrix} 0 & i_2 & 0 & 0 & i_5 \end{pmatrix}^T \text{ and } (i)_{without\_fault} = \begin{pmatrix} i_1 & 0 & i_3 & i_4 & 0 \end{pmatrix}^T.$$

In this case:

$$(Q) = \begin{pmatrix} 1 & 0 & 0 & 0 & 0 \\ 0 & 0 & 0 & 0 & 0 \\ 0 & 0 & 1 & 0 & 0 \\ 0 & 0 & 0 & 1 & 0 \\ 0 & 0 & 0 & 0 & 0 \end{pmatrix} \text{ and } (Q') = \begin{pmatrix} 0 & 0 & 0 & 0 & 0 \\ 0 & 1 & 0 & 0 & 0 \\ 0 & 0 & 0 & 0 & 0 \\ 0 & 0 & 0 & 0 & 0 \\ 0 & 0 & 0 & 0 & 1 \end{pmatrix} \quad [3.13]$$

The perturbator torque C_p generated by the faulty phases is:

$$C_p = \frac{1}{\Omega} \cdot (e)^T \cdot (i)_{fault} \quad [3.14]$$

Torque C_{em_stat} can be written as:

$$C_{em\_stat} = \frac{1}{\Omega} \cdot (e)^T \cdot (i) = \frac{1}{\Omega} \cdot (e)^T \cdot (i)_{fault} + \frac{1}{\Omega} \cdot (e)^T \cdot (i)_{without\_fault} \quad [3.15]$$

The total electromagnetic torque desired (*des*), written as C_{em_des} must fulfill:

$$C_{em\_des} = C_p + C_d + \frac{1}{\Omega} \cdot (e)^T \cdot (i)_{without\_fault} \quad [3.16]$$

The problem of the optimal supply amounts is the same as determining $(i)_{without\_fault}$ in the controllable phases, which gives C_{em_des} while compensating for the cogging torque and the perturbation torque.

3.6.2. *A heuristic approach to the solution*

When phase *k* is the only faulty phase, properly supplying the machine in the presence of a fault amounts to choosing the current i_j for each phase of index *j* and $j \neq k$ so that the total electromagnetic torque is the desired torque C_{em_des}. The equation to solve is therefore:

$$C_{em\_des} = i_k \cdot \frac{e_k}{\Omega} + C_d + \sum_{\substack{j=1 \\ j \neq k}}^{n} i_j \cdot \frac{e_j}{\Omega} \quad [3.17]$$

For this solution, we use index $opt-0$, namely:

$$(i) = (i)_{opt-0} = (i)_{fault} + (i)_{without\_fault\_opt-0} \qquad [3.18]$$

Torque $C_p = i_k \cdot \dfrac{e_k}{\Omega}$ is then the perturbator torque that contains the effect of phase k in fault and is added to the cogging torque. To simplify the presentation, we will show the case where the perturbator torque is due to e_k as a whole (there can only be a short circuit on a few turns) but we will give an experimental method in section 3.7 that allows us to deal with all cases. We propose to build a vector of stator currents $(i)_{without\_fault}$, whose j_{th} component is expressed as (it is a heuristic choice *a priori*, initially presented by [CLE 93] in the case of machines operating in normal mode):

$$i_{opt-0,j} = A(\theta) \cdot \sin\big(p \cdot (\theta - (j-1) \cdot \gamma)\big) \text{ and } j \neq k \qquad [3.19]$$

The goal is to determine coefficient A so that the currents answer the problem. By replacing the currents of [formula 3.19] in formula [3.17], the total electromagnetic torque is written as:

$$C_{em} = A(\theta) \cdot \sum_{\substack{j=1 \\ j \neq k}}^{n} \dfrac{e_j(p \cdot \theta) \cdot \sin\big(p \cdot (\theta - (j-1) \cdot \gamma)\big)}{\Omega} + \dfrac{e_k(p \cdot \theta) \cdot i_k}{\Omega} + C_d \qquad [3.20]$$

Thus, we find the current of phase m, which corresponds to a phase without fault:

$$i_{opt-0,m \atop m \neq k} = \dfrac{(C_{em\_des} - C_d) \cdot \Omega - e_k(p \cdot \theta) \cdot i_k}{\sum_{\substack{j=1 \\ j \neq k}}^{n} e_j(p \cdot \theta) \cdot \sin\big(p \cdot (\theta - (j-1) \cdot \gamma)\big)} \cdot \sin\big(p \cdot (\theta - (m-1) \cdot \gamma)\big) \qquad [3.21]$$

A variant of equation [3.21] is obtained by replacing the sinus with the fundamental component of the *back emf* of phase m, which gives a more general formula and with weaker ohmic losses.

$$i_{opt-0,j} = \frac{\left(C_{em\_des} - C_d\right) \cdot \Omega - e_k(p \cdot \theta) \cdot i_k}{\sum_{\substack{j=1 \\ j \neq k}}^{n} e_j(p \cdot \theta) \cdot e_j^1(p \cdot \theta)} \cdot e_j^1(p \cdot \theta) \quad [3.22]$$

$$(i)_{without\_fault\_opt-0} = (Q) \cdot (i)_{opt-0}$$

$$= \frac{\left(C_{em\_des} - C_d\right) \cdot \Omega - e_k \cdot i_k}{(e)^T \cdot (Q) \cdot (e^1)} \cdot (Q) \cdot (e^1) \quad [3.23]$$

RULE 3.1.- For the general case with several faulty phases (with at least two non-faulty phases) we obtain:

$$(i)_{without\_fault\_opt-0} = \frac{\left(C_{em\_des} - C_d - C_p\right) \cdot \Omega}{(e)^T \cdot (Q) \cdot (e^1)} \cdot (Q) \cdot (e^1) \quad [3.24]$$

Supplying the machine with these currents generates the desired torque and compensates for both the cogging torque and the perturbator torque generated by the faulty phases. We have a solution, from among all the possible solutions, and we will see later that there are countless ways of solving this problem.

An extension of this method is presented in [FLI 08]; it consists of determining experimental term $A(\theta)$ of formula [3.20] by using its development in Fourier series where the coefficients are identified by a neural network. This is developed in section 3.7. However these currents are not optimal in terms of ohmic losses; the search for currents minimizing these losses will be the topic of the following sections.

3.6.3. *First optimization of ohmic losses without constraint on the homopolar current*

When k is the only faulty phase, let us again take the previous problem and look for the stator currents of index $opt-1$, minimizing the ohmic losses and generating the total electromagnetic torque desired. We will write:

$$(i) = (i)_{opt-1} = (i)_{fault} + (i)_{without\_fault\_opt-1} \quad [3.25]$$

Synchronous Machines in Degraded Mode 83

The equation to solve is then:

$$C_{em\_des} \cdot \Omega = e_k \cdot i_k + C_d \cdot \Omega + (e)^T \cdot (i)_{without\_fault\_opt-1} \qquad [3.26]$$

The solution $(i)_{without\_fault\_opt-1}$ must also minimize the criterion:

$$J_1 = \sum_{\substack{j=1 \\ j \neq k}}^{n} i_j^2 \qquad [3.27]$$

We are in the presence of an optimization problem without constraint, which can be solved, for instance, using the Lagrange formulation. We will therefore write the Lagrangian:

$$L_1 = \sum_{\substack{j=1 \\ j \neq k}}^{n} i_j^2 + \lambda_1 \cdot \left[(e)^T \cdot (Q) \cdot (i) + e_k \cdot i_k - \left(C_{em\_des} + C_d \right) \cdot \Omega \right] \qquad [3.28]$$

where we make multiplier λ_1 appear. The optimization problem is comprehensively described by equations [3.26] to [3.28]. We then have to differentiate the Lagrangian with respect to each current except for phase k, and we obtain $n - 1$ equations of the type:

$$2 \cdot i_j + \lambda_1 \cdot e_j = 0, \text{ with } j = 1 \ldots n \text{ and } j \neq k \qquad [3.29]$$

This allows us to write:

$$\left(C_{em\_des} - C_d \right) \cdot \Omega - e_k \cdot i_k = (e)^T \cdot (Q) \cdot (i) = -\frac{1}{2} \cdot \lambda_1 \cdot (e)^T \cdot (Q) \cdot (e) \qquad [3.30]$$

which gives an expression of the Lagrange multiplier λ_1:

$$\lambda_1 = \frac{-2 \cdot \left(\left(C_{em\_des} - C_d \right) \Omega - e_k \cdot i_k \right)}{(e)^T \cdot (Q) \cdot (e)} \qquad [3.31]$$

Thus, we obtain the first optimal solution for the currents (without constraint on the homopolar current):

$$(Q) \cdot (i)_{opt-1} = \frac{(C_{em\_des} - C_d) \cdot \Omega - e_k \cdot i_k}{(e)^t \cdot (Q) \cdot (e)} \cdot (Q) \cdot (e) \quad [3.32]$$

EXAMPLE 3.1.– let's write the three-phase case ($n = 3$ and find again that the usual notations a, b and c are the indices of phases one, two and three). When phase a is faulty, it appears in the denominator of the square of the amplitude of *back emfs* of non-faulty phases: $\|e^2\| = e_b^2 + e_c^2$; and we obtain the expression of optimal currents:

$$(i)_{sans\_défaut\_opt-1} = \begin{pmatrix} 0 \\ i_{opt\_1,b} \\ i_{opt\_1,c} \end{pmatrix} = \frac{1}{e_b^2 + e_c^2} \cdot \begin{pmatrix} 0 \\ ((C_{em\_des} - C_d) \cdot \Omega - i_a \cdot e_a) \cdot e_b \\ ((C_{em\_des} - C_d) \cdot \Omega - i_a \cdot e_a) \cdot e_c \end{pmatrix} \quad [3.33]$$

In this case, $(Q) = \begin{pmatrix} 0 & 0 & 0 \\ 0 & 1 & 0 \\ 0 & 0 & 1 \end{pmatrix}$ and $(Q') = \begin{pmatrix} 1 & 0 & 0 \\ 0 & 0 & 0 \\ 0 & 0 & 0 \end{pmatrix}$. [3.34]

REMARK 3.1.– as $(e)^T \cdot (Q) \cdot (e) = \sum_{\substack{j=1 \\ j \neq k}}^{n} e_j^2$, we have the current of phase m:

$$i_{without\_fault\_opt-1,m \atop m \neq k} = \frac{(C_{em\_des} - C_d) \cdot \Omega - e_k \cdot i_k}{\sum_{\substack{j=1 \\ j \neq k}}^{n} e_j^2} \cdot e_m \quad [3.35]$$

RULE 3.2.– for the general case (with at least two non-faulty phases):

$$(i)_{without\_fault\_opt-1} = \frac{(C_{em\_des} - C_d - C_p) \cdot \Omega}{(e)^T \cdot (Q) \cdot (e)} \cdot (Q) \cdot (e) \quad [3.36]$$

or:

$$(i)_{without\_fault\_opt-1} = \frac{(C_{em\_des} - C_d - C_p) \cdot \Omega}{(e')^T \cdot (e')} \cdot (e') \quad [3.37]$$

with:

$$(e') = (Q) \cdot (e) \quad [3.38]$$

In the particular case of a machine with sinusoidal *back emfs* in the form:

$$e_j = p \cdot \phi_{f0} \cdot \Omega \cdot \sin(p \cdot (\theta - (j-1) \cdot \gamma)) \text{ with } j = 1...n \quad [3.39]$$

the optimal current of phase *m*, inferred from [3.32], is expressed:

$$i_{opt-1,m \atop m \neq k} = \frac{(C_{em\_des} - C_d - C_p) \cdot \Omega}{\sum_{\substack{j=1 \\ j \neq k}}^{n} e_j(p \cdot \theta) \cdot \sin(p \cdot (\theta - (j-1) \cdot \gamma))} \cdot \sin(p \cdot (\theta - (m-1) \cdot \gamma)) \quad [3.40]$$

which is the result given by the first heuristic approach (see formula [3.21]).

RULE 3.3.– when *back emfs* are sinusoidal $(i)_{opt-1} = (i)_{opt-0}$

It is interesting to note that formula [3.32] gives the optimal currents in normal mode when there is no fault [FLI 11]; in this case, matrix (Q) is the identity matrix and the optimal currents are given by:

$$(i)_{opt-1} = \frac{(C_{em\_des} - C_d) \cdot \Omega}{(e)^T \cdot (e)} \cdot (e) \quad [3.41]$$

This same approach allows us to compensate for periodical perturbator torques generated by unbalances or errors in alignment.

3.6.3.1. *Application to a three-phase machine with sinusoidal back emf*

We are going to show simulations for a three-phase machine that is initially sinusoidal, then trapezoidal for two types of fault studied separately: phase *a* in open circuit (where the current in phase *a* is 0); and saturation at $\pm I_{sat}$ on the currents, each current successively reaching saturation. The following simulations are done using $C_{em-1-des} = 1.5 N.m$.

Figure 3.5 illustrates the case of a machine with sinusoidal *back emfs* in normal functioning supplied with current, $(i)_{opt-1} = (i)_{opt-0}$. The homopolar current is 0 and the machine can be star-coupled. When phase *a* is in open circuit – the situation represented in Figure 3.6 with the same supply strategy (only (Q) changes) – the currents of phases *b* and *c* are no longer sinusoidal and have a greater amplitude.

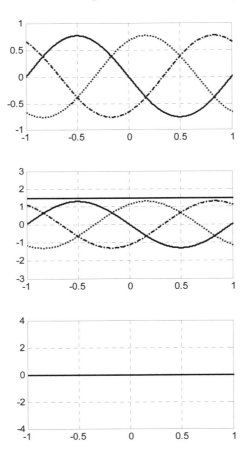

Figure 3.5. *Three-phase machine with sinusoidal back emfs without fault or cogging torque. From top to bottom: normalized back emfs, currents and torque, and homopolar current. Solution* $(i)_{opt-1}$

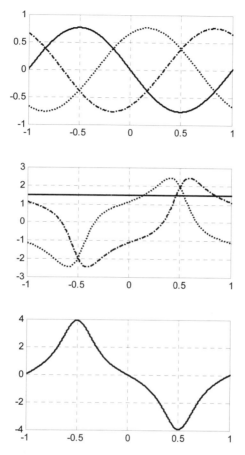

Figure 3.6. *Three-phase machine with sinusoidal back emfs with phase a in open circuit. From top to bottom: normalized back emfs, currents and torque, homopolar current. Solution* $(i)_{opt-1}$

When the currents are limited to $\pm I_{sat}$, Figure 3.7 is obtained, while two phases never saturate simultaneously. For instance, the currents of phases *b* and *c* are no longer sinusoidal when the current of phase *a* is saturated. The homopolar current is no longer 0 when one of the phases is saturated; the machine can therefore no longer be supplied in star formation.

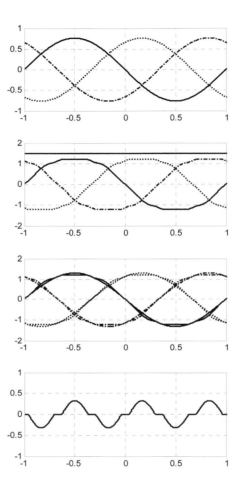

Figure 3.7. *Three phase sinusoidal machine and without cogging torque. Solution $(i)_{opt-1}$. From top to bottom: normalized back emfs, saturated currents and torque, comparison with sinusoidal currents without saturation, and homopolar current*

The limit allowed by this strategy is shown in Figure 3.8 and corresponds to $I_{sat.} = \dfrac{C_{em\_des}}{\sqrt{3}p_1\phi_{f0}}$, beyond which two currents simultaneously become saturated: in this case or with another fault, for instance with currents a and c, current b is given theoretically by:

$$i_b = \frac{\left(C_{em\_des} - C_d\right) \cdot \Omega - e_a \cdot i_a - e_c \cdot i_c}{e_b} \quad [3.42]$$

It is no longer possible to obtain a constant torque for all the positions because when e_b cancels out, the current i_b is also led to saturation.

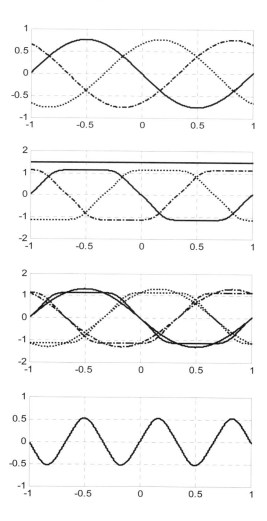

Figure 3.8. *Three-phase sinusoidal machine and without cogging torque. Solution* $(i)_{opt-1}$.
From top to bottom: normalized back emfs, saturated currents and torque, comparison with sinusoidal currents without saturation, and homopolar current

3.6.3.2. *Application to a three-phase machine with trapezoidal back emfs*

For a machine with trapezoidal *back emfs* in normal functioning supplied with current $(i)_{opt-1}$, we obtain the results shown in Figure 3.9 and Figure 3.10. The homopolar current is not 0 and the current amplitudes are also greater, with one faulty phase.

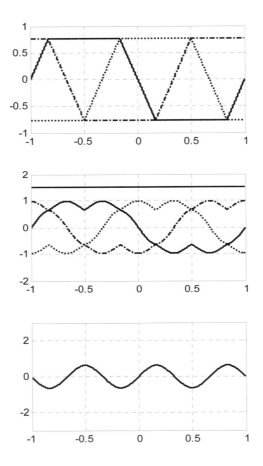

Figure 3.9. *Trapezoidal machine: machine without fault. Solution* $(i)_{opt-1}$. *From top to bottom: normalized back emfs, currents and torque, and homopolar current*

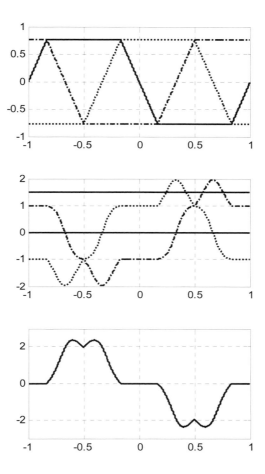

Figure 3.10. *Trapezoidal machine. Cut-out of phase a. Solution $(i)_{opt-1}$. From top to bottom: normalized back emfs, currents and torque, and homopolar current*

3.6.4. *Second optimization of ohmic losses with the sum of currents of non-faulty phases being zero*

In this section, we initially consider a star-coupled machine with faulty phase k, before generalizing to the situation where n_d phases are faulty. We are interested in the case where the current of phase k is 0 or when the short circuit of phase k in fault does not go through the converter but only through the wires of the faulty phase. As the machine is star-coupled, the homopolar current (being 0) concerns only the healthy phases. Let us look for the stator currents of index opt_2, minimizing the ohmic losses and generating the desired total electromagnetic torque. We will then write:

$$(i) = (i)_{opt\_2} = (i)_{fault} + (i)_{without\_fault\_opt\_2} \qquad [3.43]$$

The constraint imposed by the homopolar component is written:

$$\sum_{\substack{j=1 \\ j \neq k}}^{n} i_j = (w_1)^T \cdot (i) = 0 \qquad [3.44]$$

with:

$$(w_1) = \begin{pmatrix} 1 & 1 & \cdots & \underset{k^{th}}{0} & \cdots & 1 \end{pmatrix}^T \text{ and } \|w_1\|^2 = (w_1)^T \cdot (w_1) = n-1 \qquad [3.45]$$

The constraint on the torque remains unchanged:

$$\begin{aligned}(e)^T \cdot (Q) \cdot (i) &= (e')^T \cdot (i) \\ &= (e)^T \cdot (i)_{without\_fault\_opt-2} = (C_{em\_des} - C_d) \cdot \Omega - e_k \cdot i_k \end{aligned} \qquad [3.46]$$

As the currents of faulty phases are not free, the ohmic losses to be minimized only concern the non-faulty phases. We remind ourselves that index k represents the faulty phase. The criterion to be minimized is therefore:

$$J = \sum_{\substack{j=1 \\ j \neq k}}^{n} i_j^2 = (i)^T \cdot (Q) \cdot (i) \qquad [3.47]$$

We notice that $(e)^T \cdot (i)_{without\_fault-opt\_2} = ((Q).(e))^T \cdot (i)_{without\_fault-opt\_2}$. The problem amounts to finding $(i)_{without\_fault\_opt\_2}$, which minimizes the ohmic losses while not having components according to (w_1) so that the scalar product with $(e') = (Q).(e)$ gives $(C_{em\_des} - C_d) \cdot \Omega - e_k \cdot i_k$. It is possible to produce a geometrical representation in a $n - n_d$ dimensional space: Figure 3.12 emphasizes the geometrical properties of the different optimal currents. The set of points M of coordinates equal to $(Q) \cdot (i)_{opt-1} = (i)_{without\_fault\_opt-1}$ fulfilling:

$$C_{em\_des} \cdot \Omega = (e)^T \cdot (Q) \cdot (i) + (C_P + C_d) \cdot \Omega \qquad [3.48]$$

is situated on the hyperplan (P) of equation $(C_{em\_des} - C_p - C_d) \cdot \Omega = (e)^T \cdot (Q) \cdot (i)$. Vector $(Q) \cdot (e) = (K) \cdot \Omega$ is therefore a vector normal to hyperplan (P). Hyperplan (P) is shown in dark gray in Figure 3.12. Point M_{opt-1} is the point of coordinates $(Q) \cdot (i)_{opt-1}$ belonging to (P) and minimizing the norm of the vector $(Q) \cdot (i)$. Current $(i)_{without\_fault\_opt-1}$ is such that its scalar product with vector $(e') = (Q) \cdot (e)$ is equal to $(C_{em\_des} - C_d - C_p) \cdot \Omega$, thus:

$$((Q) \cdot (e))^T \cdot (Q) \cdot (i) = (e)^T \cdot (Q) \cdot (i) = (C_{em\_des} - C_d - C_p) \cdot \Omega \qquad [3.49]$$

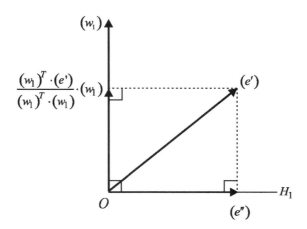

Figure 3.11. *Geometrical representation of optimal solutions opt-0, opt-1 and opt-2 in a $n - n_d$ dimensional space*

Vector $(i)_{without\_fault\_opt-1}$ of the minimum modulus is proportional to $(Q) \cdot (e)$ so that: $(i)_{without\_fault\_opt-1} = \alpha_1 \cdot (Q) \cdot (e)$ or by replacing:

$$(i)_{without\_fault\_opt-1} = \frac{(C_{em\_des} - C_d - C_p) \cdot \Omega}{(e)^T \cdot (Q) \cdot (e)} \cdot (Q) \cdot (e) \quad [3.50]$$

we again find formula [3.51]. On the same figure, point M_{opt-0} is the point of coordinates $(Q) \cdot (i)_{opt-0}$. We check that:

$$(i)_{opt-0}{}^T \cdot (Q) \cdot (e) = (i)_{opt-1}{}^T \cdot (Q) \cdot (e) \quad [3.51]$$

which gives us another way to obtain $(i)_{opt-1}$ from $(i)_{opt-0}$.

The set of points M fulfilling the constraint linked to the homopolar on $n-1$ phases: $\sum_{\substack{j=1 \\ j \neq k}}^{n} i_j = (w_1)^T \cdot (Q) \cdot (i) = 0$, is situtated in the hyperplan (H_1) of equation $(w_1)^T \cdot (Q) \cdot (i) = 0$. Vector (w_1) is therefore a vector normal to (H_1). Hyperplan (H_1) is shown in light gray in Figure 3.12. Point M_{opt-2} is the point of coordinates $(Q) \cdot (i)_{opt-2}$ belonging to (H_1) and (P) so that the norm of (i) is minimized (with constraint on the homopolar current). It is therefore the point that is the closest to O at the intersection of (H_1) and (P). As a result, vector $(Q) \cdot (i)_{opt-2}$ does not have any component according to (w_1).

Let $(e")$ be the vector $(e') = (Q).(e)$ deprived of its component according to (w_1) (see) and let's define vector (K_2). It comes:

$$(i)_{without\_fault\_opt-1} = \frac{(C_{em\_des} - C_d - C_p) \cdot \Omega}{(e)^T \cdot (Q) \cdot (e)} \cdot (Q) \cdot (e) \quad [3.52]$$

The optimal solution $(Q) \cdot (i)_{opt-2} = (i)_{without\_fault\_opt-2}$ is then written:

$$(e") = (Q) \cdot (e) - (w_1)^T \cdot (Q) \cdot (e) \cdot \frac{(w_1)}{\|(w_1)\|^2} = (K_2) \cdot \Omega \quad [3.53]$$

Synchronous Machines in Degraded Mode 95

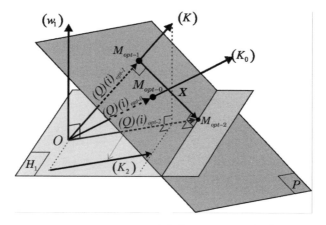

Figure 3.12. *Breaking down of (e') according to (w_1) and H_1*

This gives:

$$(i)_{without\_fault\_opt-2} = \frac{(C_{em\_des} - C_d) \cdot \Omega - e_k \cdot i_k}{((Q) \cdot (e))^T \cdot (e'')} \cdot (e'') \quad [3.54]$$

moreover:

$$((Q) \cdot (e))^T \cdot (e'') = (e'')^T \cdot (e'') \quad [3.55]$$

Pythagorus theorem gives:

$$(e'')^T \cdot (e'') = (e)^T \cdot (Q) \cdot (e) - \frac{((w_1)^T \cdot (Q) \cdot (e))^2}{\|(w_1)\|^2} = \sum_{\substack{j=1 \\ j \neq k}}^{n} e_j^2 - \frac{1}{n-1}\left(\sum_{\substack{j=1 \\ j \neq k}}^{n} e_j\right)^2 \quad [3.56]$$

from where we get:

$$(i)_{without\_fault\_opt-2} = \frac{(C_{em\_des} - C_d) \cdot \Omega - e_k \cdot i_k}{\sum_{\substack{j=1 \\ j \neq k}}^{n} e_j^2 - \frac{1}{n-1}\left(\sum_{\substack{j=1 \\ j \neq k}}^{n} e_j\right)^2} \cdot (e'') \quad [3.57]$$

96   Control of Non-conventional Synchronous Motors

RULE 3.4.– the general formulation with several faulty phases (with at least three non-faulty phases) gives:

$$(i)_{without\_fault\_opt-2} = \frac{\left(C_{em\_des} - C_d - C_p\right) \cdot \Omega}{(e)^T \cdot (Q) \cdot (e) - \frac{1}{(w_1)^T \cdot (w_1)} \left((w_1)^T \cdot (e)\right)^2} \cdot (e'')$$

[3.58]

3.6.5. *Third optimization of ohmic losses with a homopolar current of zero (in all phases)*

In this section we initially consider a star-coupled machine with a faulty phase k, before generalizing our calculations to the situation where n_d phases are faulty. We are interested in the case where the phase current is not 0 and goes through the converter. As the machine is star-coupled, the homopolar current at 0 concerns all the phases of the machine including the faulty ones. Let us look for the stator currents of index opt_3, minimizing the ohmic losses and generating the total electromagnetic torque desired. We will then write:

$$(i) = (i)_{opt-3} = (i)_{fault} + (i)_{without\_fault\_opt-3}$$

[3.59]

The constraint imposed by the homopolar component is written:

$$\sum_{j=1}^{n} i_j = (u_1)^T \cdot (i) = 0$$

[3.60]

Namely:

$$\sum_{\substack{j=1 \\ j \neq k}}^{n} i_j = (w_1)^T \cdot (i) = -i_k$$

[3.61]

The set of points M fulfilling the constraint linked to the homopolar on n phases: $\sum_{\substack{j=1 \\ j \neq k}}^{n} i_j = (w_1)^T \cdot (i) = (w_1)^T \cdot (Q) \cdot (i) = -i_k$, is situated in hyperplan (H'_1) of equation $(w_1)^T \cdot (Q) \cdot (i) = -i_k$. Vector (w_1) is hence a vector normal to (H'_1), which is itself parallel to hyperplan (H_1) (represented in dark gray in Figure 3.13). Point M_{opt-3} is the point of coordinates $(Q) \cdot (i)_{opt-3}$ belonging to (H'_1) and (P) so that the distance to O is minimized. This point is the same hyperplan as O, M_{opt-1} and M_{opt-2}. The distance between the two hyperplans is equal to $-i_k$ (or $-(w'_1) \cdot (i)_{fault}$ in the general case with several faulty phases) so that (H'_1) and (H_1) overlap when the fault current is 0. In this case, M_{opt-3} and M_{opt-2} also overlap and the optimal solutions $(Q) \cdot (i)_{opt-3}$ and $(Q) \cdot (i)_{opt-2}$ are the same.

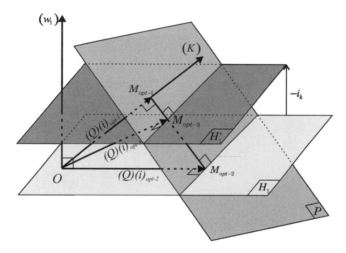

Figure 3.13. *Geometrical representation of optimal solutions opt-1, opt-2 and opt-3 in a $n - n_d$ dimensional space*

RULE 3.5.– when the fault currents are 0, $(i)_{opt-2} = (i)_{opt-3}$, solutions $(Q) \cdot (i)_{opt-2}$ or $(Q) \cdot (i)_{opt-3}$ only exist if (H'_1) and (P) are not parallel, otherwise some components of $(Q) \cdot (i)_{opt-2}$ or $(Q) \cdot (i)_{opt-3}$ would tend to infinity. In practice it needs to be $n - n_d \geq 3$ so that $(Q) \cdot (i)_{opt-2}$ or $(Q) \cdot (i)_{opt-3}$ exists for all the rotor positions.

98   Control of Non-conventional Synchronous Motors

The equation of hyperplan P is given by:

$$(e')^T \cdot (i) = \left(C_{em\_des} - C_d\right) \cdot \Omega - e_k \cdot i_k \qquad [3.62]$$

The equation of hyperplan H_1' is defined by:

$$(w_1)^T \cdot (i) = -i_k \qquad [3.63]$$

The equation of hyperplan H_1 is defined by:

$$(w_1)^T \cdot (i) = 0 \qquad [3.64]$$

Optimal current $(Q) \cdot (i)_{opt-1}$ is;

$$(Q) \cdot (i)_{opt-1} = (i)_{without\_fault\_opt-1} = \frac{\left(C_{em\_des} - C_d - C_p\right) \cdot \Omega}{(e')^T \cdot (e')} \cdot (e') \quad [3.65]$$

Optimal current $(Q) \cdot (i)_{opt-2}$ is:

$$(Q) \cdot (i)_{opt-2} = (i)_{without\_fault\_opt-2} = \frac{\left(C_{em\_des} - C_d - C_p\right) \cdot \Omega}{(e'')^T \cdot (e'')} \cdot (e'') \quad [3.66]$$

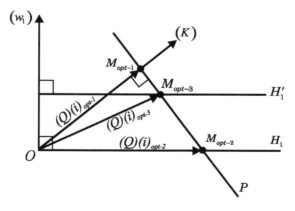

Figure 3.14. *Optimal opt-1, opt-2 and opt-3 solutions in actual size*

Optimal current $(Q)\cdot(i)_{opt-3}$ is a linear combination of the two others:

$$(Q)\cdot(i)_{opt-3} = (Q)\cdot(i)_{opt-2} + \alpha_3 \cdot \left((Q)\cdot(i)_{opt-1} - (Q)\cdot(i)_{opt-2}\right) \quad [3.67]$$

as:

$$(w_1)^T \cdot (Q)\cdot(i)_{opt-2} = 0 \quad [3.68]$$

we have:

$$(w_1)^T \cdot (i)_{opt-3} = \alpha_3 \cdot (w_1)^T \cdot (i)_{opt-1} = -i_k \quad [3.69]$$

We deduce α_3 by:

$$\alpha_3 = \frac{-i_k}{(w_1)^T \cdot (i)_{opt-1}} \quad [3.70]$$

Let us replace [3.70] in [3.67]. We then have:

$$(i)_{without\_fault\_opt-3} = \frac{\left(C_{em\_des} - C_d - C_p\right)\cdot \Omega}{(e'')^T \cdot (e'')} \cdot (e'') - \frac{i_k \cdot \left((i)_{opt-1} - (i)_{opt-2}\right)}{(w_1)^T \cdot (i)_{opt-1}} \quad [3.71]$$

We also have:

$$(w_1)^T \cdot (i)_{opt-1} = \frac{\left(C_{em\_des} - C_d - C_p\right)\cdot \Omega \cdot \sum_{\substack{j=1 \\ j\neq k}}^{n} e_j}{\sum_{\substack{j=1 \\ j\neq k}}^{n} e_j^2} \quad [3.72]$$

Let:

$$(X) = \frac{(i)_{opt-1} - (i)_{opt-2}}{(w_1)^T \cdot (i)_{opt-1}} \quad [3.73]$$

For the *m*th component of X (which represents a non-faulty phase):

$$X_m = \frac{-e_m \cdot \frac{1}{(w_1)^T \cdot (w_1)} \left(\sum_{\substack{j=1 \\ j \neq k}}^{n} e_j \right) + \frac{1}{(w_1)^T \cdot (w_1)} \cdot \sum_{\substack{j=1 \\ j \neq k}}^{n} e_j^2}{\left(\sum_{\substack{j=1 \\ j \neq k}}^{n} e_j^2 - \frac{1}{(w_1)^T \cdot (w_1)} \left(\sum_{\substack{j=1 \\ j \neq k}}^{n} e_j \right)^2 \right)} \quad [3.74]$$

This gives component m of the current:

$$i_m_{m \neq k} = \frac{\left((C_{em\_des} - C_d) \cdot \Omega - e_k \cdot i_k \right) \cdot e"_m}{(e")^T (e")} - \frac{1}{n-1} \cdot \frac{\sum_{\substack{j=1 \\ j \neq k}}^{n} e_j^2 - e_m \sum_{\substack{j=1 \\ j \neq k}}^{n} e_j}{(e")^T (e")} \cdot i_k \quad [3.75]$$

RULE 3.6.– General formulation with several faulty phases (with at least 3 non-faulty phases):

$$(i)_{without\_fault\_opt-3} = \frac{(C_{em\_des} - C_d - C_p) \cdot \Omega}{(e")^T (e")} (e")$$

$$- \frac{(w'_1)^T \cdot (i)_{fault}}{(w_1)^T (w_1)} \cdot \frac{\left((e')^T (e') \right) \cdot (w_1) - \left((w_1)^T (e') \right) \cdot (e')}{(e")^T (e")} \quad [3.76]$$

This result completes the previous studies of that where the fault currents were 0 when the homopolar was taken into account, in particular in [ATA 03] and [DAW 07].

Table 3.2 summarizes the different solutions possible according to the number n of phases and number n_d of faulty phases. For a constant torque we need at least two non-faulty phases (solutions $(Q) \cdot (i)_{opt-0}$ and $(Q) \cdot (i)_{opt-1}$). To have an additional constraint on the sum of currents, we need at least three non-faulty phases (solutions $(Q) \cdot (i)_{opt-2}$ and $(Q) \cdot (i)_{opt-3}$).

Synchronous Machines in Degraded Mode   101

Number of faulty phases \ Number of phases	2	3	4	5	......	n
0	$(i)_{opt-0}$ $(i)_{opt-1}$	$(i)_{opt-0}$ $(i)_{opt-1}$ $(i)_{opt-2}$	$(i)_{opt-0}$ $(i)_{opt-1}$ $(i)_{opt-2}$	$(i)_{opt-0}$ $(i)_{opt-1}$ $(i)_{opt-2}$	$(i)_{opt-0}$ $(i)_{opt-1}$ $(i)_{opt-2}$	$(i)_{opt-0}$ $(i)_{opt-1}$ $(i)_{opt-2}$
1		$(i)_{opt-0}$ $(i)_{opt-1}$	$(i)_{opt-0}$ $(i)_{opt-1}$ $(i)_{opt-2}$ $(i)_{opt-3}$	$(i)_{opt-0}$ $(i)_{opt-1}$ $(i)_{opt-2}$ $(i)_{opt-3}$	$(i)_{opt-0}$ $(i)_{opt-1}$ $(i)_{opt-2}$ $(i)_{opt-3}$	$(i)_{opt-0}$ $(i)_{opt-1}$ $(i)_{opt-2}$ $(i)_{opt-3}$
2			$(i)_{opt-0}$ $(i)_{opt-1}$	$(i)_{opt-0}$ $(i)_{opt-1}$ $(i)_{opt-2}$ $(i)_{opt-3}$	$(i)_{opt-0}$ $(i)_{opt-1}$ $(i)_{opt-2}$ $(i)_{opt-3}$	$(i)_{opt-0}$ $(i)_{opt-1}$ $(i)_{opt-2}$ $(i)_{opt-3}$
3				$(i)_{opt-0}$ $(i)_{opt-1}$	$(i)_{opt-0}$ $(i)_{opt-1}$ $(i)_{opt-2}$ $(i)_{opt-3}$	$(i)_{opt-0}$ $(i)_{opt-1}$ $(i)_{opt-2}$ $(i)_{opt-3}$
$\vdots$	$\vdots$	$\vdots$	$\vdots$	$\vdots$	$\vdots$	$\vdots$
$n_d = n-2$						$(i)_{opt-0}$ $(i)_{opt-1}$

Table 3.2. *The different possible solutions*

3.6.6. *Global formulations*

The diverse constraints acting on the current concern the n_d faulty phases (the nature of the fault imposes constraints on the faulty phases), the possible relationship with the homopolar current that needs to be fulfilled and that on the desired torque.

These constraints can be written in matrix form:

$$(B) \cdot (i) - (C) = (0) \qquad [3.77]$$

If the number of constraints is equal to the number n of currents, the problem amounts to a solution that is imposed by constraints:

$$(i) = (B)^{-1} \cdot (C) \qquad [3.78]$$

In practice, this case is only theoretically possible for $n = 3$, but particular positions appear where matrix B is no longer invertible. This is the case, for instance, if the current of phase a is faulty; when the *back emf* of phases b and c are equal then currents i_b and i_c diverge.

When n is > 3, there is an infinity of solutions. We can then introduce the criterion to minimize the n phases (although some phases are faulty):

$$J = \sum_{j=1}^{n} i_j^2 = (i)^T \cdot (i) \qquad [3.79]$$

The associated Lagrangian can be written:

$$L = J + (\Lambda) \cdot \left((B) \cdot (i) - (C)\right) \qquad [3.80]$$

and (Λ) is a line vector containing as many multipliers as there are constraints. The Lagrange conditions give:

$$(\Lambda)^T = -2\left((B) \cdot (B)^T\right)^{-1} \cdot (C) \qquad [3.81]$$

and

$$(i)_{opt} = (B)^T \cdot \left((B) \cdot (B)^T\right)^{-1} \cdot (C) \qquad [3.82]$$

The three possible cases are:

– a fault current with independent phases;

– a fault current that is independent from the other phase currents, wth a homopolar current of 0 on $n-1$ phases;

– a fault current independent of the other phase currents, with a homopolar current of 0 on n phases.

3.6.6.1. *Case 1: a fault current with independent phases*

The constraints are written as follows:

– the phase current of index k is imposed by the fault, namely:

$$\begin{pmatrix} 0 & 0 & \cdots & 1 & \cdots & 0 \end{pmatrix}.(i) = i_k \qquad [3.83]$$

– the desired torque imposes:

$$(e)^T \cdot (i) = \left(C_{em\_des} - C_d\right) \cdot \Omega \qquad [3.84]$$

namely:

$$(B_1) = \begin{bmatrix} 0 & 0 & \cdots & 1 & \cdots & 0 \\ e_1 & e_2 & \cdots & e_k & \cdots & e_n \end{bmatrix} \text{ and } (C_1) = \begin{bmatrix} i_k \\ \left(C_{em\_des} - C_d\right) \cdot \Omega \end{bmatrix} \qquad [3.85]$$

Then:

$$(i)_{opt\_1} = (B_1)^T \cdot \left((B_1) \cdot (B_1)^T\right)^{-1} \cdot (C_1) \qquad [3.86]$$

which is equivalent to:

$$(Q) \cdot (i)_{opt-1} = \frac{(C_{em\_des} - C_d) \cdot \Omega - e_k \cdot i_k}{(e')^T \cdot (e')} \cdot (e') \qquad [3.87]$$

3.6.6.2. Case 2: a fault current that is independent of other phase currents, with a homopolar current of 0 on n − 1 phases

The sum of currents of $n - 1$ non-faulty phases must be 0. A third constraint is added that concerns the homopolar current on $n - 1$ phases:

$$\sum_{\substack{j=1 \\ j \neq k}}^{n} i_j = (w_1)^T \cdot (i) = 0$$

[3.88]

We then get:

$$(B_2) = \begin{bmatrix} 0 & 0 & \ldots & 1 & \ldots & 0 \\ 1 & 1 & \ldots & 0 & \ldots & 1 \\ e_1 & e_2 & \ldots & e_k & \ldots & e_n \end{bmatrix} \text{ and } (C_2) = \begin{bmatrix} i_k \\ 0 \\ (C_{em\_des} - C_d) \cdot \Omega \end{bmatrix} \qquad [3.89]$$

Then:

$$(i)_{opt-2} = (B_2)^T \cdot \left((B_2) \cdot (B_2)^T \right)^{-1} \cdot (C_2) \qquad [3.90]$$

which is equivalent to:

$$(Q) \cdot (i)_{opt-2} = \frac{(C_{em\_des} - C_d) \cdot \Omega - e_k \cdot i_k}{(e'')^T \cdot (e'')} \cdot (e'') \qquad [3.91]$$

3.6.6.3. *Case 3: a fault current independent of the other phase currents, with homopolar current of 0 on n phases*

In this case:

$$\sum_{j=1}^{n} i_j = (u_1)^T \cdot (i) = 0 \qquad [3.92]$$

We then get:

$$(B_3) = \begin{bmatrix} 0 & 0 & \dots & 1 & \dots & 0 \\ 1 & 1 & \dots & 1 & \dots & 1 \\ e_1 & e_2 & \dots & e_k & \dots & e_n \end{bmatrix} \text{ and } (C_3) = \begin{bmatrix} i_k \\ 0 \\ (C_{em\_des} - C_d) \cdot \Omega \end{bmatrix} \qquad [3.93]$$

then:

$$(i)_{opt-3} = (B_3)^T \cdot \left((B_3) \cdot (B_3)^T\right)^{-1} \cdot (C_3) \qquad [3.94]$$

which is equivalent to:

$$(Q) \cdot (i)_{opt-3} = \frac{(C_{em\_des} - C_d - C_p) \cdot \Omega}{(e")^T \cdot (e")} \cdot (e") - \frac{i_k \cdot \left((i)_{opt-1} - (i)_{opt-2}\right)}{(w_1)^T \cdot (i)_{opt-1}} \qquad [3.95]$$

3.6.6.4. *Application to a five-phase machine and independent phases with two phases in open circuit*

Let us consider the example of a five-phase machine with sinusoidal *back emfs* and two faulty phases and currents of 0. In we check that solutions $(i)_{opt-0}$ and $(i)_{opt-1}$ overlap in accordance with rule 3.3. We check in that solutions $(i)_{opt-2}$ and $(i)_{opt-3}$ overlap in accordance with rule 3.5. Figure 3.17 allows us to compare the currents of phase *a* obtained using different solutions on the same graph. We also check that the best ohmic losses are obtained for $(i)_{opt-1}$ in accordance with Figure 3.13.

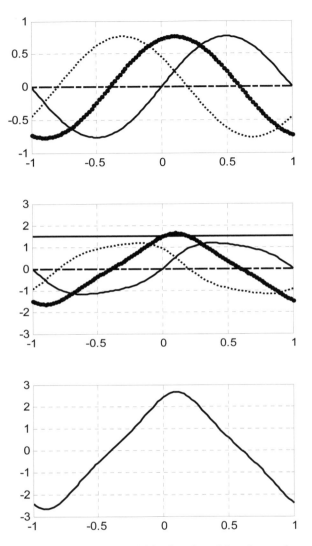

Figure 3.15. *Machine with sinusoidal back emfs and five phases, the two phases b and c are in open circuit; strategy* $(i)_{opt-0} = (i)_{opt-1}$

Synchronous Machines in Degraded Mode 107

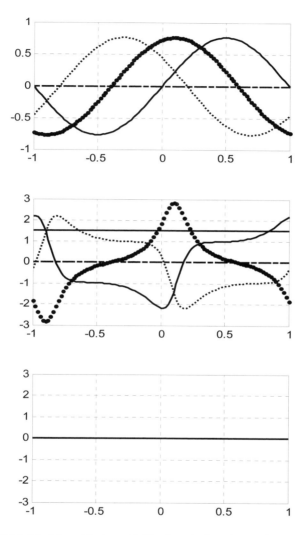

Figure 3.16. *Machine with sinusoidal back emfs, five phases, the two phases b and c are in open circuit; strategy* $(i)_{opt-2} = (i)_{opt-3}$

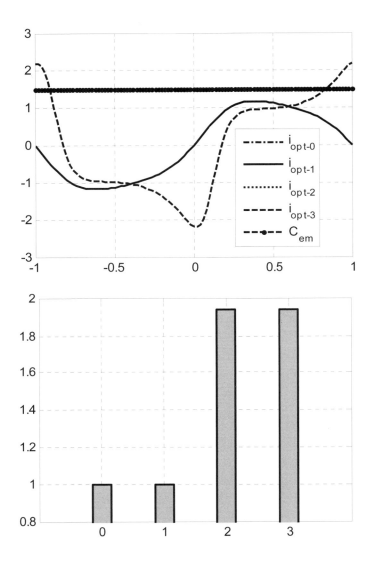

Figure 3.17. *Machine with sinusoidal back emfs with five phases, where phases b and c are in an open circuit. Top: comparison of the current obtained (phase a) for the different approaches, $(i)_{opt-0} = (i)_{opt-1}$, $(i)_{opt-2} = (i)_{opt-3}$. Bottom: comparison of ohmic losses*

3.6.6.5. *Application to a non-sinusoidal five-phase machine with a phase in open circuit*

Let us consider the example of a five-phase machine with non-sinusoidal *back emfs* and with a current of 0 on one phase. We will show in Figure 3.18 that solutions $(i)_{opt-2}$ and $(i)_{opt-3}$ overlap in accordance with rule 3.5 and give identical ohmic losses. Solution $(i)_{opt-1}$ is the one that gives the best ohmic losses in accordance with Figure 3.13.

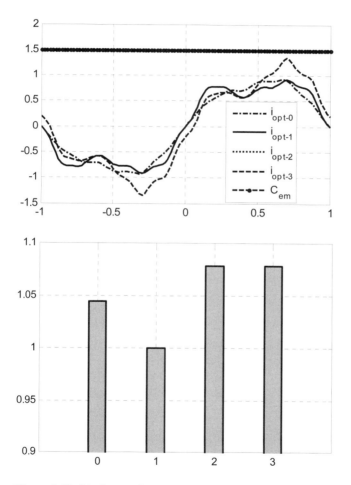

Figure 3.18. *Machine with non-sinusoidal back emfs and five phases, with phase b in open circuit*

3.6.6.6. *Application to a sinusoidal five-phase machine in the presence of saturation*

Let us consider the example of a five-phase machine with sinusoidal *back emfs* with a saturation at $\pm I_{sat}$. We can see that solutions $(i)_{opt-0}$ and $(i)_{opt-1}$ overlap, in accordance with rule 3.3, and give identical ohmic losses. Solution $(i)_{opt-3}$ allows us to obtain a homopolar current of 0 (see Figure 3.20). $(i)_{opt-1}$ is the solution that gives the best ohmic losses (see Figure 3.21) in accordance with Figure 3.13.

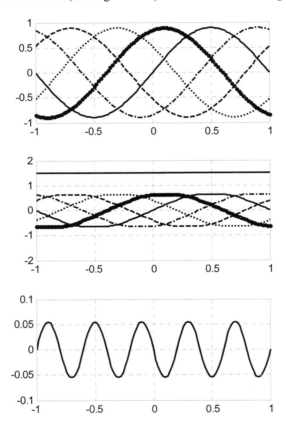

Figure 3.19. *Five-phase machine with sinusoidal back emfs, with saturation. From top to bottom: normalized back emfs, currents and torque, homopolar current. Solution* $(i)_{opt-0} = (i)_{opt-1}$

Synchronous Machines in Degraded Mode   111

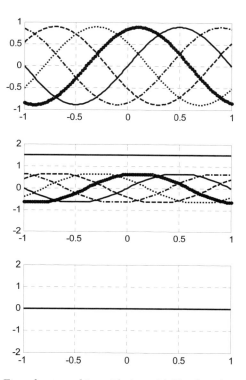

Figure 3.20. *Five-phase machine with sinusoidal back emfs, with saturation. From top to bottom: normalized back emfs, currents and torque, homopolar current. Solution* $(i)_{opt-3}$

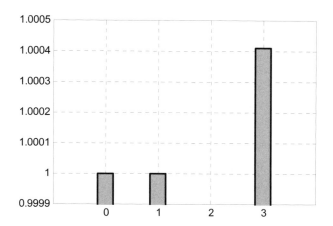

Figure 3.21. *Comparison of the ohmic losses for solutions opt-0, opt-1 and opt-3*

112   Control of Non-conventional Synchronous Motors

3.6.6.7. *Application to a five-phase machine with sinusoidal back emfs and phase a in short circuit*

Let us consider the example of a five-phase machine with sinusoidal *back emfs* with phase a in short circuit and a unipolar current (blocking of a converter diode). Figure 3.22 shows that solutions $(i)_{opt-0}$ and $(i)_{opt-1}$ overlap in accordance with rule 3.3 and give identical ohmic losses. $(i)_{opt-1}$ is the solution that gives the best ohmic losses (see Figure 3.23) in accordance with Figure 3.13.

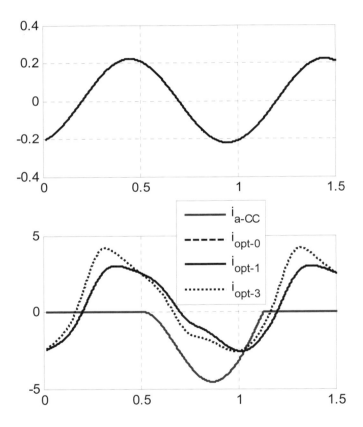

Figure 3.22. *Five-phase machine with phase a in short-circuit: a comparison of solutions*

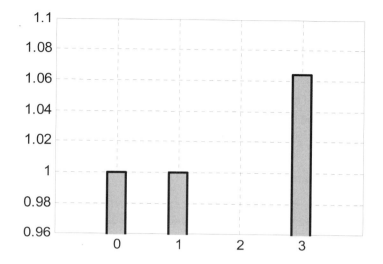

Figure 3.23. *Five-phase machine with phase a in short-circuit: a comparison of the ohmic losses*

3.7. Experimental learning strategy in closed loop to obtain optimal currents in all cases

The main ideas for control in closed loop come from the careful observation of formulae of the optimized currents *opt-0*, *opt-1* and *opt-2*. We remind ourselves of them here:

$$(i)_{without\_fault\_opt-0} = \frac{\left(C_{em\_des} - C_d - C_p\right) \cdot \Omega}{(e)^T \cdot (Q) \cdot (e^1)} \cdot (Q) \cdot (e^1) \qquad [3.96]$$

$$(i)_{without\_fault\_opt-1} = \frac{\left(C_{em\_des} - C_d - C_p\right) \cdot \Omega}{(e)^T \cdot (Q) \cdot (e)} \cdot (Q) \cdot (e) \qquad [3.97]$$

$$(i)_{without\_fault\_opt-2} = \frac{\left(C_{em\_des} - C_d - C_p\right) \cdot \Omega}{(e")^T \cdot (e")} \cdot (e") \qquad [3.98]$$

114   Control of Non-conventional Synchronous Motors

These three solutions depend on the perturbator torque and the cogging torque and are respectively proportional to (e^1), (e) and (e''); the coefficient of proportionality depends on the fault. In practice it is very difficult to exactly determine the perturbator torque C_p. For instance, in a case of short circuit between the turns of a same phase, C_p is going to depend on the number of turns in short circuit and on their position. The main idea is therefore to make the control learn these proportionality coefficients.

We can then rewrite $(i)_{opt-0} = k_{opt-0}(p\theta) \cdot (Q) \cdot \dfrac{(e^1)}{\Omega}$, where (e^1) is the vector of fundamental components of *back emfs*. For each non-faulty phase of the machine, the corresponding optimal current is the product of a scalar $k_{opt-0}(p\theta)$ and the fundamental component of the derivative of the inductor flux in the corresponding phase.

Similarly: $(i)_{opt-1} = \dfrac{(C_{em\_des} - C_d - C_p) \cdot \Omega}{(e)^T \cdot (Q) \cdot (e)} \cdot (Q) \cdot (e)$, which can also be written:

$(i)_{opt-1} = k_{opt-1}(p\theta) \cdot (Q) \cdot \dfrac{(e)}{\Omega}$, where (e) is the vector of *back emfs*. For each non-faulty phase of the machine, the corresponding optimal current is the product of a scalar k_{opt-1} and the derivative of the inductor flux in the corresponding phase.

We also have: $(i)_{opt-2} = \dfrac{(C_{em\_des} - C_d - C_p) \cdot \Omega}{(e'')^T \cdot (e'')} \cdot (e'')$, which can also be written:

$(i)_{opt-2} = k_{opt-2}(p\theta) \cdot \dfrac{(e'')}{\Omega}$, where (e'') is the vector of *back emfs* of non-faulty phases deprived of their homopolar component. For instance, for each non-faulty phase of a n phase machine, the corresponding optimal current is the product of a scalar $k_{opt-2}(p\theta)$ and of the derivative of the inductor flux in the corresponding phase deprived of its harmonic components or rank multiple of n. Figure 3.24 shows the simulation results of three proportionality coefficients: $k_{opt-0}(p\theta)$, $k_{opt-1}(p\theta)$ and $k_{opt-2}(p\theta)$. It appears that these coefficients are periodical, have a non-zero mean value and can easily be broken down in Fourier series.

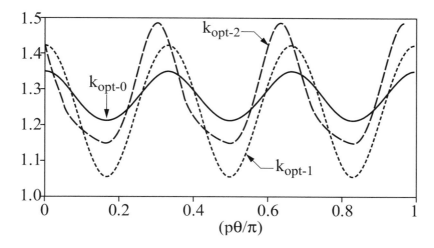

Figure 3.24. *Three optimal functions $k_{opt-i}(p\theta)$, i=1, 2 and 3 for a machine with trapezoidal back emfs*

Figure 3.25 of the neural control of the torque of the three-phase synchronous machine comes from these observations. According to the strategy chosen (and so to the possibilities of the inverter and machine coupling), functions $k_{opt-0}(p\theta)$, $k_{opt-1}(p\theta)$ or $k_{opt-2}(p\theta)$ will be learnt and synthesized by the Adaline network of the torque controller if the switch is in position 1.

There are three possible strategies:

– Strategy 0: only the fundamental component of the derivative of the flux is used. This strategy can be used for all machines and all couplings with or without connected neutral.

– Strategy 1: the derivative of the flux is used completely, with all its harmonic components. This strategy can be used for star-coupled machines without connection of the neutral, which does not have harmonic components that are multiples of three in their *back emfs*, or all the machines that support a homopolar current.

– Strategy 2: the derivative of the flux is used without its harmonic components of rank multiples of three. This strategy can be used for all star-coupled machines without connection of the neutral.

In Figure 3.25, the torque controller acts as an integrator (LMS algorithm) to supply a value of $k_{opt-i}(p\theta)$, $i = 0,1,2$ (according to the strategy adopted) necessary to generate the desired torque (the cogging torque is taken into account in the torque calculation). Besides the current, servo-control errors are compensated for by the neural controller. Details of the implementation are given in [NGU 10].

If there is one switch in speed regulation (switch in position 0), the speed ripples resulting from the torque ripples (taking into account the cogging torque and the perturbation torque) are reflected by a PLL locked on the signal of the incremental sensor of position operating as a synchronous demodulator. The Adaline network acts so that the speed measured is always equal to the reference speed. In this case, the torque will be the exact torque necessary to drive the machine with the criterion resulting from the chosen strategy.

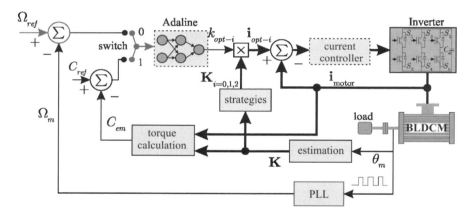

Figure 3.25. *Neural diagram to obtain optimal currents*

3.8. Simulation results

Figure 3.26 shows the case of a three-phase machine with trapezoidal *back emf* without fault, with cogging torque, in steady state for the three strategies 0, 1 and 2. Figure 3.27 shows the same machine but in the presence of a fault in phase *c*, the control convergence. Few periods of the current can reach the optimal currents and adapt to the fault. This shows the great interest in self-adaptive control.

Synchronous Machines in Degraded Mode 117

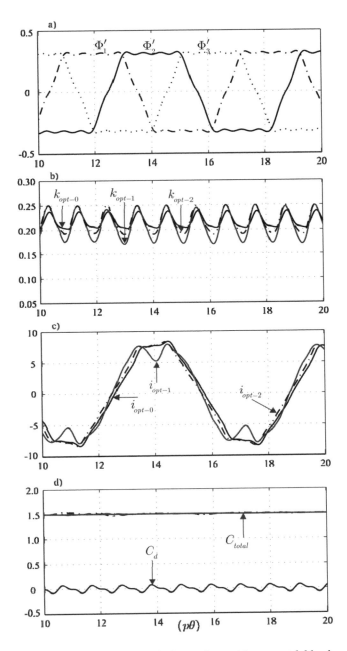

Figure 3.26. *Results obtained with a non-faulty machine with trapezoidal back emfs in steady state. The cogging torque is compensated for*

118   Control of Non-conventional Synchronous Motors

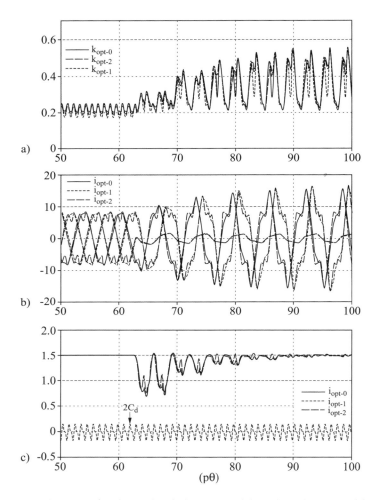

Figure 3.27. *Results obtained with the trapezoidal machine: learning of the Adaline network with a fault in phase c*

3.9. General conclusion

In this chapter we initially showed that the failures of a converter-synchronous machine set mainly concerned the static converter and, to a lesser extent, the coils of the machine. Thus, the main faults that we see mostly concern short circuit and open circuits, the first often leading to the second. We have also tried to establish supply modes allowing us to produce the desired torque on the machine shaft, to compensate for the perturbator torque due to the faulty phases and the cogging torque, while minimizing the ohmic losses. We have established the formulae for currents that allow us to answer the different issues and this has distinguished the possibilities offered by the coupling of the machine's coils and the nature of the fault. Conditions on the number of non-faulty phases either allow or disallow these different strategies. In particular, in a star-coupled machine, if the fault is a short circuit on one phase inside the motor, a homopolar current of 0 will only concern the non-faulty phases. On the other hand, if the short circuit is at the level of the static converter the homopolar current of 0 will concern all the phases of the machine, including the faulty one. These theoretical formulae require knowledge of the perturbator torque and the cogging torque, and are therefore difficult to determine experimentally. The closed loop strategy, based on the online learning of the artificial neural network, is thus proposed to obtain the machine's supply currents. Results from simulations have allowed us to emphasize the main properties of these supply currents, and this is for different examples of faulty synchronous machines.

3.10. Glossary

p : number of pole pairs

n : number of phases

θ : instantaneous position of the rotor

Ω : mechanical rotor speed

t : time

$(i) = \begin{pmatrix} i_1 & i_2 & \cdots & i_n \end{pmatrix}^T$: stator currents

$(i) = (i)_{fault} + (i)_{without\_fault}$

$(i)_{fault} = (Q') \cdot (i)$

$(i)_{without\_fault} = (Q) \cdot (i)$

$(i)_{opt-i}$: optimal currents of the ith strategy

k_{opt-i} : optimal function of the ith strategy

$(e) = \begin{pmatrix} e_1 & e_2 & \cdots & e_k & \cdots & e_n \end{pmatrix}^T$: back electromotive forces (*back emfs*)

120 Control of Non-conventional Synchronous Motors

$(e^1) = \begin{pmatrix} e_1^1 & e_2^1 & \cdots & e_k^1 & \cdots & e_n^1 \end{pmatrix}^T$: vector gathering the fundamental components of *back emfs*

$(e') = \begin{pmatrix} e_1 & e_2 & \cdots & 0 & \cdots & e_n \end{pmatrix}^T = (Q) \cdot (e)$: *back emfs* of non-faulty phases

$(e'') = (e') - \dfrac{(w_1)^T \cdot (e')}{(w_1)^T \cdot (w_1)} \cdot (w_1)$

$(u_1) = \begin{pmatrix} 1 & 1 & \cdots & 1 & \cdots & 1 \end{pmatrix}^T$: unit vector

$(w_1) = \begin{pmatrix} 1 & 1 & \cdots & 0 & \cdots & 1 \end{pmatrix}^T = (Q) \cdot (u_1)$: vector with 1 as the component for the non-faulty phases and 0 for faulty phases; it has n_d zeros

$(w'_1) = \begin{pmatrix} 0 & 0 & \cdots & 1 & \cdots & 0 \end{pmatrix}^T = (Q') \cdot (u_1)$: vector with 0 as the component for non-faulty phases and 1 for faulty phases

$n_d = (w'_1)^T \cdot (w'_1) = (u_1)^T \cdot (Q') \cdot (u_1)$: number of faulty phases

ψ_{jf} : stator flux of phase j due to the rotor

(Ψ_f) : stator flux vector of the phase due to the rotor

i_k : current of phase k

C_d : cogging torque

$C_p \cdot \Omega = (e_f)^T \cdot (Q') \cdot (i)_{fault} = (e_f)^T \cdot (i)_{fault}$

I_{ref} I_{sat} : reference current and saturation current

$C_{em-stat}$, $C_{em-des} = C_{ref}$: stator electromagnetic torques, desired electromagnetic torque and reference electromagnetic torque

L_{jk} : mutual inductance between phases j and k

J_1 : function representing the stator ohmic losses

L_1 : Lagrange function to be minimized

λ_1 : Lagrange multiplier

(P): hyperplan defining the desired torque

(H_1): hyperplan defining the neutral current od 0

(H_1'): hyperplan defining the neutral current of 0 with the faulty phases

M_{opt-i}: optimal points of coordinates equal to $(i)_{opt-i}$

3.11. Bibliography

[AGH 00] AGHILI F., BUEHLER M., HOLLERBACH J.M., "Optimal commutation laws in the frequency domain for PM synchronous direct-drive motors", *IEEE Transactions on Power Electronics*, vol. 15, no. 6, 2000.

[AGH 08] AGHILI F., "Adaptive reshaping of excitation currents for accurate torque control of brushless motors", *IEEE Transactions on Control Systems Technology*, vol. 16, no. 2, 2008.

[ATA 03] ATALLAH K., WANG J., HOWE D., "Torque ripple minimization in modular permanent-magnet brushless machines", *IEEE Transactions on Industry Applications*, vol. 39, no. 6, 2003.

[BIR 96] BIROLINI A., *Zuverlässigkeit von Geräten und Systemen*, Springer, Germany, 1996.

[BOL 00] BOLOGNANI S., ZORDAN M., ZIGLIOTTO M., "Experimental fault-tolerant control of a PMSM drive", *IEEE Transactions on Industrial Electronics*, vol. 47, no. 5, 2000.

[CHA 99] CHAPMAN P. L. SUDHOFF S., "Optimal control of permanent-magnet AC machine drives with a novel multiple reference frame estimator/regulator", *34th Annual Conference on Industry Applications in Society*, 3rd -7th October, Phoenix, USA,1999.

[CLE 93] CLENET S., LEFEVRE Y., SADOWSKI N., ASTIER S., LAJOIE-MAZENC M., "Compensation of permanent magnet motors torque ripple by means of current supply waveshapes control determined by finite element method", *IEEE Transactions on Magnetics*, vol. 29, no. 2, 1993.

[DOD 95] US Department of Defense, Military Handbook – Reliability Prediction of Electronic Equipment, MIL-HDBK-217F, US DoD, February 1995.

[DWA 07] DWARI S., PARSA L., LIPO T. A., "Optimal control of a five-phase integrated modular permanent magnet motor under normal and open-circuit fault conditions", *Power Electronics Specialists Conference (PESC 07)*, pp. 1639-1644, 2007.

[DWA 08] DWARI S., L. PARSA, "An optimal control technique for multiphase PM machine under open-circuit faults", *The IEEE Transaction on Industrial Electronics*, vol. 55, no. 5, 2008.

[DWA 11] DWARI S., PARSA L., "Fault tolerant control of five-phase permanent-magnet motors with trapezoidal back EMF", *IEEE Transactions on Industrial Electronics*, vol. 58, no. 2, 2011.

[FAV 90] FAVRE E., JUFER M., "Cogging torque suppression by a current control", *Procedings of the International Conference on ElectricMachines*, Cambridge, MA, pp. 601-606, 1990.

[FAV 93] FAVRE E., CARDOLETTI L., JUFER M., "Permanent-magnet synchronous motors: a comprehensive approach to cogging torque suppression", *IEEE Trans. Ind. Applicat.*, vol. 29, pp. 1141-1149,1993.

[FLI 08] FLIELLER D., GRESSIER J., STURTZER G., OULD ABDESLAM D., WIRA P., "Optimal currents based on Adalines to control a permanent magnet synchronous machine", 34th *Annual Conference of IEEE, (IECON 2008)*, p . 2702-2707, November 10-13, 2008.

[FLI 11] FLIELLER D., LOUIS J.-P., STURTZER G., NGUYEN N.K., "Optimal Supply and Synchronous Motors Torque Control : Designs in the a-b-c Reference Frame", in: *Control of Synchronous Motors*, J.-P. LOUIS (ed.), ISTE Ltd., London and John Wiley & Sons, New York, 2011.

[HAY 99] HAYLOCK J.A., MECROW B.C., JACK A.G., ATKINSON D.J., "Operation of fault tolerant machines with winding failures", *IEEE Trans. Energy Conv.*, vol. 14 , pp 1490-1495, 1999.

[HUN 92] HUNG J.Y., DING Z., "Minimization of torque ripple in permanent magnet motors – A closed form solution", *International Conference on Industrial Electronics, Control, Instrumentation, and Automation*, vol. 1, pp. 459-463, 1992.

[HUN 93] HUNG J.Y., DING Z., "Design of currents to reduce torque ripple in brushless permanent magnet motors", *IEE Proceedings-B*, vol. 140, no. 4, 1993.

[HUN 94] HUNG J.Y., "Design of the most efficient excitation for a class of electric motor", *IEEE Transactions on Circuits and Systems-I: Fundamental Theory and Applications*, vol. 41, no. 4, 1994.

[JOK 98] JOKSIMOVIC G., PENMAN J., "The detection of interturn short circuits in the stator windings of operating motors", *Proc. IEEE Conf. IECON '98*, vol. 4, pp. 1974-1979, 1998.

[KES 09] KESTELYN X., SEMAIL E., CREVITS Y., "Generation of on-line optimal current references for multi-phase permanent magnet machines with open-circuited phases", *IEEE International Electric Machines and Drives Conference (IEMDC 09)*, pp. 689-694, 2009.

[KOG 03] KOGURE H., SHINOHARA K., NONAKA A., "Magnet configurations and current control for high torque to current ratio in interior permanent magnet synchronous motors", *The IEEE International on Electric Machines and Drives Conference*, vol. 1, pp. 353-259, 2003.

[KRU 03] KRÜGER H., Betriebsdiagnose bei Elektronikmotoren, doctoral thesis, Technische Universität Carolo-Wilhelmina zu Braunschweig, Bosch, 2003.

[LAJ 95] LAJOIE-MAZENC M., VIAROUGE P., "Alimentation des machines synchrones, traité de génie electriques", D3630 and D 3631, *Techniques de l'Ingénieur*, 1991.

[LEB 99] LEBEY T., CAMBRONE J.P., "CASTELAN P., "Problèmes posés par l'utilisation de variateurs de vitesse à l'isolation des machines tournantes basse tension", *Proc. EF'99 Lilles*, March 1999.

[LI 09] LI L., JI H., ZHANG L., SUN H., "Study on torque ripple attenuation for BLDCM based on vector control method", *2009 2nd International Conference on Intelligent Network and Intelligent Systems*, pp. 605-608, 2009.

[LU 08] LU H., ZHANG L., QU W., "A new torque control method for torque ripple minimization of BLDC motors with un-ideal back EMF", *IEEE Transactions on Power Electronics*, vol. 23, no. 2, 2008.

[NGU 10] NGUYEN N-K., Approche neuromimétique pour l'identification et la commande des systèmes électriques: application au filtrage actif et aux actionneurs synchrones, Doctoral thesis, UHA, 2010.

[PEN 94] PENMAN J., SEDDING H.G., FINK W.T., "Detection and location of interturn short circuits in the stator windings of operating motors", *Trans. Energy Conversion*, vol. 9, pp. 652-658, 1994.

[SCH 03] SCHWAB H., KLÖNNE A., RECK S., RAMESOHL I., STRUTZER G., KEITH B., "Reliability evaluation of a permanent magnet synchronous motor drive for an automotive application", *Proc. Conf. Power Electr. Applicat. EPE 2003*, 2003.

[SCH 04] SCHWAB H., Stratégies de commande d'actionneurs synchrones à aimants permanents intégrant la sureté de fonctionnement, Doctoral thesis, UHA, 2004.

[SHA 08] SHAMSI-NEJAD M. A., NAHID-MOBARAKEH B., PIERFEDERICI S., MEIBODY-TABAR F., "Fault tolerant and minimum loss control of double-star synchronous machines under open phase conditions", *IEEE Transactions on Industrial Electronics*, vol. 55, no. 5, 2008.

[STU 02] STURTZER G., FLIELLER D., LOUIS J.-P., "Extension of the Park's transformation applied to non-sinusoidal saturated synchronous motors", *EPE Journal*, vol. 12, no. 3, pp. 16-20, 2002.

[STU 03] STURTZER G., FLIELLER D., LOUIS J.-P., "Mathematical and experimental method to obtain the inverse modelling of nonsinusoidal and saturated synchronous reluctance motors", *IEEE Transactions on Energy Conversion*, vol. 18, no. 4, pp. 494-500, 2003.

[WAL 88] WALLACE A., SPEE R., "The simulation of brushless DC drive failures", *Proc. PESC '88*, Kyoto, Japan, April 1988.

[WAN 03] WANG J., ATALLAH K., HOWE D., "Optimal torque control of fault-tolerant permanent magnet brushless machines", *IEEE Transactions on Magnetics*, vol. 39, no. 5, 2003.

[WU 03] WU A.P., CHAPMAN P.L., "Cancellation of torque ripple due to non-idealities of permanent magnet synchronous machine drives", *34th Annual Conference of Power Electronics Specialist PESC '03*, vol. 1, pp. 256- 261, 2003.

[WU 05] WU A.P., CHAPMAN P.L., "Simple expressions for optimal current waveforms for permanent-magnet synchronous machines drive", *IEEE Transactions on Energy Conversion*, vol. 20, no. 1, 2005.

[XU 08] XU F., LI T., TANG P., "A low cost drive strategy for BLDC motor with low torque ripples", *3rd IEEE Conference on Industrial Electronics and Applications (ICIEA 2008)*, pp. 2499-2502, 2008.

[ZHU 09] ZHU J., ERTUGRUL N., SOONG W.L., "Fault remedial strategies in a fault-tolerant brushless permanent magnet AC motor drive with redundancy", *6th International Conference on Power Electronics and Motion Control (IPEMC 09)*, pp. 423-427, 2009.

Chapter 4

Control of the Double-star Synchronous Machine Supplied by PWM Inverters

4.1. Introduction

In high-power applications, such as naval propulsion, the constraints of power segmentation, reliability and safety of functioning have led to the development of specific electrical actuators created by associating machines with a high number of phases with static converters. Thus, during the 1980s, the double-star synchronous machine with two current commutators was used for the electrical propulsion systems of ships [BEN 98, KET 95, KHE 95, KOT 96]. This supply mode, with thyristor current commutators, ensured the simplicity and reliability of the electrical actuator. However the supply of the machine with pulses of currents rich in harmonics led to ripples in the electromagnetic torque. A shift of 30° between the two stars of the machine stator reduced these torque ripples [KHE 95, WER 84]. The use of the smoothing inductance necessary to manufacture the direct current (DC) source supplying the commutator, and thyristor type semi-conductor components decreased the performance of the electrical actuator in dynamic state.

Since the end of the 1990s, thanks to progress made in the field of power electronics and industrial computer science, a new power electronic converter – the double star synchronous machine – is being investigated. The supply of double star synchronous machines with PWM inverters produces sinusoidal currents in the machine windings [ABB 84, NEL 73, SIN 02] that considerably decrease the torque ripples. Furthermore, the vectorial control algorithms that are well known for the

---

Chapter written by Mohamed Fouad BENKHORIS.

three phase synchronous machine [LEO 96, LOU 95, VAS 94] can be extended to the double-star synchronous machine to improve the performance in static and dynamic states.

This chapter is devoted to the vectorial control of the double-star synchronous machine supplied with two PWM inverters. The control strategy is similar to that of a three-phase synchronous machine; an internal loop to control the torque and an external loop to control the speed. Usually for a three-phase machine, control of the torque requires control of currents in reference frame dq linked to the rotor and the management of specific problems, such as the coupling between axes d and q [LOU 10a, LOU 10b]. As for the three-phase synchronous machine, the control algorithm of the double-star synchronous machine is closely linked to the dynamic model of the machine. Two modeling approaches can be distinguished. The first considers the machine as two three-phase machines that are magnetically coupled [TER 00]. In this case, we apply the classic Concordia and then Park transformations a [LOU 04a, LOU 04b] to each star and manage the magnetic coupling between the two stars. The control algorithm of the total torque controls the torques produced by the two stars in a way similar to that of a classical three-phase machine and manages the new coupling introduced by the two stars. The second approach consists of managing the coupling at the modeling level [MER 05]. In this context, the machine is described on an orthonormal basis whose dimension is identical to that of the initial basis. The modeling principle requires us to generalize the vectorial formalism developed for polyphase machines with equally distributed phases [SEM 00, SEM 10] in the case of double-star or even multi-star machines [BEN 04]. Hence, the current vector is projected in an orthonormal space with six dimensions. All currents are decoupled. The first two components of current vector are equivalent to those of the classical triphase machine in Concordia's frame. Then, the application of the usual Park's transformation introduces the classical coupling. The other four currents are non-sequential ones and remain decoupled from the two currents in Park's frame.

Briefly, the problem with the control of a double-star synchronous machine control is linked to the dynamic modeling of the machine. Thus, after a quick description of the electrical actuator and a reminder of the basic matricial electrical equations of the machine, we explain the dynamic models of the machine. The last part of this chapter is dedicated to the specific problems linked to control, such as the management of decoupling and the generation of references in relation to control.

4.2. Description of the electrical actuator

Each star of the machine is supplied with its own three-phase PWM inverter. The two inverters can be supplied with the same DC bus or with two different DC buses. In this study, we consider the most general case, illustrated in Figure 4.1. It consists of supplying each inverter with its own DC voltage source. We assume that the two DC buses have the same amplitude.

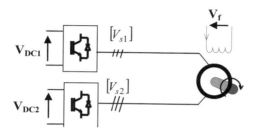

Figure 4.1. *Supply structure of the double-star synchronous machine*

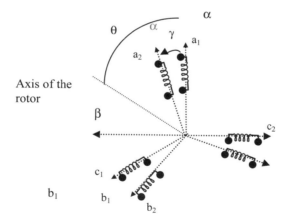

Figure 4.2. *Distribution of statoric phases of the double-star synchronous machine*

Like every rotating electrical machine, the double-star synchronous machine consists of a stator and a rotor. The machine considered is a salient pole synchronous machine. The stator illustrated in Figure 4.2 has two three-phase windings shifted by an angle γ generally equal to 30°. The rotor is identical to a classical synchronous machine. It can be with permanent magnets or a coiled rotor. In the latter case it includes an inductor winding supplied with DC and damper windings modeled by two windings in short-circuit: one on the direct axis of the rotor; and the other on the quadrature axis. The inductor winding is supplied with a DC voltage V_f via a current-controlled chopper.

4.3. Basic equations

4.3.1. *Voltage equations*

The models developed from now on are based on the following hypotheses:

– the two three-phase windings making the stator are identical, electrically insulated and shifted by an angle γ;

– the magnetomotive forces have a sinusoidal distribution;

– only the first space harmonics are considered and the mutual inductances are characterized by their fundamental component alone; and

– the magnetic saturation is disregarded.

The electrical equations of the machine are written in the following general matricial form:

$$\begin{bmatrix}[V_{s1}]\\ [V_{s2}]\\ [V_R]\end{bmatrix} = [R]\begin{bmatrix}[I_{s1}]\\ [I_{s2}]\\ [I_R]\end{bmatrix} + \frac{d}{dt}[L]\begin{bmatrix}[I_{s1}]\\ [I_{s2}]\\ [I_R]\end{bmatrix} \qquad [4.1]$$

The current and voltage vectors are defined as follows:

$$\begin{cases}[V_{s1}]=\begin{bmatrix}v_{a1}&v_{b1}&v_{c1}\end{bmatrix}^t\\ [V_{s2}]=\begin{bmatrix}v_{a2}&v_{b2}&v_{c2}\end{bmatrix}^t\end{cases}, \quad \begin{cases}[I_{s1}]=\begin{bmatrix}i_{a1}&i_{b1}&i_{c1}\end{bmatrix}^t\\ [I_{s2}]=\begin{bmatrix}i_{a2}&i_{b2}&i_{c2}\end{bmatrix}^t\end{cases}$$

$$\begin{cases}[V_R]=\begin{bmatrix}v_F&0&0\end{bmatrix}^t\\ [I_R]=\begin{bmatrix}i_F&i_D&i_Q\end{bmatrix}^t\end{cases}$$

The resistance matrix is diagonal and the terms in it are the values of the resistances of the different windings:

$$[R]=\begin{bmatrix}r_s[1]_3 & [0]_{3\times3} & [0]_{3\times3}\\ [0]_{3\times3} & r_s[1]_3 & [0]_{3\times3}\\ [0]_{3\times3} & [0]_{3\times3} & [R_R]_{3\times3}\end{bmatrix}; \quad [R_R]=\begin{bmatrix}r_F & 0 & 0\\ 0 & r_D & 0\\ 0 & 0 & r_Q\end{bmatrix} \qquad [4.2]$$

The inductance matrix can be broken down on its diagonal according to the self-inductance matrices of the different winding systems: the first star, the second star and the rotor. The other inductance matrices are introduced by the magnetic coupling between the winding systems:

$$[L] = \begin{bmatrix} [L_{s1}] & [M_{s1s2}] & [M_{s1R}] \\ [M_{s1s2}]^t & [L_{s2}] & [M_{s2R}] \\ [M_{s1R}]^t & [M_{s2R}]^t & [L_R] \end{bmatrix}$$
[4.3]

Matrices $[L_{s1}]$ and $[L_{s2}]$ represent the inductance matrices of the first and second star, respectively. Each of them is broken down as the sum of a constant matrix and of a matrix varying as a function of the rotor position reflecting the saliency of the machine:

$$[L_{sx}] = [L_{ss0}] + [L_{ss}(\beta_x)]$$
[4.4]

with:

$$[L_{ss0}] = \begin{bmatrix} L_s & M_{ss}\cos(\frac{2\pi}{3}) & M_{ss}\cos(\frac{2\pi}{3}) \\ M_{ss}\cos(\frac{2\pi}{3}) & L_s & M_{ss}\cos(\frac{2\pi}{3}) \\ M_{ss}\cos(\frac{2\pi}{3}) & M_{ss}\cos(\frac{2\pi}{3}) & L_s \end{bmatrix}$$

$$[L_{ss}(\beta_x)] = M_{sfm} \begin{bmatrix} \cos(2\beta_x) & \cos(2\beta_x - \frac{2\pi}{3}) & \cos(2\beta_x + \frac{2\pi}{3}) \\ \cos(2\beta_x - \frac{2\pi}{3}) & \cos(2\beta_x + \frac{2\pi}{3}) & \cos(2\beta_x) \\ \cos(2\beta_x + \frac{2\pi}{3}) & \cos(2\beta_x) & \cos(2\beta_x - \frac{2\pi}{3}) \end{bmatrix}$$

Angle β_x depends on electric position θ and on the given star ($x = 1, 2$):

$$\beta_x = \theta - (x-1)\frac{\pi}{6}$$
[4.5]

The magnetic coupling between the two stars of the stator is characterized by inductance matrix $[M_{s1s2}]$, which is put in the following form:

$$[M_{s1s2}] = [M_{ss0}] + [M_{ss}(\theta)]$$
[4.6]

with:

$$[M_{ss0}] = M_{ss} \begin{bmatrix} \cos(\frac{\pi}{6}) & \cos(\frac{\pi}{6}+2\frac{\pi}{3}) & \cos(\frac{\pi}{6}-2\frac{\pi}{3}) \\ \cos(\frac{\pi}{6}-2\frac{\pi}{3}) & \cos(\frac{\pi}{6}) & \cos(\frac{\pi}{6}+2\frac{\pi}{3}) \\ \cos(\frac{\pi}{6}+2\frac{\pi}{3}) & \cos(\frac{\pi}{6}-2\frac{\pi}{3}) & \cos(\frac{\pi}{6}) \end{bmatrix}$$

$$[M_{ss}(\theta)] = M_{sfm} \begin{bmatrix} \cos(2\theta-\frac{\pi}{6}) & \cos(2\theta-\frac{\pi}{6}-2\frac{\pi}{3}) & \cos(2\theta-\frac{\pi}{6}+2\frac{\pi}{3}) \\ \cos(2\theta-\frac{\pi}{6}-2\frac{\pi}{3}) & \cos(2\theta-\frac{\pi}{6}+2\frac{\pi}{3}) & \cos(2\theta-\frac{\pi}{6}) \\ \cos(2\theta-\frac{\pi}{6}+2\frac{\pi}{3}) & \cos(2\theta-\frac{\pi}{6}) & \cos(2\theta-\frac{\pi}{6}-2\frac{\pi}{3}) \end{bmatrix}$$

Matrix $[M_{ss0}]$ is a constant matrix; it depends on the shift angle of the two stars equal to $\pi/6$. Matrix $[M_{ss}(\theta)]$ cancels out in the case of a machine with smooth poles.

Self-inductance L_s can be written as the sum of an inductance of statoric leaks and inductance M_{ss}.

$$L_s = l_{fs} + M_{ss} \qquad [4.7]$$

The coupling between the stator and rotor is characterized by two matrices $[M_{s1r}]$ and $[M_{s2r}]$:

$$[M_{sxR}] = [M_{sR}(\beta_x)] = \begin{bmatrix} M_{sF}\cos(\beta_x) & M_{sD}\cos(\beta_x) & -M_{sQ}\sin(\beta_x) \\ M_{sF}\cos(\beta_x - \frac{2\pi}{3}) & M_{sD}\cos(\beta_x - \frac{2\pi}{3}) & -M_{sQ}\sin(\beta_x - \frac{2\pi}{3}) \\ M_{sF}\cos(\beta_x + \frac{2\pi}{3}) & M_{sD}\cos(\beta_x + \frac{2\pi}{3}) & -M_{sQ}\sin(\beta_x + \frac{2\pi}{3}) \end{bmatrix} \qquad [4.8]$$

with: $x = 1, 2$

Since the rotor is identical to that of a classic three-phase machine, the rotor inductance matrix remains unchanged. It is written as:

$$[L_R] = \begin{bmatrix} L_F & M_{FD} & 0 \\ M_{FD} & L_D & 0 \\ 0 & 0 & L_Q \end{bmatrix} \qquad [4.9]$$

4.3.2. Equation of the electromagnetic torque

The general expression of the torque can be inferred from the derivative of the coenergy:

$$\Gamma = \frac{1}{2} \cdot p \cdot [i]^t \left(\frac{\partial}{\partial \theta} [L] \right) \cdot [i] \quad\quad [4.10]$$

from which:

$$\Gamma = \frac{1}{2} \cdot p \cdot \begin{bmatrix} [I_{s1}]^t & [I_{s2}]^t & [I_R]^t \end{bmatrix} \left(\frac{\partial}{\partial \theta} [L] \right) \cdot \begin{bmatrix} [I_{s1}] \\ [I_{s2}] \\ [I_R] \end{bmatrix} \quad\quad [4.11]$$

4.4. Dynamic models of the double-star synchronous machine

With the natural basis, as shown by the voltage equations of the double-star synchronous machine in equation [4.1], the inductance matrix of the stator (equation [4.3]) is full. When we are interested in establishing a model to control or study transient states, it is usual to look for a referential in which the inductance matrix is diagonal and the state variables are continuous variables. To achieve this, in the case of classic three-phase machines, electrical engineers have defined transformation matrices to change the referential. These transformations are the Concordia and Park matrices. The aim of this section, as in the case of three-phase machines, is to look for a new basis and define transformations to switch from the natural basis to this new basis. These transformations, which are in fact changes of reference point, form the basic tools for establishing dynamic models that aim to control double star-machines. Two modeling approaches can be distinguished. The first considers the machine as two three-phase systems. The machine is represented by two coupled submachines in the Park reference system. The second approach takes advantage of the symmetry property of the inductance matrix and considers the double-star machine to be a polyphase machine. It is important to underline, however, that contrary to a polyphase winding, with three-phase double winding the six phases are identical but are not equally distributed.

4.4.1. Dynamic model in referential $d_1q_1d_2q_2$

4.4.1.1. Definition of the transformations

As shown in Figure 4.2, the three-phase double winding constituting the stator consists of two three-phase windings shifted by an angle equal to $\pi/6$.

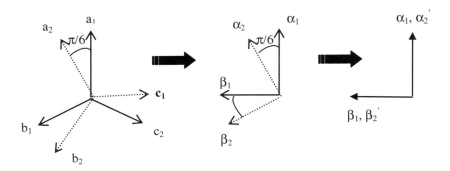

Figure 4.3. *Transformation of the three-phase double system into a two-phase double system*

The transformation from a three-phase double system to two two-phase orthonormal systems [MOU 99, TER 00] is illustrated in Figure 4.3. It consists of performing the following two steps:

a) applying the Concordia transformation to each three-phase system forming a star. This operation allows us to transform the two three-phase systems into two two-phase systems:

$$\begin{bmatrix} X_{\alpha 1} \\ X_{\beta 1} \\ X_{01} \end{bmatrix} = [T_{33}]^t \cdot \begin{bmatrix} X_{a1} \\ X_{b1} \\ X_{c1} \end{bmatrix} \quad \begin{bmatrix} X_{\alpha 2} \\ X_{\beta 2} \\ X_{02} \end{bmatrix} = [T_{33}]^t \cdot \begin{bmatrix} X_{a2} \\ X_{b2} \\ X_{c2} \end{bmatrix} \quad [T_{33}] = [[T_{32}] \ [T_{31}]] \quad [4.12]$$

with:

$$[T_{33}] = [[T_{32}] \ [T_{31}]] \quad ; \quad [T_{32}] = \sqrt{\frac{2}{3}} \cdot \begin{bmatrix} 1 & 0 \\ -\frac{1}{2} & -\frac{\sqrt{3}}{2} \\ -\frac{1}{2} & \frac{\sqrt{3}}{2} \end{bmatrix} \quad ; \quad [T_{31}] = \begin{bmatrix} \frac{1}{\sqrt{3}} \\ \frac{1}{\sqrt{3}} \\ \frac{1}{\sqrt{3}} \end{bmatrix}$$

b) apply a rotation of angle $\pi/6$ for the second two-phase system. This operation allows us to minimize the coupling between the two two-phase systems. In fact only the equivalent phases according to the same α or β axis show a magnetic coupling.

$$\begin{bmatrix} X'_{\alpha 2} \\ X'_{\beta 2} \\ X_{02} \end{bmatrix} = \begin{bmatrix} [T_{32}] \cdot P(\frac{\pi}{6}) \\ [T_{31}]^t \end{bmatrix}^t \cdot \begin{bmatrix} X_{a2} \\ X_{b2} \\ X_{c2} \end{bmatrix} = \begin{bmatrix} [T^*_{32}] \\ [T_{31}]^t \end{bmatrix} \cdot \begin{bmatrix} X_{a2} \\ X_{b2} \\ X_{c2} \end{bmatrix} \quad [4.13]$$

with:

$$P(\frac{\pi}{6}) = \begin{bmatrix} \cos\frac{\pi}{6} & \sin\frac{\pi}{6} \\ -\sin\frac{\pi}{6} & \cos\frac{\pi}{6} \end{bmatrix}$$

By assuming that the neutral points of the two stars are not linked to the sources supplying the inverters, we suppress the homopolar components and keep just the variables following referentials (α_1, β_1) and (α'_2, β'_2). In the same way as the three-phase synchronous machine, we apply the rotation of angle θ to the two two-phase systems:

$$\begin{bmatrix} [X_{dq1}] \\ [X_{dq2}] \end{bmatrix} = \begin{bmatrix} [P(\theta)]^t & [0]_{2x2} \\ [0]_{2x2} & [P(\theta)]^t \end{bmatrix} \begin{bmatrix} [X_{\alpha\beta1}] \\ [X'_{\alpha\beta2}] \end{bmatrix} \quad [4.14]$$

The model of the machine in Park referential (d,q) linked to the rotor is shown in Figure 4.4:

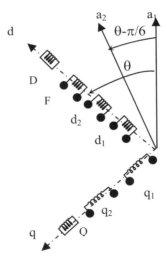

Figure 4.4. *Representation of the machine in referential dq*

4.4.1.2. *Voltage equations in referential $d_1q_1d_2q_2$*

By applying the transformations defined in equations [4.12], [4.13] and [4.14] to voltage equations [4.1], we can infer the electrical model of the synchronous machine with salient poles written in referential dq, or more precisely in referential $d_1q_1d_2q_2$:

$$\begin{bmatrix}[V_{dq1}]\\ [V_{dq2}]\\ [V_R]\end{bmatrix}=\begin{bmatrix}[R_{dq}] & [R_{dqm}] & [R_{dqS}]\\ [R_{dqm}] & [R_{dq}] & [R_{dqS}]\\ [0]_{3x2} & [0]_{3x2} & [R_{sG}]\end{bmatrix}\cdot\begin{bmatrix}[I_{dq1}]\\ [I_{dq2}]\\ [I_R]\end{bmatrix}+\begin{bmatrix}[L_{dq}] & [M_{dqm}] & [L_{dqS}]\\ [M_{dqm}] & [L_{dq}] & [L_{dqS}]\\ [L_{dqS}] & [L_{dqS}] & [L_R]\end{bmatrix}\cdot\frac{d}{dt}\begin{bmatrix}[I_{dq1}]\\ [I_{dq2}]\\ [I_R]\end{bmatrix}\quad [4.15]$$

where:

$$[V_{dqi}]=\begin{bmatrix}V_{di}\\ V_{qi}\end{bmatrix};\ [I_{dqi}]=\begin{bmatrix}I_{di}\\ I_{qi}\end{bmatrix}$$

Each matrix is defined as follows:

$$[R_{dq}]=\begin{bmatrix}r_s & -\omega L_q\\ \omega L_d & r_s\end{bmatrix}\ [R_{dqm}]=\omega\begin{bmatrix}0 & -L_{qm}\\ L_{dm} & 0\end{bmatrix},$$

$$[R_{dqS}]=\omega\begin{bmatrix}0 & 0 & -M_{qQ}\\ M_{dF} & M_{dD} & 0\end{bmatrix}\ [R_{sG}]=\begin{bmatrix}r_F & & \\ & r_D & \\ & & r_Q\end{bmatrix}\quad [4.16]$$

$$[L_{dq}]=\begin{bmatrix}L_d & 0\\ 0 & L_q\end{bmatrix}[M_{dqm}]=\begin{bmatrix}L_{dm} & 0\\ 0 & L_{qm}\end{bmatrix}[L_{dqS}]=\begin{bmatrix}M_{dF} & M_{dD} & 0\\ 0 & 0 & M_{qQ}\end{bmatrix}$$

with:

$$\begin{cases}L_d=L_s+\dfrac{M_{ss}}{2}+\dfrac{3}{2}M_{sfm}\\ L_q=L_s+\dfrac{M_{ss}}{2}-\dfrac{3}{2}M_{sfm}\end{cases}\text{ and }\begin{cases}L_{dm}=\dfrac{3}{2}M_{ss}+\dfrac{3}{2}M_{sfm}\\ L_{qm}=\dfrac{3}{2}M_{ss}-\dfrac{3}{2}M_{sfm}\end{cases}\quad [4.17]$$

$$M_{dF}=\sqrt{\dfrac{3}{2}}M_{SF};\quad M_{dD}=\sqrt{\dfrac{3}{2}}M_{SD};\quad M_{qQ}=\sqrt{\dfrac{3}{2}}M_{SQ}$$

By emphasizing the leakage inductances, the expressions of inductances given above become:

$$\begin{cases} L_d = l_{fs} + \frac{3}{2} M_{ss} + \frac{3}{2} M_{sfm} \\ L_q = l_{fs} + \frac{3}{2} M_{ss} - \frac{3}{2} M_{sfm} \end{cases} \quad [4.18]$$

By emphasizing the coupling inductances between the stars in their respective Park referentials, we can rewrite the inductances according to axes d and q in the following form:

$$\begin{cases} L_d = l_{fs} + L_{dm} \\ L_q = l_{fs} + L_{qm} \end{cases} \quad [4.19]$$

4.4.1.3. *Expression of the electromagnetic torque in referential $d_1q_1d_2q_2$*

By applying the transformations defined in equations [4.12], [4.13] and [4.14] to the general expression of torque [4.11], we infer the expression of torque in this referential. It is put in the form:

$$\Gamma = P\left\{ \left[M_{dF}i_F + M_{dD}I_D + \frac{1}{2}\left[(L_d - L_q) + (L_{dm} - L_{qm})\right](i_{d1} + i_{d2}) \right](i_{q1} + i_{q2}) - M_{qQ}I_Q(i_{d1} + i_{d2}) \right\} \quad [4.20]$$

By emphasizing the contribution of each star and the interaction between the two, expression [4.20] of torque can be rewritten as the sum of three terms:

$$\Gamma = \Gamma_1 + \Gamma_2 + \Gamma_{12} \quad [4.21]$$

with:

$$\begin{cases} \Gamma_1 = P\left(M_{dF}i_F i_{q1} + \frac{1}{2}\left[(L_d - L_q) + (L_{dm} - L_{qm})\right]i_{d1}i_{q1} + M_{dD}I_D i_{q1} - M_{qQ}I_Q i_{d1} \right) \\ \Gamma_2 = P\left(M_{dF}i_F i_{q2} + \frac{1}{2}\left[(L_d - L_q) + (L_{dm} - L_{qm})\right]i_{d2}i_{q2} + M_{dD}I_D i_{q2} - M_{qQ}I_Q i_{d2} \right) \\ \Gamma_{12} = \frac{P}{2}\left[(L_d - L_q) + (L_{dm} - L_{qm})\right](i_{d1}i_{q2} + i_{d2}i_{q1}) \end{cases}$$

In the case of a machine with smooth poles and without damper, the expression of torque is directly inferred from expression [4.21]:

$$\Gamma = P.M_{dF}.i_F.(i_{q1} + i_{q2}) \qquad [4.22]$$

We get a result similar to that of a three-phase synchronous machine with smooth poles. Only the currents following the quadrature axis generate torque. In fact, by replacing the sum of the two currents following axis q with the quadrature current in the case of a three-phase machine, we obtain the classic expression of torque in a three-phase machine in the Park referential.

4.4.2. Dynamic model in referential $dqz_1z_2\,z_3z_4$

In this context, we do not separate the two stars and attempt to establish a completely decoupled dynamic model. This model can be obtained in two different ways: either by taking advantage of the analysis and then using the dynamic model developed in section 4.4.1.2, or by rewriting the model on an orthonormal basis. In the latter case, we first need to establish the change of basis matrices.

4.4.2.1. Voltage equations in referential $d^+q^+d^-q^-$

Relationship [4.20] shows that the electromagnetic torque depends on the sum of currents in axis d or q [MER 05, MOU 99]. This remark leads us to define new state variables that are the sum of currents following the same axis ($i_{d1}+i_{d2}$, $i_{q1}+i_{q2}$). For reasons of bijectivity and to keep the initial dimension of the state vector, the differences between the respective currents following the same axis must appear in the state vector. From now on, the new state vector becomes:

$$[i_{d1}+i_{d2} \quad i_{q1}+i_{q2} \quad i_{d1}-i_{d2} \quad i_{q1}-i_{q2}]^t \qquad [4.23]$$

The dynamic model of the machine will then be expressed in a new referential, written (d^+, q^+, O^+, d^-, q^-, O^-). It is obtained from the model by respectively calculating the sums and differences of equations according to the different axes. In order to keep the norm of the current vector, we use the following normalized transformation:

$$\begin{bmatrix} x_d^+ \\ x_q^+ \\ x_d^- \\ x_q^- \end{bmatrix} = \frac{1}{\sqrt{2}} \begin{bmatrix} 1 & 0 & 1 & 0 \\ 0 & 1 & 0 & 1 \\ 1 & 0 & -1 & 0 \\ 0 & 1 & 0 & -1 \end{bmatrix} \begin{bmatrix} x_{d1} \\ x_{q1} \\ x_{d2} \\ x_{q2} \end{bmatrix} \qquad [4.24]$$

Starting from voltage equation [4.15], by introducing the variable change defined above we establish the new dynamic model:

$$\begin{bmatrix} V_d^+ \\ V_q^+ \end{bmatrix} = \begin{bmatrix} R_s & 0 \\ 0 & R_s \end{bmatrix} \begin{bmatrix} i_d^+ \\ i_q^+ \end{bmatrix} + \omega \begin{bmatrix} 0 & -L_q^+ \\ L_d^+ & 0 \end{bmatrix} \begin{bmatrix} i_d^+ \\ i_q^+ \end{bmatrix} + \begin{bmatrix} L_d^+ & 0 \\ 0 & L_q^+ \end{bmatrix} \frac{d}{dt} \begin{bmatrix} i_d^+ \\ i_q^+ \end{bmatrix}$$

$$+ \sqrt{2}.\omega \left[R_{dqs} \right].[I_R] + \sqrt{2}.\left[L_{dqs} \right] \frac{d}{dt}.[I_R] \qquad [4.25]$$

$$\begin{bmatrix} V_d^- \\ V_q^- \end{bmatrix} = \begin{bmatrix} R_s & 0 \\ 0 & R_s \end{bmatrix} \begin{bmatrix} i_d^- \\ i_q^- \end{bmatrix} + \begin{bmatrix} l_{fs} & 0 \\ 0 & l_{fs} \end{bmatrix} \begin{bmatrix} i_d^- \\ i_q^- \end{bmatrix} + \omega \begin{bmatrix} 0 & -l_{fs} \\ l_{fs} & 0 \end{bmatrix} \begin{bmatrix} i_d^- \\ i_q^- \end{bmatrix}$$

with:

$$L_d^+ = l_{fs} + 3.M_{ss} + 3.M_{sfm} \qquad L_q^+ = l_{fs} + 3.M_{ss} - 3.M_{sfm}$$

This dynamic model (equation [4.25]) consists of two completely decoupled sub-systems (d^+q^+) and (d^-q^-). The first sub-system (d^+q^+) is similar to that of a three-phase machine described in the Park referential (dq). The second sub-system (d^-q^-) is described by two decoupled passive circuits whose inductance is that of leakage. It is important to underline that in the framework of the sinusoidal distribution hypotheses, only the first system contributes to the electromechanical conversion of energy.

4.4.2.2. *Voltage equations in referential* $\alpha\beta z_1 z_2 z_3 z_4$

Using the approach developed above, it takes several steps to obtain a friendly work referential d^+,q^+,d^-,q^-. After applying the classic Concordia and Park transformations to each three-phase star of the double-star machine, the equations of the machine were still coupled. An analysis of the model obtained allowed us to define a third simple transformation (equation [4.24]) defining a new basis by which to describe the double-star machine using a dynamic model. Another modeling approach based on generalizing the vectorial formalism established for polyphase systems or the projection of different vectors in a six-dimensional orthonormal space leads to a decoupled dynamic model.

4.4.2.2.1. Generalization of the vectorial formalism

This approach exploits the symmetry property of inductance matrices of the windings that make up the stator. This property [BAS 95] implies that the inductance matrix is diagonalizable and that an orthogonal basis for eigenvectors exists. As a consequence, we know that there is a referential in which the equations of the machine are decoupled.

In [SEM 00], this basis has been obtained in the case of a machine whose phases are regularly distributed (the statoric windings are dephased by $2\pi/n$ for a winding with n phases). Yet, in the case of a double-star machine, as shown in Figure 4.2, the six phases are not regularly distributed. It is necessary to extrapolate the vectorial formalism to this type of machine. The dephasing between the two stars is 30°, as shown in Figure 4.5, so the equivalent polyphase machine has 12 phases that are regularly distributed [MAD 04].

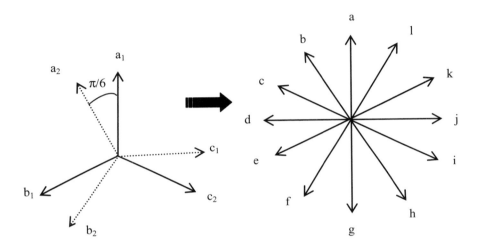

Figure 4.5. *Transformation of double-star winding into a 12-phase winding*

Knowing that the currents going through real phases (i_{a1}, i_{a2}, i_{b1}, i_{b2}, i_{c1}, i_{c2}) are linear combinations of the currents going through the fictitious phases (i_a, i_b, i_c, i_d, i_e, i_f, i_g, i_h, i_i, i_j), we infer matrix $[C_{12 \times 6}]$, allowing the transformation of a double-star system into an equivalent 12-phase system:

$$[i_a \; i_b \; i_c \; i_d \; i_e \; i_f \; i_g \; i_h \; i_i \; i_j \; i_k \; i_l]^t = [C_{12 \times 6}] \begin{bmatrix} i_{a1} \\ i_{a2} \\ i_{b1} \\ i_{b2} \\ i_{c1} \\ i_{c2} \end{bmatrix} \quad [4.26]$$

with:

$$[C_{12x6}] = \frac{1}{2}\begin{bmatrix} 1 & 0 & 0 & 0 & 0 & 0 & -1 & 0 & 0 & 0 & 0 & 0 \\ 0 & 1 & 0 & 0 & 0 & 0 & 0 & -1 & 0 & 0 & 0 & 0 \\ 0 & 0 & 0 & 0 & 1 & 0 & 0 & 0 & 0 & -1 & 0 \\ 0 & 0 & 0 & 0 & 0 & 1 & 0 & 0 & 0 & 0 & 0 & -1 \\ 0 & 0 & -1 & 0 & 0 & 0 & 0 & 1 & 0 & 0 & 0 \\ 0 & 0 & 0 & -1 & 0 & 0 & 0 & 0 & 1 & 0 & 0 \end{bmatrix}^t$$

Thanks to this matrix, $[C_{12x6}]$, we can transform the six-dimensional natural basis into a 12-dimensional basis. The matrix of inverse transformation is given by the following relationship:

$$[C_{6x12}] = 2[C_{12x6}]^t \qquad [4.27]$$

We refer to $[T_{12}]$ as the generalized Concordia matrix, allowing us to transform this new 12-dimensional basis into an orthonormal basis of the same dimension [BEN 04]:

$$[T_{12}]^t = \begin{bmatrix} 1 & \cos(\frac{2\pi}{12}) & \cos(2\frac{2\pi}{12}) & \cdots & \cdots & \cos(11\frac{2\pi}{12}) \\ 0 & \sin(\frac{2\pi}{12}) & \sin(2\frac{2\pi}{12}) & \cdots & \cdots & \sin(11\frac{2\pi}{12}) \\ \vdots & \vdots & \vdots & \vdots & \vdots & \vdots \\ \vdots & \vdots & \vdots & \vdots & \vdots & \vdots \\ 1 & \cos(5\frac{2\pi}{12}) & \cos(2x5\frac{2\pi}{12}) & \cdots & \cdots & \cos(2x11\frac{2\pi}{12}) \\ 0 & \sin(5\frac{2\pi}{12}) & \sin(2x5\frac{2\pi}{12}) & \cdots & \cdots & \sin(2x11\frac{2\pi}{12}) \\ \frac{1}{\sqrt{2}} & \frac{1}{\sqrt{2}} & \frac{1}{\sqrt{2}} & \cdots & \cdots & \frac{1}{\sqrt{2}} \\ \frac{1}{\sqrt{2}} & -\frac{1}{\sqrt{2}} & \frac{1}{\sqrt{2}} & \cdots & \cdots & -\frac{1}{\sqrt{2}} \end{bmatrix} \qquad [4.28]$$

The first two lines of $[T_{12}]^t$ correspond to the first two lines of the generalized Concordia matrix. They allow us to obtain the usual components according to axes α and β. The last two lines correspond to the homopolar and quasi-homopolar components, respectively. The other components are called non-sequential components. These are perpendicular to the first components.

140   Control of Non-conventional Synchronous Motors

Finally, the matrix transforming the initial six-dimensional basis into an orthonormal basis is done via the $[T_{12,6}]^t$ matrix, which is equal to the product of the two matrices, $[T_{12}]^t$ and $[c_{12x6}]$. If we keep the non-zero vectors resulting from this matricial product [BEN 04], the dimension of the transformation matrix is reduced to six.

Given the fact that the matrix resulting from the product of matrices $[C_{6x12}]$ and $[C_{12x6}]$ is not normalized, the transformation matrix is not normalized either.

By exploiting these different properties, and by taking into account the normalization factor of the matricial product $[C_{6x12}]$ and $[C_{12x6}]$, we can infer the matrix of the transformation of the initial double-star basis into the six-dimensional orthonormal basis:

$$[T_{DE,30}]^t = \sqrt{\frac{1}{3}} \begin{bmatrix} 1 & \frac{\sqrt{3}}{2} & -\frac{1}{2} & -\frac{\sqrt{3}}{2} & -\frac{1}{2} & 0 \\ 0 & \frac{1}{2} & \frac{\sqrt{3}}{2} & \frac{1}{2} & -\frac{\sqrt{3}}{2} & -1 \\ 1 & -\frac{\sqrt{3}}{2} & -\frac{1}{2} & \frac{\sqrt{3}}{2} & -\frac{1}{2} & 0 \\ 0 & \frac{1}{2} & -\frac{\sqrt{3}}{2} & \frac{1}{2} & \frac{\sqrt{3}}{2} & -1 \\ 1 & 0 & 1 & 0 & 1 & 0 \\ 0 & 1 & 0 & 1 & 0 & 1 \end{bmatrix} \quad [4.29]$$

This matrix diagonalizes the inductance matrix of a double three-phase winding where the two windings are shifted by 30°.

4.4.2.2.2. *Projection of vectors on a six-dimensional orthonormal basis*

It is clear that there is no single approach to follow to define transformation matrix [4.29]. Lipo and his co-authors [ZHA 96] have proposed an algorithm to establish this diagonalization matrix of the inductance matrix. Other similar algorithms have been established by other authors [ABB 84, BEN 03, HAD 01]. With this goal in mind, they define an orthonormal work basis. The corresponding work space is broken down in two orthogonal subspaces.

The first two-dimensional subspace is known by plane (α,β) of Concordia. It corresponds to the plane of electromechanical conversion of energy for electrical machines with sinusoidal distribution. As shown in Figure 4.2, axis α is collinear

Control of the Double-star Synchronous Machine 141

with the axis of phase a_1. The components of physical variable x in this plane are obtained by the projection of the vector components of this variable according to these axes α and β:

$$\begin{bmatrix} x_\alpha \\ x_\beta \end{bmatrix} = \begin{bmatrix} 1 & \cos(30°) & \cos(120°) & \cos(150°) & \cos(240°) & \cos(270°) \\ 0 & \sin(30°) & \sin(120°) & \sin(150°) & \cos(240°) & \cos(270°) \end{bmatrix} \begin{bmatrix} x_{a1} \\ x_{a2} \\ x_{b1} \\ x_{b2} \\ x_{c1} \\ x_{c2} \end{bmatrix}$$

[4.30]

The second subspace is four-dimensional. The different axes defining it (z_1, z_2, z_3, z_4) are orthogonal to each other and to the Concordia plane. A trivial solution takes the two homopolar vectors of the two stars as a solution for components z_3 and z_4. The other two components, z_1 and z_2, called non-sequential components, are inferred from solving the following system of equations:

$$\begin{cases} \alpha^t \beta = \alpha^t z_1 = \alpha^t z_2 = \alpha^t z_3 = \alpha^t z_4 = 0 \\ \beta^t z_1 = \beta^t z_2 = \beta^t z_3 = \beta^t z_4 = 0 \\ z_1^t z_2 = z_1^t z_3 = z_1^t z_4 = 0 \\ z_2^t z_3 = z_2^t z_4 = 0 \\ z_3^t z_4 = 0 \end{cases}$$

[4.31]

Thus, after solving the system of equations [4.31] and taking the normalization criterion into account [MER 05], we get the transform matrix of the original initial basis into the new orthonormal basis $(\alpha \ \beta \ z_1 \ z_2 \ z_3 \ z_4)$:

$$[T_6]^t = \sqrt{\frac{2}{6}} \begin{bmatrix} 1 & \cos(\frac{\pi}{6}) & \cos(2\frac{\pi}{3}) & \cos(\frac{\pi}{6}+2\frac{\pi}{3}) & \cos(4\frac{\pi}{3}) & \cos(\frac{\pi}{6}+4\frac{\pi}{3}) \\ 0 & \sin(\frac{\pi}{6}) & \sin(2\frac{\pi}{3}) & \sin(\frac{\pi}{6}+2\frac{\pi}{3}) & \sin(4\frac{\pi}{3}) & \sin(\frac{\pi}{6}+4\frac{\pi}{3}) \\ 1 & \cos(\frac{\pi}{6}+2\frac{\pi}{3}) & \cos(4\frac{\pi}{3}) & \cos(\frac{\pi}{6}) & \cos(2\frac{\pi}{3}) & \cos(\frac{\pi}{6}+4\frac{\pi}{3}) \\ 0 & \sin(\frac{\pi}{6}+2\frac{\pi}{3}) & \sin(4\frac{\pi}{3}) & \sin(\frac{\pi}{6}) & \sin(2\frac{\pi}{3}) & \sin(\frac{\pi}{6}+4\frac{\pi}{3}) \\ 1 & 0 & 1 & 0 & 1 & 0 \\ 0 & 1 & 0 & 1 & 0 & 1 \end{bmatrix}$$

[4.32]

It is important to underline that this solution is not unique.

4.4.2.2.3. Comments

Admittedly, the two approaches shown above lead to similar transformation matrices. The two transformations $[T_6]^t$ and $[T_{DE,30}]^t$ are identical. The first approach is based on linear algebra arguments. The expressions of eigenvalues can easily be generalized. However, the first approach requires the use of the equivalent polyphase machine with phases that are regularly distributed and disregard the variable reluctance of the machine in the first step. The second approach is easier. It is based on the projection of initial vectors on a new orthonormal basis. It is not based on restrictive hypotheses, such as geometrical isotropy and regularity of the phase distribution. It does, however, require the calculation of eigenvalues on a case-by-case basis and solving the [4.31] system of equations is not straightforward.

4.4.2.2.4. Voltage equations in space $\alpha\beta z_1 z_2 z_3 z_4$

Establishing voltage equations in the $\alpha\beta z_1 z_2 z_3 z_4$ space requires us to consider the stator winding as a six-phase winding rather than as two three-phase windings. The basic matricial voltage equation [4.1] remains identical but they need to be reordered:

$$\begin{bmatrix}[V_s]\\[V_R]\end{bmatrix} = [R]\begin{bmatrix}[I_s]\\[I_R]\end{bmatrix} + \frac{d}{dt}\left[[L]\begin{bmatrix}[I_s]\\[I_R]\end{bmatrix}\right] \qquad [4.33]$$

with:

$$[V_s] = \begin{bmatrix} v_{a1} & v_{a2} & v_{b1} & v_{b2} & v_{c1} & v_{c2} \end{bmatrix}^t$$

$$[I_s] = \begin{bmatrix} i_{a1} & i_{a2} & i_{b1} & i_{b2} & i_{c1} & i_{c2} \end{bmatrix}^t$$

The inductance matrix $[L]$ is written in the form:

$$[L] = \begin{bmatrix} [L_{ss}] & [M_{sR}] \\ [M_{sR}]^t & [L_R] \end{bmatrix} \qquad [4.34]$$

In this case, the inductance matrix of the stator is defined as follows:

$$[L_{ss}] = l_{fs}.[I_{6 \times 6}] + M_{ss}.[M_0] + M_{sfm}[M_2] \qquad [4.35]$$

Control of the Double-star Synchronous Machine   143

where:

$$[M_0] = \begin{bmatrix} \cos(0) & \cos(\frac{\pi}{6}) & \cos(\frac{2\pi}{3}) & \cos(\frac{2\pi}{3}+\frac{\pi}{6}) & \cos(\frac{4\pi}{3}) & \cos(\frac{4\pi}{3}+\frac{\pi}{6}) \\ \cos(\frac{\pi}{6}) & \cos(0) & \cos(\frac{4\pi}{3}+\frac{\pi}{6}) & \cos(\frac{2\pi}{3}) & \cos(\frac{2\pi}{3}+\frac{\pi}{6}) & \cos(\frac{4\pi}{3}) \\ \cos(\frac{2\pi}{3}) & \cos(\frac{4\pi}{3}+\frac{\pi}{6}) & \cos(0) & \cos(\frac{\pi}{6}) & \cos(\frac{2\pi}{3}) & \cos(\frac{2\pi}{3}+\frac{\pi}{6}) \\ \cos(\frac{2\pi}{3}+\frac{\pi}{6}) & \cos(\frac{2\pi}{3}) & \cos(\frac{\pi}{6}) & \cos(0) & \cos(\frac{4\pi}{3}+\frac{\pi}{6}) & \cos(\frac{4\pi}{3}) \\ \cos(\frac{4\pi}{3}) & \cos(\frac{2\pi}{3}+\frac{\pi}{6}) & \cos(\frac{2\pi}{3}) & \cos(\frac{4\pi}{3}+\frac{\pi}{6}) & \cos(0) & \cos(\frac{\pi}{6}) \\ \cos(\frac{4\pi}{3}+\frac{\pi}{6}) & \cos(\frac{4\pi}{3}) & \cos(\frac{2\pi}{3}+\frac{\pi}{6}) & \cos(\frac{4\pi}{3}) & \cos(\frac{\pi}{6}) & \cos(0) \end{bmatrix}$$

$$[M_2] = \begin{bmatrix} \cos(2\theta) & \cos\left(2\theta-\frac{\pi}{6}\right) & \cos\left(2\theta-\frac{2\pi}{3}\right) & \cos\left(2\theta-\frac{\pi}{6}-\frac{2\pi}{3}\right) & \cos\left(2\theta+\frac{2\pi}{3}\right) & \cos\left(2\theta-\frac{\pi}{6}+\frac{2\pi}{3}\right) \\ \cos\left(2\theta-\frac{\pi}{6}\right) & \cos\left(2\theta-2\frac{\pi}{6}\right) & \cos\left(2\theta-\frac{\pi}{6}-\frac{2\pi}{3}\right) & \cos\left(2\theta-2\frac{\pi}{6}-\frac{2\pi}{3}\right) & \cos\left(2\theta-\frac{\pi}{6}+\frac{2\pi}{3}\right) & \cos\left(2\theta-2\frac{\pi}{6}+\frac{2\pi}{3}\right) \\ \cos\left(2\theta-\frac{2\pi}{3}\right) & \cos\left(2\theta-\frac{\pi}{6}-\frac{2\pi}{3}\right) & \cos\left(2\theta-\frac{\pi}{6}-\frac{2\pi}{3}\right) & \cos\left(2\theta-\frac{\pi}{6}+\frac{2\pi}{3}\right) & \cos(2\theta) & \cos\left(2\theta-\frac{\pi}{6}\right) \\ \cos\left(2\theta-\frac{\pi}{6}-\frac{2\pi}{3}\right) & \cos\left(2\theta-2\frac{\pi}{6}-\frac{2\pi}{3}\right) & \cos\left(2\theta-\frac{\pi}{6}-\frac{2\pi}{3}\right) & \cos\left(2\theta-2\frac{\pi}{6}-\frac{2\pi}{3}\right) & \cos\left(2\theta-\frac{\pi}{6}\right) & \cos\left(2\theta-2\frac{\pi}{6}\right) \\ \cos\left(2\theta+\frac{2\pi}{3}\right) & \cos\left(2\theta-\frac{\pi}{6}+\frac{2\pi}{3}\right) & \cos(2\theta) & \cos\left(2\theta-\frac{\pi}{6}\right) & \cos\left(2\theta-\frac{2\pi}{3}\right) & \cos\left(2\theta-\frac{\pi}{6}-\frac{2\pi}{3}\right) \\ \cos\left(2\theta-\frac{\pi}{6}+\frac{2\pi}{3}\right) & \cos\left(2\theta-2\frac{\pi}{6}+\frac{2\pi}{3}\right) & \cos\left(2\theta-\frac{\pi}{6}\right) & \cos\left(2\theta-2\frac{\pi}{6}\right) & \cos\left(2\theta-\frac{\pi}{6}-\frac{2\pi}{3}\right) & \cos\left(2\theta-2\frac{\pi}{6}-\frac{2\pi}{3}\right) \end{bmatrix}$$

The stator-rotor mutual inductance matrix is given by:

$$[M_{sr}] = \begin{bmatrix} M_{sF}\cos(\theta) & M_{sD}\cos(\theta) & -M_{sQ}\sin(\theta) \\ M_{sF}\cos(\theta-\frac{\pi}{6}) & M_{sD}\cos(\theta-\frac{\pi}{6}) & -M_{sQ}\sin(\theta-\frac{\pi}{6}) \\ M_{sF}\cos(\theta-\frac{2\pi}{3}) & M_{sD}\cos(\theta-\frac{2\pi}{3}) & -M_{sQ}\sin(\theta-\frac{2\pi}{3}) \\ M_{sF}\cos(\theta-\frac{\pi}{6}-\frac{2\pi}{3}) & M_{sD}\cos(\theta-\frac{\pi}{6}-\frac{2\pi}{3}) & -M_{sQ}\sin(\theta-\frac{\pi}{6}-\frac{2\pi}{3}) \\ M_{sF}\cos(\theta-\frac{4\pi}{3}) & M_{sD}\cos(\theta-\frac{4\pi}{3}) & -M_{sQ}\sin(\theta-\frac{4\pi}{3}) \\ M_{sF}\cos(\theta-\frac{\pi}{6}-\frac{4\pi}{3}) & M_{sD}\cos(\theta-\frac{\pi}{6}-\frac{4\pi}{3}) & -M_{sQ}\sin(\theta-\frac{\pi}{6}-\frac{4\pi}{3}) \end{bmatrix} \quad [4.36]$$

By applying the basic transformation defined by relationship [4.32] to the systems of equations [4.33], we can infer the voltage equations in orthonormal basis ($\alpha\beta z_1 z_2\ z_3 z_4$):

$$\begin{bmatrix} \begin{bmatrix} V_\alpha \\ V_\beta \end{bmatrix} \\ [V_{z_1 z_2 z_3 z_4}] \end{bmatrix} = r_s \begin{bmatrix} \begin{bmatrix} i_\alpha \\ i_\beta \end{bmatrix} \\ [i_{z_1 z_2 z_3 z_4}] \end{bmatrix} + \frac{d}{dt} \left(\begin{bmatrix} (l_{fs} + 3M_{ss})[I]_{2x2} & [0]_{2x4} \\ [0]_{4x2} & l_{fs}[I]_{4x4} \end{bmatrix} \begin{bmatrix} \begin{bmatrix} i_\alpha \\ i_\beta \end{bmatrix} \\ [i_{z_1 z_2 z_3 z_4}] \end{bmatrix} \right) +$$

$$M_{sfm} \frac{d}{dt} \left(\begin{bmatrix} \begin{bmatrix} 3\cos(2\theta) & 3\sin(2\theta) \\ 3\sin(2\theta) & -3\cos(2\theta) \end{bmatrix} & [0]_{2x4} \\ [0]_{4x2} & [0]_{4x4} \end{bmatrix} \begin{bmatrix} \begin{bmatrix} i_\alpha \\ i_\beta \end{bmatrix} \\ [i_{z_1 z_2 z_3 z_4}] \end{bmatrix} \right) + \frac{d}{dt}([T_6][M_{SR}][I_R])$$ [4.37]

$$[V_R] = [R_R][I_R] + \frac{d}{dt}[L_R][I_R] + \frac{d}{dt}\left(([T_6][M_{SR}])^t \begin{bmatrix} \begin{bmatrix} i_\alpha \\ i_\beta \end{bmatrix} \\ [i_{z_1 z_2 z_3 z_4}] \end{bmatrix} \right)$$

where:

$$[T_6][M_{SR}] = \sqrt{3} \begin{bmatrix} M_{sF} \cos(\theta) & M_{sD} \cos(\theta) & -M_{sQ} \sin(\theta) \\ M_{sF} \sin(\theta) & M_{sD} \sin(\theta) & M_{sQ} \cos(\theta) \\ & [0]_{4x3} & \end{bmatrix}$$

System of equations [4.37] shows that the electrical equations describing the stator are completely decoupled. In fact, the equations according to plane (α, β) are completely decoupled from the equations according to space $(z_1 z_2 z_3 z_4)$. And the four equations in space $z_1 z_2 z_3 z_4$ are decoupled.

4.4.2.2.5. Voltage equations in space $dq z_1 z_2 z_3 z_4$

The voltage equations of the double-star machine in space $\alpha\beta z_1 z_2 z_3 z_4$ are equivalent to that of the three-phase machine in space $\alpha\beta z$. We can apply the classic Park rotation transformation to the statoric variables according to plane (α, β) to infer the dynamic model of the machine in plane dq. This model is put in the following form:

$$\begin{bmatrix} V_d \\ V_q \end{bmatrix} = r_s \begin{bmatrix} i_d \\ i_q \end{bmatrix} + \begin{bmatrix} L_d^* & 0 \\ 0 & L_q^* \end{bmatrix} \frac{d}{dt} \begin{bmatrix} i_d \\ i_q \end{bmatrix} + \omega \begin{bmatrix} 0 & -L_q^* \\ L_d^* & 0 \end{bmatrix} \frac{d}{dt} \begin{bmatrix} i_d \\ i_q \end{bmatrix} +$$
$$\sqrt{3}\omega \begin{bmatrix} 0 & 0 & -M_{sQ} \\ M_{sF} & M_{sD} & 0 \end{bmatrix} [I_R] + \sqrt{3} \begin{bmatrix} M_{sF} & M_{sD} & 0 \\ 0 & 0 & M_{sQ} \end{bmatrix} \frac{d}{dt} [I_R]$$

[4.38]

$$[V_{z1z2z3z4}] = r_s [i_{z1z2z3z4}] + l_{fs} \frac{d}{dt} [i_{z1z2z3z4}]$$

$$[V_R] = [R_R][I_R] + [L_R]\frac{d}{dt}[I_R] + \sqrt{3} \begin{bmatrix} M_{sF} & M_{sD} & 0 \\ 0 & 0 & M_{sQ} \end{bmatrix}^t \frac{d}{dt} \begin{bmatrix} i_d \\ i_q \end{bmatrix}$$

with:

$$L_d^* = l_{fs} + 3Mss + 3M_{sfm}$$

$$L_q^* = l_{fs} + 3Mss - 3M_{sfm}$$

Comparison of relationships [4.25] and [4.38] shows that the two dynamic models are identical. Referential d^+q^+ is the classic Park referential dq.

4.4.2.3. *Expression of electromagnetic torque in referential $dqz_1z_2\,z_3z_4$*

Starting from general torque expression [4.11] and applying the matrix of referential change from the initial basis to the orthonormal basis [4.32], then Park rotation, we get the expression of the torque:

$$\Gamma = P\left[\left(L_d^* - L_q^* \right) \cdot i_d i_q + M_{dF}^* i_f i_q + M_{dD}^* i_D i_q - M_{qQ}^* i_Q i_d \right]$$

[4.39]

with:

$$L_d^* = l_{fs} + 3.M_{ss} + 3.M_{sfm} \quad L_q^* = l_{fs} + 3.M_{ss} - 3.M_{sfm}$$
$$M_{dF}^* = \sqrt{3}M_{sf} \; ; \; M_{dD}^* = \sqrt{3}M_{sD} \; ; \; M_{qQ}^* = \sqrt{3}M_{sQ}$$

The same result can be obtained quickly by analyzing the system of equations [4.38]. In fact, the electromotive forces (emfs) according to space $z_1z_2z_3z_4$ are 0. The only currents in the Park referential therefore contribute to the electromechanical conversion of energy. We notice that according to plane (d,q), the model of the machine is equivalent to that of the three-phase synchronous machine in plane

(d,q), provided that we replace the different and mutual inductances and their counterparts with stars.

4.5. Control of the double-star synchronous machine

In this context, we assume that the excitation current is kept constant and that it is not disrupted by the other currents. We disregard the effect of the dampers. Figures 4.6 and 4.7 illustrate the principle diagrams for control of the double-star synchronous machine in referentials $d_1q_1d_2q_2$ and $dqz_1z_2z_3z_4$, respectively. The difference between the two lies chiefly in the generation of reference currents from the reference torque generated by the speed regulator, and the control loops of reference currents. The latter referentials are broken down into two: the control blocks of the currents and the decoupling blocks. In the rest of this section we will focus on problems specific to the double-star synchronous machine. These are the decoupling algorithms in each of the two referentials and the generation of reference currents from a reference torque.

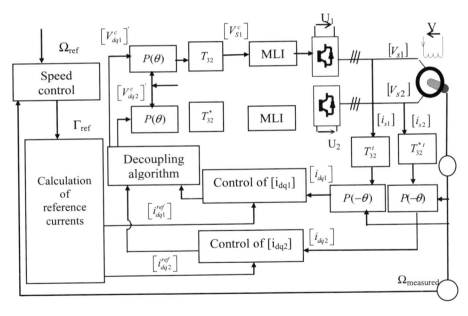

Figure 4.6. *Control structure of double-star synchrone machine (DSSM) in referential $d_1q_1d_2q_2$*

Control of the Double-star Synchronous Machine   147

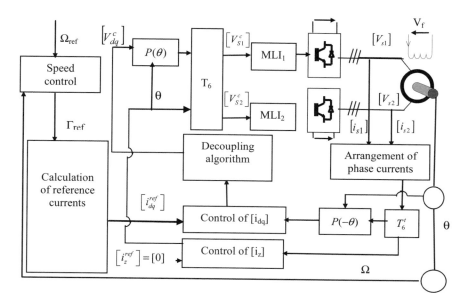

Figure 4.7. *Control structure of double-star synchrone machine in referential $dqz_1z_2z_3z_4$*

4.5.1. *Control in referential $d_1q_1d_2q_2$*

4.5.1.1. *Decoupling algorithm*

4.5.1.1.1. Principle of the decoupling approach

By disregarding the effect of dampers and assuming that the excitation current is regulated and kept constant, the machine, as shown in Figure 4.8, is replaced with an equivalent with four windings distributed two-by-two according to axes d and q in Park referential.

By arranging the components of the current vector according to the adequate order $[I_{d1}, I_{d2}, I_{q1}, I_{q2}]$, the electrical equations given by the relationships [4.15] can be rewritten in the following matricial form:

$$[V] = r_s[I] + \omega[M_{CR}][I] + [M_{CD}]\frac{d}{dt}[I] + [E] \qquad [4.40]$$

with:

$$[V] = \begin{bmatrix} [V_{d1d2}]^t & [V_{q1q2}]^t \end{bmatrix}^t \qquad [I] = \begin{bmatrix} [I_{d1d2}]^t & [I_{q1q2}]^t \end{bmatrix}^t \; ; \; [E] = \begin{bmatrix} 0 & 0 & [E_{q1q2}]^t \end{bmatrix}^t$$

$$[V_{d1d2}] = [V_{d1} \quad V_{d2}]^t \quad [V_{q1q2}] = [V_{q1} \quad V_{q2}]^t \quad [I_{d1d2}] = [I_{d1} \quad I_{d2}]^t \quad [I_{q1q2}] = [I_{q1} \quad I_{q2}]^t$$

$$[E_{q1q2}] = [\omega M_{dF} i_F \quad \omega M_{dF} i_F]^t$$

$$[M_{CR}] = \begin{bmatrix} [0]_{2 \times 2} & -[C_q] \\ [C_d] & [0]_{2 \times 2} \end{bmatrix} \quad [M_{CD}] = \begin{bmatrix} [C_d] & [0]_{2 \times 2} \\ [0]_{2 \times 2} & [C_q] \end{bmatrix} \quad [C_d] = \begin{bmatrix} L_d & L_{dm} \\ L_{dm} & L_d \end{bmatrix}$$

$$[C_q] = \begin{bmatrix} L_q & L_{qm} \\ L_{qm} & L_q \end{bmatrix}$$

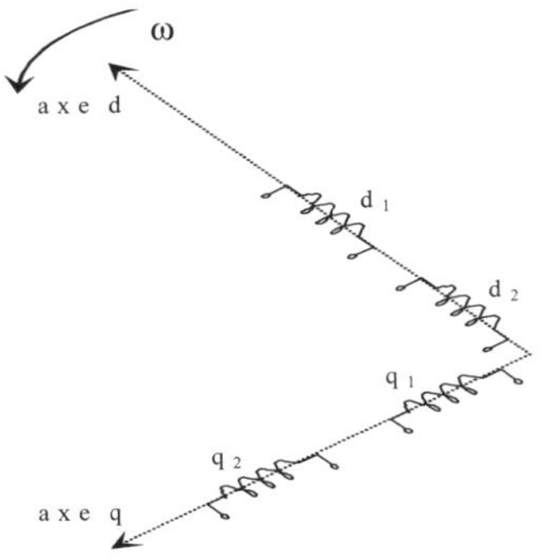

Figure 4.8. *Equivalent windings in referential (d, q).*

Equation [4.40] shows that there are two types of coupling between the state variables. The first coupling is between the state variables of axes d and q due to the rotation term introduced by the Park transformation. The second is due to the mutual inductance effect (L_{dm} and L_{qm}) between the windings of the same axis.

The perturbation terms introduced by the components of emf vector $[E]$ can be compensated for in a traditional way, as in the case of a DC motor or a three-phase synchronous motor.

The decoupling principle consists of bringing back the real system, representing the electrical actuator initially with fourth-order multivariables in an equivalent fictitious system where each state variable is decoupled with respect to the others. Initial input vector $[V]$ becomes $[V']$ by means of a decoupling algorithm. As shown in Figure 4.9, the new system is equivalent to four independent monovariable subsystems from the point of view of the regulators. As a consequence, all the problems now lie in the establishment of a decoupling algorithm [BEN 04, TER 00].

Control of the Double-star Synchronous Machine   149

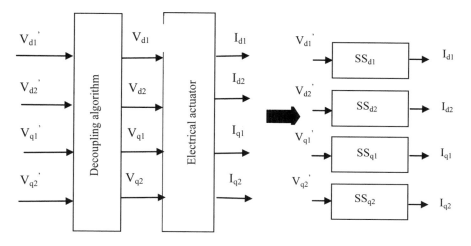

Figure 4.9. *Decoupling principle*

After factorization and introduction of Laplace operator p, equation [4.40] is put in the following form:

$$(r_s[I_d]_{4\times 4} + p[M_{CD}] + \omega[M_{CR}]) \, [I] = [U] \qquad [4.41]$$

where:

$[U] = [V] - [E]$ and $[I_d]_{4\times 4}$ is the fourth-order identity matrix.

We define a transfer matrix H, linking input vector $[U]$ to state vector $[I]$:

$$[I] = [H][U] \qquad [4.42]$$

with:

$$H = (r_s[I]_{4\times 4} + p[M_{CD}] + \omega[M_{CR}])^{-1} \qquad [4.43]$$

Given the existence of coupling between the state variables, matrix $[H]$ is not diagonal. Furthermore, since this matrix depends on the speed, the system is then in non-steady state. As a consequence, without being cautious, every attempt to establish the transfer matrix from Laplace operator p would be wrong.

150   Control of Non-conventional Synchronous Motors

However, for such a motor, the dynamic of the mechanical mode is much slower than that of the electrical mode. Thus, the speed evolution is so weak with respect to electrical variables that the speed can be considered constant piecewise, making the diagonalization of the transfer matrix possible.

Two solutions can be considered according to the evolution of the speed. In the case of steady-state systems, the rotational speed of the motor can be assumed to be constant. The problem amounts to diagonalizing the transfer matrix. In the case of quasi-steady-state systems, the speed evolves but with very weak dynamic with respect to the electrical variables. The solution considered consists of combining the method of diagonalization and static compensation of the terms introduced by the rotational speed. The first method is referred to as complete diagonalization and the second as partial diagonalization.

4.5.1.1.2. Complete diagonalization

In accordance with Figure 4.9, control vector [U] is then replaced with a new vector, [U']. The system described by relationship [4.42] becomes:

$$[I] = [H_{diag}][U'] \qquad [4.44]$$

Where $[H_{diag}]$ is a diagonal matrix equal to the product of the desired decoupling matrix and the transfer matrix:

$$[H_{diag}] = [H][M] \qquad [4.45]$$

In this new system, the control vector is:

$$[U'] = [M]^{-1}[U] \qquad [4.46]$$

The matrix of change of basis [M] representing the decoupling algorithm depends on the transfer matrix "imposed" by the system and on the diagonalization matrix $[H_{diag}]$ chosen:

$$[M] = [H]^{-1}[H_{diag}] \qquad [4.47]$$

From now on, the solution is not unique. The heart of the problem lies in the judicious choice of transfer matrix $[H_{diag}]$. The former allows us to impose the dynamics of the system in open loop.

Control of the Double-star Synchronous Machine 151

The evolution of each state variable linked to a winding d_i or q_i is ruled by an equation in the form:

$$V_x = r_s i_x + L_x \frac{di_x}{dt} + e_x \quad [4.48]$$

with:

$$x = d_i, q_i \quad i = 1,2$$

e_x is a perturbation term for state variable i_x. Generally it reproduces the coupling terms with the other windings for axis d or q as well as it does the windings of quadrature axis q or d.

Given that the aim of decoupling is to make the evolution of each state variable i_x independent of the others, it appears to be relevant for each variable i_x to keep its own time constant:

$$\tau_x = L_x / r_s \quad [4.49]$$

By taking into account these different criteria, we can define the diagonalization matrix:

$$[H_{diag}] = \begin{bmatrix} \frac{1}{r_s + L_d p}[I_d]_{2x2} & [0]_{2x2} \\ [0]_{2x2} & \frac{1}{r_s + L_q p}[I_d]_{2x2} \end{bmatrix} \quad [4.50]$$

By replacing $[H_{diag}]$ with its expression in relationship [4.47], we get the matrix of change of basis:

$$[M] = \begin{bmatrix} 1 & \frac{L_{dm}p}{r_s + L_d p} & \frac{-\omega L_q}{r_s + L_d p} & \frac{-\omega L_{qm}}{r_s + L_d p} \\ \frac{L_{dm}p}{r_s + L_d p} & 1 & \frac{-\omega L_{qm}}{r_s + L_d p} & \frac{-\omega L_q}{r_s + L_d p} \\ \frac{L_d p}{r_s + L_d p} & \frac{L_{dm}p}{r_s + L_d p} & 1 & \frac{-\omega L_{qm}}{r_s + L_d p} \\ \frac{L_{dm}p}{r_s + L_d p} & \frac{L_d p}{r_s + L_d p} & \frac{-\omega L_{qm}}{r_s + L_d p} & 1 \end{bmatrix} \quad [4.51]$$

152   Control of Non-conventional Synchronous Motors

Figure 4.10 illustrates the final structure of the block describing the decoupling algorithm. The inputs of this block are the outputs of the regulators of current vectors $[i_{dq1}]$ and $[i_{dq2}]$, and the outputs of the block are the control voltage vectors in referential $d_1q_1d_2q_2$ $[V^c_{dq1}]$ and $[V^c_{dq2}]$. The two blocks of demultiplexing and multiplexing only allow us to reorder the input and output vectors:

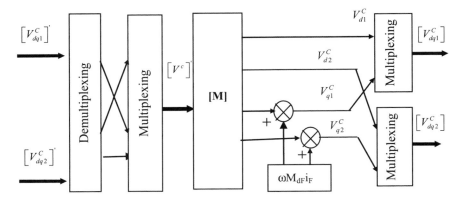

Figure 4.10. *Decoupling structure following the complete diagonalization approach*

4.5.1.1.3. Partial diagonalization

As shown in by relationship [4.51], the matrix of the change of basis M depends on the rotational speed. In order to free ourselves from the presence of rotational speed ω in the transfer matrix, the two axes are initially decoupled by compensation of the terms depending on ω. It leads to two independent subsystems – one for each axis. The two subsystems obtained are linear and steady state, so it is possible to diagonalize the transfer matrix of each by introducing two matrices of change of basis $[M_D]$ and $[M_Q]$, respectively, for axes d and q. This decoupling process, combining the compensation and diagonalization methods of each axis, is called the partial diagonalization method.

Electrical equations [4.40] can be put in the form of two subsystems of equations:

$$[U_d] = [V_{d1d2}] + \omega[C_q \, I_{q1q2}] = r_s [I_{d1d2}] + [C_d] \frac{d}{dt}[I_{d1d2}] \qquad [4.52]$$

$$[U_q] = [V_{q1q2}] - \omega[C_d \, I_{d1d2}] - [E_{q1q2}] = r_s [I_{q1q2}] + [C_q] \frac{d}{dt}[I_{q1q2}] \qquad [4.53]$$

We define the two decoupling matrices $[M_D]$ and $[M_Q]$ for the two subsystems. By following the same approach as in the previous section, we get:

$$[M_D] = \begin{bmatrix} 1 & \dfrac{L_{dm}p}{r_s + L_d p} \\ \dfrac{L_{dm}p}{r_s + L_d p} & 1 \end{bmatrix} \quad [4.54]$$

$$[M_Q] = \begin{bmatrix} 1 & \dfrac{L_{qm}p}{r_s + L_q p} \\ \dfrac{L_{qm}p}{r_s + L_q p} & 1 \end{bmatrix} \quad [4.55]$$

The new structure describing the decoupling algorithm is given in Figure 4.11.

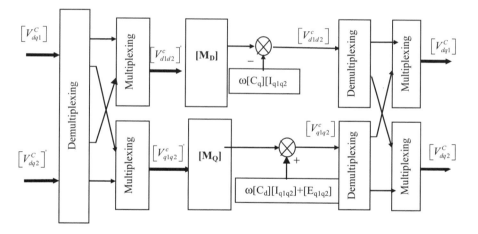

Figure 4.11. *Decoupling structure following the partial diagonalization approach*

4.5.1.1.4. Validation of decoupling algorithms [TER 00]

As we have seen previously, the model in referential $d_1q_1d_2q_2$ is characterized by two types of coupling: one between the state variables according to the same axis, d or q; and the other between variables according to axis d and state variables according to axis q.

When stationary, only the couplings between the state variables according to the same axis remain. Figures 4.12 and 4.13 show the efficiency of the two decoupling algorithms, complete and partial diagonalization. A step variation of one current

154   Control of Non-conventional Synchronous Motors

(i_{d1} or i_{d2}) along the *d* axis introduced a very weak disturbance of 0.5% on the other current of the same axis (i_{d2} or i_{d1}). The same remarks can be said when applying a step variation along the *q* axis. Since the speed is 0, the state variables according to the other axis are unaffected.

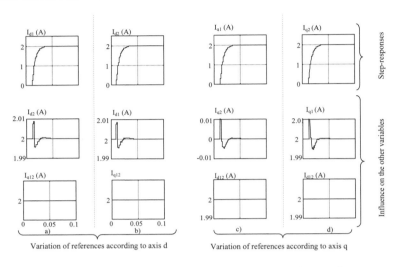

Figure 4.12. *Validation of the decoupling algorithm by complete diagonalization when stationary*

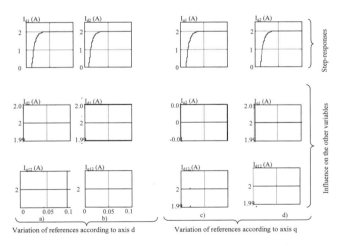

Figure 4.13. *Validation of the decoupling algorithm by partial diagonalization when stationary*

Figures 4.14 and 4.15 show the decoupling efficiency at non-zero speed (when the machine is in motion). In fact, at a rotational speed of 100 rpm, two tests have been performed for the two decoupling algorithms. The first test (left column) varies the reference currents according to axis q, a level of iq_1 at 0.5 s and a level of iq_2 at 0.15 s. The second test (right column) measures the reference current variations according to axis d, a level of id_1 at 0.5 s and a level of id_2 at 0.15 s. The results obtained confirm those measured when stationary and show the very weak perturbation due to a state variable on axis (d or q) compared to the variables of the other axis (q or d).

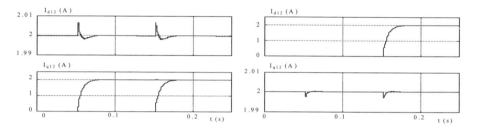

Figure 4.14. *Validation of the decoupling algorithm by complete diagonalization at a speed of 100 rpm*

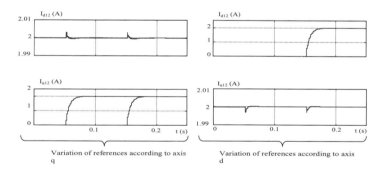

Figure 4.15. *Validation of the decoupling algorithm by partial diagonalization at a speed of 100 rpm*

4.5.1.2. *Generation of current references*

If we do not take the effects of the dampers into account, the expression of torque [4.20] becomes:

156   Control of Non-conventional Synchronous Motors

$$\Gamma = P\left[M_{dF}i_F + \frac{1}{2}\left[(L_d - L_q) + (L_{dm} - L_{qm})\right](i_{d1} + i_{d2}) \right](i_{q1} + i_{q2})$$ [4.56]

In the case of the machine with smooth poles and without dampers, the expression of torque is directly inferred from expression [4.21]:

$$\Gamma = P.M_{dF}.i_F.(i_{q1} + i_{q2})$$ [4.57]

Relationship [4.56] shows that the torque in referential $d_1q_1d_2q_2$ depends on currents i_{di} and i_{qi} ($i = 1,2$) and the inductor current. Relationship [4.57] shows that in the case of a machine with smooth poles, the expression of torque is simplified and no longer depends on the two currents according to axis q. The reference currents according to axis d are kept constant and equal to 0. In the most general case, by referring to the case of a DC machine, to have a torque that is directly proportional to the currents according to axis q, we set the currents to 0 according to axis d:

$$I_{d1}^{ref} = I_{d2}^{ref} = 0$$ [4.58]

Since the two stars are identical, the reference torque is equally distributed between the two stars. The two reference currents according to axis q are equal and defined by the following relationship:

$$I_{q1}^{ref} = I_{q2}^{ref} = \frac{\Gamma_{ref}}{2PM_{dF}i_F}$$ [4.59]

We notice that we again find the results known for torque control of a three-phase synchronous machine. Each star is equivalent to a three-phase machine.

4.5.2. Control in referential $dqz_1z_2z_3z_4$

4.5.2.1. Decoupling algorithm

As shown by the voltage equations in this referential [4.38], the dynamic model of the double-star synchronous machine is made of two systems, dq and $z_1z_2z_3z_4$, that are completely decoupled. The equations describing subsystem dq are equivalent to that of a three-phase synchronous machine in Park referential. The control algorithm is then identical to that of a three-phase machine. Sub-system $z_1z_2z_3z_4$ is described by four identical and completely decoupled first-order equations whose current control is easy.

4.5.2.2. *Generation of current references*

If we do not take the effects of dampers into account, the expression of torque [4.39] becomes:

$$\Gamma = P(L_d^* - L_q^*)i_d i_q + PM_{dF}^* i_f i_q \qquad [4.60a]$$

In the case of the machine with smooth poles, the expression of torque is simplified and becomes:

$$\Gamma = PM_{dF}^* i_f i_q \qquad [4.60b]$$

The former relationship [4.60a] shows that within the framework of our hypotheses, which consider the sinusoidal distribution machine, the torque in plane $dqz_1z_2z_3z_4$ does not depend on sequential currents (i_{z1}, i_{z2}, i_{z3} and i_{z4}). Knowing that i_{z3} and i_{z4} correspond to homopolar currents of the machine and that each star has a non-linked neutral point, these two currents have a value of 0. Furthermore, since currents i_{z1} and i_{z2} do not contribute to the development of torque, they are set to 0:

$$i_{z1}^{ref} = i_{z2}^{ref} = 0 \qquad [4.61]$$

We notice that the expression of torque for the machine in this referential is equivalent to that of the three-phase synchronous machine in Park referential. We therefore adopt the same torque control strategy as that of a three-phase machine [LOU 10b]. Thus, in the case of the machine with smooth poles, the reference current according to axis *d* is kept constant and equal to 0. In the most general case, by referring to the DC machine to have a torque that is directly proportional to the current according to axis *q* we keep the current at 0 according to axis *d*. We again find the expressions of reference currents in the Park referential of a three-phase machine:

$$I_d^{ref} = 0 \qquad [4.62]$$

$$I_q^{ref} = \frac{\Gamma_{ref}}{PM_{dF}^* i_F} \qquad [4.63]$$

4.6. Bibliography

[ABB 84] ABBAS M.A., CHRISTEN R., JAHNS T.M., "Six-phase voltage source inverter driven induction motor", *IEEE Transactions on Industry Applications*, vol. 1A-20, no. 5, 1984.

[BAS 95] BASILI B., PESKINE C. "Algèbre", in: *Bibliothèque des Sciences*, Diderot Editeur Paris, 1995.

[BEN 03] BENKHORIS M.F., MERABTÈNE M., MEIBODYTABAR F., DAVAT B., SEMAIL E., "Approches de modélisation de la MSDE alimentée par des onduleurs de tension en vue de la commande" *Revue Internationale de Génie Electrique RS série RIGE*, vol. 6, no. 5, pp 579-608, 2003.

[BEN 04] BENKHORIS M.F., Modélisation dynamique et commande des systèmes convertisseurs - machines complexes, habilitation à diriger des recherches, University of Nantes, March 2004.

[HAD 01] HADIOUCHE D., Contribution à l'étude de la machine asynchrone double étoile : modélisation, alimentation et structure, PhD thesis, Henri Poincaré university, Nancy1, December 2001.

[KET 95] KETTELER K.H., "Multisystem propulsion concept on the double star circuit", *EPE, 1995*, pp.159-166, 1995.

[KHE 95] KHELOUI A., MEIBODY-TABAR F., DAVAT B., "Current commutation analysis in self-controlled double star synchronous machine taking into account saturation effect" *Electric Machines and Power Systems*, vol. 23, issue 5, pp. 557-569, 1995.

[KOT 96] KOTNY J.L., ROGER D., ROMARY R., "Analytical determination of the double star synchronous machine commutating reactance", 6th *International Conference on Power Electronics and Variable Speed Drives (PEVD96), IEE*, pp. 306-310, 1996.

[LEO 96] LEONHARD W., *Control of Electrical Drives*, Springer, 1996.

[LOU 95] LOUIS J.-P., Bergmann C. "Commande numérique des machines – Evolution des commandes", *Techniques de l'Ingénieur*, D3644, vol. DAB, pp. 3640-3644, 1995.

[LOU 04a] LOUIS J.-P. (ed.), *Modélisation des machines électriques en vue de leur commande, Concepts généraux*, Traité EGEM, Série Génie Électrique, Hermès-Lavoisier, Paris, 2004.

[LOU 04b] LOUIS J.-P. (ed.), *Modèles pour la commande des actionneurs électriques*, Traité EGEM, Série Génie Électrique, Hermès-Lavoisier, Paris, 2004.

[LOU 11a] LOUIS J.-P. (ed.), *Control of Synchronous Motors*, ISTE Ltd, London and John Wiley and Sons, New York, 2011.

[LOU 11b] LOUIS J.-P., FLIELLER, D., NGUYEN N. K., STURTZER G., "Optimal supplies and synchronous motors torque controls. Design in the d-q reference frame", Chapter 3 in LOUIS J.-P., *Control of Synchronous Motors*, ISTE Ltd., London and John Wiley and Sons, New York, 2011.

[MAD 04] MADANI N., Commande à structure variable d'une machine asynchrone double étoile, alimentée par deux onduleurs MLI, modélisation dynamique, alimentation et validation expérimentale, PhD thesis, University of Nantes, Saint-Nazaire, December 2004.

[MER 05] MERABTENE M., Modélisation dynamique et commande d'une machine synchrone double étoile alimentée par des onduleurs ML. Fonctionnement en mode normal et dégradé, PhD thesis, University of Nantes, Saint-Nazaire, July 2005.

[MOU 98] MOUBAYED N., MEIBODY-TABAR F., DAVAT B., "Alimentation par des onduleurs de tension d'une machine synchrone double étoile", *Revue Internationale de Génie Electrique*, vol. 1, no. 4, pp.457-470, 1998.

[MOU 99] MOUBAYED N., Alimentation par onduleurs de tension des machines multi-étoiles, PhD thesis, INP Lorraine, Nancy, 1999.

[NEL 73] NELSON R.H., KRAUSE P.C., "Induction machine analysis for arbitrary displacement between multiple winding set", *IEEE Industry Application Society*, Milwaukee, USA, pp. 841-848, October 1973.

[SEM 00] SEMAIL E., Outils et méthodologie d'étude des systèmes électriques polyphasés. Généralisation de la méthode des vecteurs d'espace, PhD thesis, University of Lille, June 2000.

[SEM 11] SEMAIL E., KESTELYN X., "Modélisation vectorielle et commande de machines polyphasées à pôles lisses alimentées par onduleur de tension", Chapter 5 in LOUIS J.-P., *Commandes d'actionneurs électriques synchrones et spéciaux*, Traité EGEM, Série Génie Électrique, Hermès-Lavoisier, Paris, 2011.

[SIN 02] SINGH G.K., "Multi-phase induction machine drive research-a survey", *Electric Power Systems Research*, vol. 61, pp. 139-147, 2002.

[TER 99] TERRIEN F., BENKHORIS M.F., "Analysis of double star motor drives for electrical propulsion", *IEE, Elect Mach and Drives*, London, pp. 90-94, September 1-3, 1999.

[TER 00] TERRIEN F., Commande d'une machine synchrone double étoile alimentée par des onduleurs MLI, modélisation, simulation et prototype expérimental, PhD thesis, University of Nantes, Saint-Nazaire, December 2000.

[VAS 94] VAS P. *Vector Control of AC Machines*, Clarendon Press Oxford, 1994.

[WER 84] WERREN L., "Synchronous machine with 2 three-phase windings, spatially displaced by 30°el. Commutation reactance and model for converter-performance simulation", *ICEM 84*, Lausanne, Switzerland, vol. 2, pp 781-784, September 1984.

[ZHA 95] ZHAO Y., LIPO T.A., "Space vector PWM control of dual three phase induction machine using vector space decomposition", *IEEE Trans. Ind. Appl.*, vol. 31, no. 5, pp. 1100-1109, 1995.

Chapter 5

Vectorial Modeling and Control of Multiphase Machines with Non-salient Poles Supplied by an Inverter

5.1. Introduction and presentation of the electrical machines

This chapter is devoted to the modeling and control of electrical machines with at least two independent statoric currents. The star-coupled three-phase machine without a neutral terminal and the triangle-coupled three-phase machine are the most basic ones. More precisely, this chapter aims to emphasize the particularities created by a number of independent currents greater than two with respect to the classic three-phase machine.

To reach this goal, we will restrict ourselves to machines fulfilling certain hypotheses:

– constant magnetic air-gap (without variable reluctance effect);

– without magnetic saturation effect;

– built regularly, i.e. it is impossible to discriminate against phases as all of the phases are characterized by the same technological realization.

In practice, these hypotheses allow us to deal with at least two large families of machines:

– synchronous machines with surface-mounted magnets;

Chapter written by Xavier KESTELYN and Eric SEMAIL.

– squirrel-cage induction machines.

Obviously real machines do not perfectly fulfill the hypotheses, but we will assume that the phenomena induced by the non-fulfilling of these hypotheses generate second-order phenomena that can be implicitly compensated for by robust control.

For control, squirrel-cage induction machines can be distinguished from synchronous machines with magnets, mainly because the magnetization of the machine is entirely controlled by the inverter supplying the stator. In fact, for synchronous machines with magnets, part of the magnetic field within the machine is not controlled by the supply. Hence, the study of the control of synchronous machines has more constraints than that of induction machines. For this reason, we will restrict the study in this chapter to that of synchronous machines with permanent magnets. The particularity of these machines lies in the number of phases and taking both space and time harmonics into account. This problem of the impact of harmonics on control has already been tackled in the case of three-phase synchronous machines with trapezoidal electromotive forces in [GRE 94] and [LOU 10]. In this chapter, we emphasize the originality induced by increasing the number of phases.

Finally, from among the machines studied we can also distinguish between several sub-families by focusing on the connections that can be observed between the different coils of the statoric phases:

– machine without coupling between phases: each phase that has two connection terminals is generally supplied by a monophase H inverter;

– simple star machine: n coils constituting the n phases are connected by a common point, the neutral terminal of the machine;

– multi-star machine: k stars connected to n/k coils by a common point. We will therefore have k neutral points;

– machine with polygonal coupling: n coils constituting the n phases are connected in series. For the three-phase machines, we refer to this as triangle coupling.

This chapter will only deal with the control of independent phase machines and simple star machines. The five-phase machine will be the reference example allowing easy generalization to n-phase machines from the vectorial formulation chosen.

5.2. Control model of inverter-fed permanent magnet synchronous machines

The following hypotheses and notations will be used to model the machine:

– the n phases are identical and shifted by an angle $\alpha = 2\pi/n$, and p is the number of pole pairs of the machine;

– the machine has non-salient poles and is not saturated.

Figure 5.1 represents a two-pole n-phase machine in which variable g (a voltage, current, flux, etc.) is written g_k with respect to phase k.

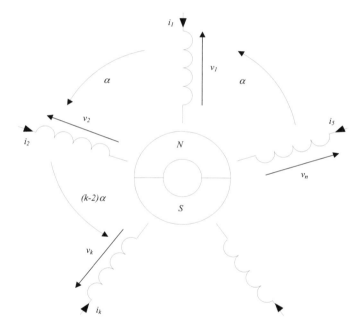

Figure 5.1. *Symbolic representation of a synchronous two-pole n-phase machine*

5.2.1. *Characteristic spaces and generalization of the notion of an equivalent two-phase machine*

5.2.1.1. *Equations in the natural basis of the stator and general vectorial expression*

A Euclidean vectorial space E^n of dimension n as well as an orthonormal basis $B^n = \{\vec{x}_1^n, \vec{x}_2^n, ..., \vec{x}_n^n\}$ is combined with the n-phase machine. This is referred to as natural, since the coordinates of vector $\vec{g}$ in this basis are measurable variables g_k of stator phases:

$$\vec{g} = g_1\vec{x}_1^n + g_2\vec{x}_2^n + \ldots + g_n\vec{x}_n^n = \sum_{k=1}^{n} g_k\vec{x}_k^n \quad [5.1]$$

Thus, the following vectors can be defined:

– voltage: $\vec{v} = v_1\vec{x}_1^n + v_2\vec{x}_2^n + \ldots + v_n\vec{x}_n^n$;

– current: $\vec{i} = i_1\vec{x}_1^n + i_2\vec{x}_2^n + \ldots + i_n\vec{x}_n^n$;

– linked flux: $\vec{\phi} = \phi_1\vec{x}_1^n + \phi_2\vec{x}_2^n + \ldots + \phi_n\vec{x}_n^n$.

By considering the resistance R_s of a stator phase, we can determine a single voltage vectorial equation that gathers the scalar voltage equations of each phase:

$$\vec{v} = R_s \vec{i} + \left(\frac{d\vec{\phi}}{dt}\right)_{/B^n} \quad [5.2]$$

From the hypothesis of non-saturation, equation [5.2] can be written as:

$$\vec{v} = R_s\vec{i} + \left(\frac{d\vec{\phi}_{ss}}{dt}\right)_{/B^n} + \left(\frac{d\vec{\phi}_{sr}}{dt}\right)_{/B^n} \quad [5.3]$$

In [5.3], the term $\left(\dfrac{d\vec{\phi}_{ss}}{dt}\right)_{/B^n}$ is due to the contribution of stator currents, while $\left(\dfrac{d\vec{\phi}_{sr}}{dt}\right)_{/B^n}$ is due to the contribution of the rotor (permanent magnets or inductor coil for a synchronous machine, and rotor coils or bars for an induction machine).

In the case of a synchronous machine with non-salient poles, equation [5.3] is more classically written as:

$$\vec{v} = R_s\vec{i} + \lambda\left(\left(\frac{d\vec{i}}{dt}\right)_{/B^n}\right) + \vec{e} \quad [5.4]$$

in which:

– λ is a linear application (or morphism) so that $\lambda(\vec{i}) = \vec{\phi}_{ss}$. This application is commonly written in natural basis B^n in the form of a symmetrical matrix ($L_{xy} = L_{yx}$) detailed by equation [5.5]. This matrix, given the hypothesis of constructive regularity that leads to all phases being equivalent, is circulating, i.e. we get line n°2 of the matrix from line n°1 by simple shifting of one rank ($L_{21} = L_{1n}$, $L_{22} = L_{11}$, $L_{23} = L_{12}$, etc.);

$$[L_s^n] = mat(\lambda, B^n) = \begin{pmatrix} L_{11} & L_{12} & \cdots & L_{1n} \\ L_{21} & L_{22} & \cdots & L_{2n} \\ \vdots & \vdots & \ddots & \vdots \\ L_{n1} & L_{n2} & \cdots & L_{nn} \end{pmatrix} \quad [5.5]$$

– $\vec{e} = \left(\dfrac{d\vec{\phi}_{sr}}{dt}\right)_{/B^n}$ is the electromotive forces (emfs) vector, which is written in the form of a function $\vec{\varepsilon}$ (speed-normalized emf) depending only on θ, the relative position of the rotor with respect to the stator coils, and on speed Ω of the rotor, see [5.6]:

$$\vec{e} = \Omega\, \vec{\varepsilon}(\theta) \quad [5.6]$$

More than allowing synthetic writing, the vectorial relationships ease the calculations of powers and torque. The instantaneous power flowing through the machine is obtained by the simple scalar product of the voltage and current vectors:

$$p = \vec{v}.\vec{i} = \sum_{k=1}^{n} v_k i_k \quad [5.7]$$

By replacing expression [5.4] of the voltage vector in equation [5.7], we get:

$$p = R_s(\vec{i})^2 + \lambda\left(\left(\dfrac{d\vec{i}}{dt}\right)_{/B^n}\right).\vec{i} + \vec{e}.\vec{i} \quad [5.8]$$

Here, we recognize the following within the framework of the hypotheses:

– $p_j = R_s(\vec{i})^2$, the stator Joule losses;

166   Control of Non-conventional Synchronous Motors

$- p_{mag} = \left(\lambda \left(\dfrac{d\vec{i}}{dt} \right)_{/B^n} \right) . \vec{i}$, the power linked to the magnetic energy stored;

$- p_{em} = \vec{e} . \vec{i}$, the electromechanical power developed by the machine that lies in the origin of the creation of an electromagnetic torque c, expressed by:

$$c = \dfrac{p_{em}}{\Omega} = \dfrac{\vec{e}.\vec{i}}{\Omega} = \vec{\varepsilon}.\vec{i} = \sum_{k=1}^{n} \varepsilon_n i_n \qquad [5.9]$$

5.2.1.2. *Determination of a decoupling basis*

If vectorial relationship $\vec{\phi}_{ss} = \lambda(\vec{i})$ between the stator flux and stator current vectors remains true, whichever the basis of space E^n chosen, it is not true when it comes to matricial relationships between the coordinates of these vectors. The vector coordinates are obtained by projecting the vectors on the generating vectors of the basis. We can then understand that if we project the vectors in another basis, the relationships between the coordinates will change. The matrix that characterizes morphism λ in the natural basis is generally a full matrix. Therefore, couplings between the different phases that appear are not appreciated within the control framework.

If there is a basis in which the matrix of morphism λ is diagonal, we prefer to work in this basis when considering control. The coordinates in this basis of voltage, current and flux vectors are no longer measurable, but fictitious.

In the case of morphisms characterized by a symmetrical matrix (which is the case of λ), we are assured of the existence of such bases that ensure decoupling of the different coordinates. Furthermore, these bases are orthogonal and the eigenvalues are real.

A basis can be determined by analyzing the inductance matrix $\left[L_s^n \right]$, the characteristic matrix of morphism λ between the stator flux and stator current vectors in the natural basis. In this basis, eigenvectors are associated with the eigenvalues Λ_k of morphism λ. We recall that eigenvalues Λ_k are the solutions to characteristic equation [5.10], in which $[I_n]$ is the identity matrix of dimension n:

$$\det\left(\Lambda [I_n] - [L_s^n] \right) = 0 \qquad [5.10]$$

The detailed calculation of these eigenvalues and the associated eigenvectors can be found in [SEM 00], and in an analogous form in [WHI 59].

In the new decoupling basis $B^d = \{\vec{x}_1^d, \vec{x}_2^d, ..., \vec{x}_n^d\}$, characteristic matrix $[L_s^d]$ of morphism λ is expressed:

$$[L_s^d] = mat(\lambda, B^d) = \begin{pmatrix} \Lambda_1 & 0 & \cdots & 0 \\ 0 & \Lambda_2 & \cdots & 0 \\ \vdots & \vdots & \ddots & \vdots \\ 0 & 0 & \cdots & \Lambda_n \end{pmatrix} \quad [5.11]$$

Matrix $[L_s^d]$ is obtained by classic basis change, as described by [5.12] with $[T_{nn}]$ the matrix of basis change. We are reminded that obtaining the coefficients of matrix $[T_{nn}]$ is easy since, by definition, each column of $[T_{nn}]$ consists of coordinates, in the natural basis, of one of the new eigenvectors that constitute the orthonormal decoupling basis.

Things being as they are, in practice matrix $[T_{nn}]$ is less useful than its inverse $[T_{nn}]^{-1}$. In fact, it is relationship [5.13] that allows us to obtain the coordinates of a vector in the new decoupled basis as a function of coordinates in the natural basis.

Although obtaining $[T_{nn}]^{-1}$ from $[T_{nn}]$ is rarely, it is in our case, since the inverse matrix is identical to the transposed matrix: $[T_{nn}]^{-1} = [T_{nn}]^t$. This property comes from the fact that matrix $[L_s^n]$ is symmetrical, which implies the orthogonality property of matrix $[T_{nn}]$.

Therefore each line of $[T_{nn}]^{-1}$, presented in [5.14], is also defined by coordinates in the natural basis and the eigenvectors that make up the new decoupling basis. We clarify that in equation [5.14]: $\alpha = 2\pi/n$; $c_k = \cos(k\alpha)$; $s_k = \sin(k\alpha)$; if n is even, $K = (n-2)\alpha/2$; if n is odd, $K = (n-1)\alpha/2$ and we will omit the last line of the transformation matrix.

The analysis that has just been carried out aims to emphasize the specificity of $[T_{nn}]$ or $[T_{nn}]^{-1}$, whose synthesis does not come from the aim to simplify calculation but directly from the analysis of the inductance matrix.

$$[L_s^d] = [T_{nn}]^{-1} [L_s^n][T_{nn}] \quad [5.12]$$

$$\begin{pmatrix} \vec{x}_1^d \\ \vec{x}_2^d \\ \vdots \\ \vec{x}_n^d \end{pmatrix} = [T_{nn}]^{-1} \begin{pmatrix} \vec{x}_1^n \\ \vec{x}_2^n \\ \vdots \\ \vec{x}_n^n \end{pmatrix}$$ [5.13]

$$[T_{nn}]^{-1} = \sqrt{\frac{2}{n}} \begin{pmatrix} 1 & c_1 & c_2 & c_3 & \cdots & c_3 & c_2 & c_1 \\ 0 & s_1 & s_2 & s_3 & \cdots & -s_3 & -s_2 & -s_1 \\ 1 & c_2 & c_4 & c_6 & \cdots & c_6 & c_4 & c_2 \\ 0 & s_2 & s_4 & s_6 & \cdots & -s_6 & -s_4 & -s_2 \\ 1 & c_3 & c_6 & c_9 & \cdots & c_9 & c_6 & c_3 \\ 0 & s_3 & s_6 & s_9 & \cdots & -s_9 & -s_6 & -s_3 \\ \vdots & \vdots & \vdots & \vdots & \ddots & \vdots & \vdots & \vdots & \cdots \\ 1 & c_K & c_{2K} & c_{3K} & \cdots & c_{3K} & c_{2K} & c_K \\ 0 & s_K & s_{2K} & s_{3K} & \cdots & -s_{3K} & -s_{2K} & -s_K \\ \frac{1}{\sqrt{2}} & \frac{1}{\sqrt{2}} & \frac{1}{\sqrt{2}} & \frac{1}{\sqrt{2}} & \cdots & \frac{1}{\sqrt{2}} & \frac{1}{\sqrt{2}} & \frac{1}{\sqrt{2}} \\ \frac{1}{\sqrt{2}} & \frac{-1}{\sqrt{2}} & \frac{1}{\sqrt{2}} & \frac{-1}{\sqrt{2}} & \cdots & \frac{-1}{\sqrt{2}} & \frac{1}{\sqrt{2}} & \frac{-1}{\sqrt{2}} \end{pmatrix}$$ [5.14]

Besides relationships [5.11] to [5.14], some properties of morphism λ are very useful. Given the circularity of inductance matrix $[L_s^n]$, the eigenvalues Λ_k of λ are generally "double", i.e. we can associate two independent eigenvectors with them. Only one of the eigenvalues is "single" (i.e. associated with a single eigenvector) in the case of an odd number of phases n; two are "single" in the case of an even number of phases n.

Thus, the latter properties allow us to justify the use of star couplings for machines with an odd number of phases and multi-star couplings for machines with an even number of phases when considering control. In fact, these couplings allow us to ensure that the currents associated with the single eigenvalues (also called homopolar currents) are rigorously kept at 0.

Finally, it is the presence of a double eigenvalue, better known as cyclical inductance, which has allowed us to introduce the notion of the equivalent two-phase machine, which is well accepted in the three-phase case.

5.2.1.3. *Equations in the decoupling basis and independent energy fluxes*

In section 5.2.1.2 we have shown that it is possible to determine an orthonormal decoupling basis in which inductance matrix $[L_s^d]$ is diagonal. Furthermore, we have assumed that the eigenvalues of this matrix were double, except for one single in the case odd n and two singles in the case even n.

The initial vectorial space can therefore be broken down into a sum of vectorial subspaces of one or two dimensions, each combined with an eigenvalue Λ_k of $[L_s^d]$, double or single. These subspaces, also referred to as eigenspaces of morphism λ are orthogonal because all the eigenvectors of morphism λ are orthogonal. A vector $\vec{g}$ belonging to vectorial space E^n can then be broken down into a unique sum of vectors of one or two dimensions, each belonging to a vectorial Eigen subspace E^{se} of one or two dimensions. The E^{se} vectors are obtained by orthogonal projection of vector $\vec{g}$ on each of subspaces E^{se}.

Applied to the voltage equation [5.4], this break down leads to the following equation, in which N is the number of subspaces, E^{se}:

$$\vec{v} = \sum_{m=1}^{N} \vec{v}_k^d = \sum_{m=1}^{N} \left(R_s \vec{i}_m^d + \Lambda_m \cdot \left(\frac{d\vec{i}_m^d}{dt}\right)_{/B^d} + \vec{e}_m^d \right) \quad [5.15]$$

If we now look for the electromagnetic torque using a power balance, by recalling that Eigen subspaces E^{se} are orthogonal, we can find:

$$p = \vec{v}.\vec{i} = \left(\sum_{m=1}^{N} \vec{v}_k^d\right) \cdot \left(\sum_{m=1}^{N} \vec{i}_k^d\right) = \sum_{m=1}^{N} \left(\vec{v}_k^d . \vec{i}_k^d\right)$$
$$= \sum_{m=1}^{N} \left(R_s \left(\vec{i}_m^d\right)^2 + \Lambda_k \cdot \left(\frac{d\vec{i}_m^d}{dt}\right)_{/B^d} \cdot \vec{i}_m^d + \vec{e}_m^d . \vec{i}_m^d \right) \quad [5.16]$$

In [5.16] a term that associates with Joule losses appears, then another term with the storage of magnetic energy and finally a term with the electromechanical conversion. By expanding this last term according to equation [5.17], it appears that the total mechanical energy of the machine is the sum of N mechanical energies, each associated with a subspace, E^{se}. It is therefore possible to consider that the torque supplied by the real machine is the sum of torques supplied by N fictitious machines. Each of these machines is characterized by its resistance, R_s, its inductance, Λ_k, and its emf vector, $\vec{e}_m^d$. According to the dimension of the eigenspace with which it is associated, the fictitious machine will be monophase or two-phase. These machines all run at the same speed. They can therefore be considered mechanically coupled.

$$c = \sum_{m=1}^{N} c_m^d = \sum_{m=1}^{N} \frac{\vec{e}_m^d \cdot \vec{i}_m^d}{\Omega} = \sum_{m=1}^{N} \vec{\varepsilon}_m^d \cdot \vec{i}_m^d \qquad [5.17]$$

Figure 5.2 illustrates the equivalence between an *n*-phase machine and a set of fictitious monophase and two-phase machines.

It has to be noted that if we can find an infinity of T_{nn} type transformations, there is one and only one break down into fictitious machines (mathematically, breaking down a vector on the eigenspaces of morphism λ). This uniqueness is a key point of the vectorial approach with respect to matricial approaches using an infinite number of transformations. Naturally this break down can also be applied to the three-phase case: the three-phase machine is *a priori* equivalent to two machines – one monophase and the other two-phase. We will see why only the two-phase machine is kept when the machine is star-coupled.

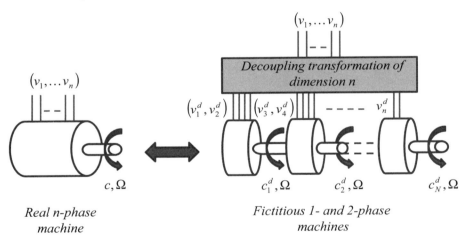

Real n-phase machine   *Fictitious 1- and 2-phase machines*

Figure 5.2. *Equivalence between an n-phase machine and a set of fictitious one- and two-phase mechanically coupled machines*

5.2.1.4. *Fundamental harmonic properties of fictitious machines*

The concept of fictitious machines allows us to transform the real machine, whose phases are magnetically coupled, into a sum of magnetically decoupled one- and two-phase fictitious machines. In this section, we show that the variables associated with a fictitious machine consist of a harmonic group of the real machine's variables. This aspect is fundamental when working out the control of a machine or during its design phase, particularly in the case of fault-tolerant machines. Each fictitious machine possesses its own features that need to be known for proper design and control.

5.2.1.4.1. Characteristic harmonics groups

Let $\vec{g}$ be a vector variable associated with the real machine. We assume that each of the vector variable coordinates has a period $2\pi/p$, i.e. it can be broken down into a Fourier series. We define θ as the mechanical angle between the rotor and the stator.

$$\vec{g} = \sum_{k=1}^{n} g_k \vec{x}_k^n \qquad [5.18]$$

with:

$$g_k = \sum_{h=1}^{\infty} g_h^{\max} \sin\left(h\left(p\theta - (k-1)\frac{2\pi}{n}\right)\right) \qquad [5.19]$$

We calculate the coordinates of the vector variables associated with fictitious machines by projecting the vector variable on the different subspaces E^{se} associated with fictitious machines. We will find the details of the calculations in [KES 03]. We only discuss the basic idea here.

Assuming that a subspace associated with fictitious machine number m ($m \in \{1, N\}$) is generated by vectors $\{\vec{x}_{2m-1}^d, \vec{x}_{2m}^d\}$ of decoupling basis B^d, defined by equation [5.13] and corresponding to the lines of matrix $[T_{nn}]^{-1}$ given by [5.14]. Equation [5.20] reminds us that projection $\vec{g}_{2m-1}^d$ of vector variable $\vec{g}$ is obtained by simple scalar products.

$$\vec{g}_{2m-1}^d = \left(\vec{g}.\vec{x}_{2m-1}^d\right) \vec{x}_{2m-1}^d + \left(\vec{g}.\vec{x}_{2m}^d\right) \vec{x}_{2m}^d = g_{2m-1}^d \vec{x}_{2m-1}^d + g_{2m}^d \vec{x}_{2m}^d \qquad [5.20]$$

In [KES 03] and [SEM 04a] it is shown that the coordinates of the vector variable written in the basis generating a subspace (or fictitious machine) consist of a harmonic group of the variables of the real machine. With the presence of harmonic families, we find a result expressed by Klingshirn [KLI 83] within the framework of the supply of a multiphase asynchronous machine with inverter in full square wave mode in steady state.

Equations [5.21] and [5.22] give the expression of the coordinates of $\vec{g}^d_{2m-1}$ in an Eigen subspace. In these equations, the harmonic ranks (assumed to be positive) that appear for m fictitious machines are $h = nl + \sigma m$, with $\sigma = \{-1, 0, +1\}$ and $l \in \mathbb{N}$. σ allows us to take into account whether a vector is homopolar $(\sigma = 0)$, rotating in clockwise (direct) direction $(\sigma = +1)$ or anticlockwise (retrograde) direction $(\sigma = -1)$.

$$g^d_{2m-1} = \sqrt{\frac{n}{2}\left(\sum_{l=0}^{\infty} g_h^{\max} \sin(hp\theta)\right)} \quad \text{with} \quad h = nl + \sigma m \qquad [5.21]$$

$$g^d_{2m} = -\sqrt{\frac{n}{2}\left(\sum_{l=0}^{\infty} \sigma g_h^{\max} \cos(hp\theta)\right)} \quad \text{with} \quad h = nl + \sigma m \qquad [5.22]$$

Table 5.1 summarizes the harmonics groups associated with three-phase, five-phase and seven-phase machines. [SEM 03a] contains the case of the six-phase double-star machine, referred to as a double-star three-phase machine. The fictitious machine is the principal machine associated with the first harmonic. That machine associated with the second harmonic is secondary. The machine associated with the third harmonic, is the tertiary machine and that associated with harmonic n is homopolar. In the case of a monophase fictitious machine $(\sigma = 0)$, we will replace coefficient $\sqrt{\frac{n}{2}}$ with $\sqrt{n}$.

	Three-phase machine	Five-phase machine	Seven-phase machine
Principal machine	$m=1$, $\sigma=\pm 1$ $h=1, 2, 4, 5, 7, \ldots$	$m=1$, $\sigma=\pm 1$ $h=1, 4, 6, 9, 11, \ldots$	$m=1$, $\sigma=\pm 1$ $h=1, 6, 8, 13, 15, \ldots$
Secondary machine	Nonexistent	$m=2$, $\sigma=\pm 1$ $h=2, 3, 7, 8, 12, \ldots$	$m=2$, $\sigma=\pm 1$ $h=2, 5, 9, 12, 16, \ldots$
Tertiary machine	Nonexistent	Nonexistent	$m=3$, $\sigma=\pm 1$ $h=3, 4, 10, 11, 17, \ldots$
Homopolar machine	$m=2$, $\sigma=0$ $h=0, 3, 6, 9, \ldots$	$m=3$, $\sigma=0$ $h=0, 5, 10, 15, \ldots$	$m=4$, $\sigma=0$ $h=0, 7, 14, 21, \ldots$

Table 5.1. *Summary table of harmonic groups associated with three-phase, five-phase and seven-phase machines*

For instance, harmonics 1 and 2 associated with a five-phase machine generate rotating vectors in the direct direction, 3 and 4 in the retrograde direction, 6 and 7 in the direct direction, etc.

We again find the well-known results for the three-phase machine. The odd harmonics of electromotive force of rank 5 (retrograde) and rank 7 (direct), by interaction with the first current harmonic, induce (direct) torque pulsations of rank 6. Similarly, the harmonics of rank 11 (retrograde) and 13 (direct) induce torque pulsations of rank 12.

For a five-phase machine, the harmonics of rank 9 (retrograde) and 11 (direct), by interaction with the first current harmonics, induce torque pulsations of rank 10. The harmonics of rank 7 (direct) and 13 (retrograde) interact with the harmonics of rank 3 (retrograde) to generate torque pulsations of rank 10.

It must be noticed that in general the harmonics of even ranks have a value of 0 except in the case of pole asymmetry. Therefore their case has not been developed.

5.2.1.4.2. *Relationship between emf harmonics and torque generated by a fictitious machine*

If we apply the "real variables to decoupled variables" transformation to the emfs of a machine possessing n phases, we can formulate different remarks:

– If the emfs of the real machine are sinusoidal; only the principal fictitious machine has an emf. In this case, according to equation [5.17], only the fictitious machine can generate a torque.

– If the number of phases of the machine is odd, the emf harmonics of the real machine that has a rank multiple of the number of phases ($h=an$, a integer) cannot generate a constant torque. In fact, these harmonics are assigned to the monophase fictitious machine referred to as being homopolar. This remark partially justifies the quasi-systematic use of star coupling between the phases, which ensures 0 current in the homopolar machine.

– If the emfs of the real machine have odd harmonics whose rank h is less than or equal to the number of phases n, the fictitious machines have a sinusoidal emf (or some are 0 if $h<n$).

– If the emfs of the real machine have more harmonics than the number of phases, there is at least one fictitious machine possessing non-sinusoidal emf.

We will notice that these remarks increase in importance when designing the control of the machine.

5.2.1.4.3. Inductances and electrical time constants of fictitious machines: the impact of harmonics

If we can restrict ourselves to estimating (with suitable precision) the value of inductances of a three-phase machine with non-salient poles, that is not saturated and is star-coupled by considering that the magnetomotive forces are sinusoidal, it is imperative to take account of the magnetomotive force harmonics when estimating the inductances of a multiphase machine. In the opposite case, there will be considerable error in the estimation of electrical time constants associated with fictitious machines, the parameters necessary for proper design of the machine's supply system and in tuning the associated current controllers.

Equation [5.23] reminds us the analytical expression of the inductance between phases j and k by considering sinusoidal magnetomotive forces [LOU 04d, WHI 59]:

$$L_{jk} = \frac{2\mu_0 (k_s N_s)^2 DL}{\pi e} \cos(\delta_{jk}) + l_{leaks} \qquad [5.23]$$

with:

- μ_0: permeability of air;
- k_s: coil coefficient;
- N_s: the number of turns of a coil (with p coils per phase);
- D: the inner diameter of the stator;
- L: the effective length of the stator;
- e: the thickness of the magnetic air-gap (air + magnets);
- $\delta_{jk} = \dfrac{2\pi(j-k)}{n}$: the angle separating phases j and k; and
- l_{leaks}: leakage inductances (generally considered zero-valued if $j \neq k$).

In this case, the calculation of inductances associated with fictitious machines leads to the following conclusions:

- the inductance associated with the principal fictitious machine (or cyclical inductance) is $\Lambda_1 = \dfrac{n}{2}L + l_{leaks}$, with L being the self-inductance of a phase ($j=k$ in equation [5.23]);

– the inductances associated with the other fictitious machines are $\Lambda_m = l_{leaks}, m \geq 2$.

Thus, we can conclude that a multiphase machine with sinusoidal magnetomotive forces possessing few leaks is not a "good" machine in the way we mean when discussing a three-phase star-coupled machine. In fact, some fictitious machines have very weak inductances (because they are only equal to the leakages inductances) and will require the machine to be supplied with inverters with a very high chopping frequency. If this condition is not fulfilled, we observe large amplitude interference currents induced by the pulse width modulation (PWM).

Similarly, modeling a multiphase machine by assuming the systematic hypothesis that magnetomotive forces are sinusoidal can lead us to make quite significant errors in the evaluation of the inductances associated with fictitious machines [LOC 06, SEM 03b]. This error leads us to considerably over-dimension the supply system and to mistune the correctors of associated currents.

If we take the magnetomotive force harmonics into account, we need to add the contribution due to the harmonics to the fundamental inductance expressed by equation [5.23]. The magnetic system being considered linear, we can apply a superposition theorem. The inductances associated with the real machine are then expressed by the following equation in which q are the magnetomotive force harmonic ranks that are kept:

$$L_{jk} = \sum_q L^q \cos(q\delta_{jk}) + l_{leaks} \qquad [5.24]$$

In the particular case of concentrated coils with diametral pitch (located in only two notches and separated by a pole pitch), the harmonic inductance L^q is expressed by:

$$L^q = \frac{1}{q^2} \frac{2\mu_0 N_s^2 DL}{\pi e}, \quad q \text{ odd} \qquad [5.25]$$

If we consider the magnetomotive force harmonics, the expression of inductance associated with m fictitious machines is given by:

$$\Lambda_m = \frac{n}{2} \sum_h L^h + l_{leakage} \qquad [5.26]$$

According to the fictitious machine, the harmonic ranks h to consider keeping are recorded in Table 5.1. In equation [5.26], we are reminded that that $h = nl + \sigma m$ with l being the integer and m the number of fictitious machines related.

176   Control of Non-conventional Synchronous Motors

According to equation [5.26], we conclude that the inductances associated with fictitious machines other than the principal fictitious machine, are not only equal to the leakage inductance but also to a particular group of harmonic inductances. Thus, the existence of magnetomotive force harmonics allows us to increase the electrical time constant of a fictitious machine and allows the use of a smaller chopping frequency when the machine is supplied by an inverter.

It is this multiharmonics approach that fundamentally distinguishes the results presented in this chapter from those presented in Chapter IX of [WHI 59], for which only a first harmonic approach was used to model multiphase machines. This first harmonic approach had as a corollary in that only the two-phase machine associated with the first harmonic was susceptible to torque generation and that the other two-phase and monophase machines were reduced to circuits characterized by the leakage inductance of the machine and the stator resistance.

We remember with the proposed approach several fictitious machines can contribute to the generation of torque (average but also pulsating), but there are design constraints so that a multiphase machine can be supplied with inverters whose PWM frequency is not too high. Here, we must take into account the harmonics whose impact is no longer a second-order phenomenon with respect to supply by an inverter [SCU 09].

5.2.1.5. *Examples*

5.2.1.5.1. Three-phase machine

In this section, we consider a classic example of a three-phase synchronous machine with sinusoidal emf. It is equipped with distributed stator coils that generate a magnetomotive force with sinusoidal spatial distribution.

We associate a vectorial three-dimensional space with the machine that is equipped with an orthonormal natural basis $B^n = \{\vec{x}_1^n, \vec{x}_2^n, \vec{x}_3^n\}$. The vectorial voltage equation of this machine is given by equation [5.4], where the inductance matrix, characteristic of linear application λ, and the emf vectors are detailed by:

$$\left[L_s^n\right] = L \begin{pmatrix} 1 + \dfrac{l_{leaks}}{L} & \cos\dfrac{2\pi}{3} & \cos\dfrac{2\pi}{3} \\ \cos\dfrac{2\pi}{3} & 1 + \dfrac{l_{leaks}}{L} & \cos\dfrac{2\pi}{3} \\ \cos\dfrac{2\pi}{3} & \cos\dfrac{2\pi}{3} & 1 + \dfrac{l_{leaks}}{L} \end{pmatrix} \quad [5.27]$$

and:

$$\vec{e} = \varepsilon_{max}\Omega\left(\sin(p\theta)\vec{x}_1^n + \sin\left(p\theta - \frac{2\pi}{3}\right)\vec{x}_2^n + \sin\left(p\theta - \frac{4\pi}{3}\right)\vec{x}_3^n\right) \quad [5.28]$$

If we apply the Concordia transformation expressed by [5.29], which is determined by equation [5.14] in the case where $n = 3$, the real machine of referenced variables $(1,2,3)$ is broken down into two fictitious machines: a two-phase machine, referred to as the principal machine, referenced (α,β); and a homopolar monophase machine, referenced z.

$$[T_{33}]^{-1} = \sqrt{\frac{2}{3}}\begin{pmatrix} 1 & \cos\frac{2\pi}{3} & \cos\frac{2\pi}{3} \\ 0 & \sin\frac{2\pi}{3} & -\sin\frac{2\pi}{3} \\ \frac{1}{\sqrt{2}} & \frac{1}{\sqrt{2}} & \frac{1}{\sqrt{2}} \end{pmatrix}, \begin{pmatrix} g_\alpha \\ g_\beta \\ g_z \end{pmatrix} = [T_{33}]^{-1}\begin{pmatrix} g_1 \\ g_2 \\ g_3 \end{pmatrix}, g = \{v,i,e\} \quad [5.29]$$

The expression of the vectors of the new basis is given as a function of the vectors of the natural basis by:

$$\begin{cases} \vec{x}_\alpha = \sqrt{\frac{2}{3}}\left(\vec{x}_1 + \cos\frac{2\pi}{3}\vec{x}_2 + \cos\frac{2\pi}{3}\vec{x}_3\right) \\ \vec{x}_\beta = \sqrt{\frac{2}{3}}\left(\sin\frac{2\pi}{3}\vec{x}_2 - \sin\frac{2\pi}{3}\vec{x}_3\right) \\ \vec{x}_z = \sqrt{\frac{1}{3}}\left(\vec{x}_1 + \vec{x}_2 + \vec{x}_3\right) \end{cases} \quad [5.30]$$

The vectorial voltage equations associated with the fictitious machines are given by:

$$\begin{cases} \vec{v}_{\alpha\beta} = R_s\vec{i}_{\alpha\beta} + \Lambda_{\alpha\beta}\left(\frac{d\vec{i}_{\alpha\beta}}{dt}\right)_{/B^n} + \vec{e}_{\alpha\beta} \\ \vec{v}_z = R_s\vec{i}_z + \Lambda_z\left(\frac{d\vec{i}_z}{dt}\right)_{/B^n} + \vec{e}_z \end{cases} \quad [5.31]$$

Finally, their characteristic variables are given by:

$$\left[L_s^d\right] = \begin{bmatrix} \Lambda_{\alpha\beta} = \frac{3}{2}L + l_{leaks} & 0 & 0 \\ 0 & \Lambda_{\alpha\beta} & 0 \\ 0 & 0 & \Lambda_z = l_{leaks} \end{bmatrix} \qquad [5.32]$$

and:

$$\begin{cases} \vec{e}_{\alpha\beta} = \sqrt{\frac{3}{2}}\varepsilon_{max}\Omega\left(\sin(p\theta)\vec{x}_\alpha - \cos(p\theta)\vec{x}_\beta\right) \\ \vec{e}_z = 0\vec{x}_z \end{cases} \qquad [5.33]$$

Analysis of the characteristic variables of fictitious machines gives us information about their specificities. Only the principal fictitious machine can generate a torque, its emf being sinusoidal and non-zero. Its electrical time constant, $\left(\frac{3}{2}L + l_{leakage}\right)/R_s$, allows us to determine the chopping frequency of the associated inverter. The homopolar fictitious machine does not possess an emf and thus cannot generate a torque. Its supply can only generate losses. If it possessed a non-zero emf linked to the presence of harmonics of rank 3, it could admittedly generate a torque of non-zero mean value [GRE 94], but due to its monophase feature it would have a significant pulsating component that would have to be compensated for by the two-phase machine. Finally, its electrical time constant, which would have a small value, l_{leaks}/R_s, that would demand supply from an inverter with a very high chopping frequency.

These conclusions bring an additional argument for a quasi-systematic coupling of three-phase machines. Such coupling will ensure the non-supply of the homopolar machine. The main economical argument is that we use three legs instead of six.

5.2.1.5.2. Five-phase machine

In this section, we illustrate the notions previously introduced, by means of a five-phase synchronous machine with surface-mounted permanent magnets and concentrated coils with a diametral pitch. Figure 5.3 shows the machine before assembly and a diagram of a cut.

We associate a five-dimensional vectorial space equipped with a natural orthonormal basis $B^n = \{\vec{x}_1^n, \vec{x}_2^n, \vec{x}_3^n, \vec{x}_4^n, \vec{x}_5^n\}$ with the machine. The vectorial voltage equation of this machine is given by [5.4], where the inductance matrix, characteristic of linear application λ, and the emf vectors are detailed by equations [5.34] and [5.35]. We note that L corresponds to the self-inductance of a phase, M_1 to the mutual inductance between two phases dephased by $2\pi/5$, and M_2 to the mutual inductance between two phases dephased by $4\pi/5$.

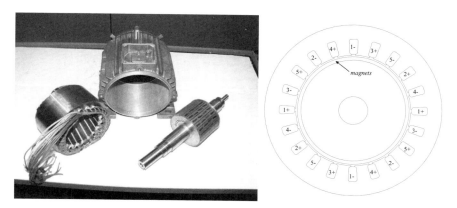

Figure 5.3. *Views of the disassembled machine and of a cut*

Finally, Table 5.2 gives the relative harmonic content of the emf of the five-phase machine considered.

$$[L_s^n] = \begin{pmatrix} L & M_1 & M_2 & M_2 & M_1 \\ M_1 & L & M_1 & M_2 & M_2 \\ M_2 & M_1 & L & M_1 & M_2 \\ M_2 & M_2 & M_1 & L & M_1 \\ M_1 & M_2 & M_2 & M_1 & L \end{pmatrix}$$ [5.34]

$$\vec{e} = \Omega . \sum_{k=1}^{n} \left(\sum_{h=1}^{\infty} \varepsilon_{max}^h \sin\left(h\left(p\theta - (k-1)\frac{2\pi}{n} \right) \right) \vec{x}_k^n \right)$$ [5.35]

Harmonics	$h=1$	$h=3$	$h=5$	$h=7$	$h=9$
Relative rate	100%	28.5%	12.4%	5.1%	1.7%

Table 5.2. *Relative harmonics content of the emf of the five-phase machine considered*

By applying the Concordia transformation of the fifth dimension to the five-phase machine, we show that the real machine is equivalent to the association of three fictitious machines: a two-phase fictitious machine referred to as the "principal", a two-phase fictitious machine referred to as "secondary" and a monophase fictitious machine referred to as "homopolar". Each of these fictitious machines is characterized by a harmonic group of variables associated with the real machine. We will find these groups with the help of Table 5.1.

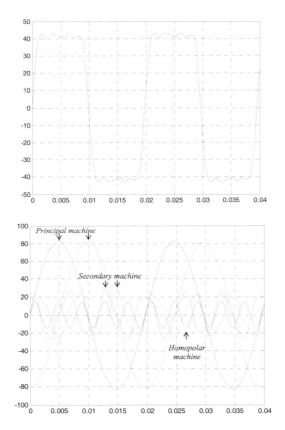

Figure 5.4. *The emf of the real machine (top) and of fictitious machines (bottom) at 1,500 rpm*

We are initially interested in the partition property of emf harmonics among fictitious machines. Figure 5.4 shows a recording of one of the emfs of the real machine and emfs of the associated fictitious machines at 1,500 rpm. We check that the emf of fictitious machines are mainly composed of:

– fundamental and harmonics of rank 9 for the principal machine,
– harmonics of ranks 3 and 7 for the secondary machine; and
– harmonic of rank 5 for the homopolar machine.

We then conclude that each of these machines can generate a torque. Obviously, for a given amplitude of current in the fictitious machines, the torque generated is greater for the principal machine than for the secondary and homopolar machine.

We are now interested in the inductances associated with the fictitious machines. Equation [5.36] gives the inductance matrix $\left[L_s^d\right]$ of the machine in the decoupling basis. Λ_p, Λ_s and Λ_z represent the inductances associated with the principal, secondary and homopolar machines, respectively.

$$\left[L_s^d\right] = \begin{pmatrix} \Lambda_p & 0 & 0 & 0 & 0 \\ 0 & \Lambda_p & 0 & 0 & 0 \\ 0 & 0 & \Lambda_s & 0 & 0 \\ 0 & 0 & 0 & \Lambda_s & 0 \\ 0 & 0 & 0 & 0 & \Lambda_z \end{pmatrix} \qquad [5.36]$$

If we only consider the fundamental component of the magnetomotive force, the inductances associated with the fictitious machines, whose calculations come from equation [5.23], are:

$$\begin{cases} \Lambda_p = 2.59mH \\ \Lambda_s = 0.348mH \\ \Lambda_z = 0.348mH \end{cases} \qquad [5.37]$$

If we now take into account harmonics 1, 3 and 5 of the stator magnetomotive force, the inductances, stemming from the calculations from equations [5.24] and [5.25] become:

$$\begin{cases} \Lambda_p = 2.59mH \\ \Lambda_s = 0.597mH \\ \Lambda_z = 0.438mH \end{cases} \qquad [5.38]$$

As shown in section 5.1.4.3, the hypothesis that only considers the fundamental component of the magnetomotive force leads to large errors in the evaluation of inductances associated with the secondary and homopolar fictitious machines. When dimensioning the chopping frequency of the inverter, these errors lead to unnecessary oversizing of the machine's supply system, in particular the apparent power of the inverter. We notice, however, that even by considering the harmonics, the ratio between the inductance of the principal machine and that of the secondary machine, $\Lambda_p / \Lambda_s = 4.3$, is still high in the case of this five-phase machine.

From another point of view, if we imagine that the machine had been designed to have low magnetic leakages and a single harmonic of magnetomotive force (and thus emf), the determination of the frequency of the PWM from the single time constant of the principal machine would lead to the observation of parasitic currents of very high amplitude in practice.

Finally, as long as the inductance linked to the homopolar machine is the smallest, the star coupling will ensure that there is no current, at least in the homopolar machine. It only remains to control the currents in the secondary machine.

5.2.2. *The inverter seen from the machine*

For the model of an *n*-leg inverter, this section will take up, the main results of Chapter 8 [KES 09a].

Let us remind ourselves that the function of the inverter is to apply the voltages calculated by the control to the electrical machine. More precisely, knowing that we are positioned within the framework of a PWM control, we will try to apply voltages whose mean values correspond to those calculated by the control. The real voltages will in fact consist of a "rolling mean" component and a "noise" component that will be filtered by the inductive circuit of the machine. We will then obtain the mean currents desired, which cause the mean electromagnetic torque developed by the machine.

According to the control strategies close to the inverter, noise voltages can be significantly different, even if the mean value of the voltage is the same. These noise voltages will induce parasitic currents whose amplitudes will be added to the "effective" mean current. When the amplitude of these noise currents becomes important, these currents become dimensioning for the choice of inverter transistor rating. For machines with more than three phases, it has been seen in section 5.2.1.4 that, contrary to the star-coupled three-phase machine without a neutral point, there are several inductive circuits (one per fictitious machine). The analysis of fictitious voltages imposed by the inverter in each fictitious machine, both for its mean component and its "noise" component, is therefore fundamental in controlling the mean currents and the parasitic currents of the real machine.

Vectorial Modeling and Control of Multiphase Machines 183

At this level, two cases have to be distinguished, which have already been seen for the three-phase machine. In other words, depending on the way the phases of the machine are coupled, a fictitious machine can be dependent or not dependent on the inverter. For this reason we will show the case of the three-leg inverter using an approach that is easy to generalize to the case of a n-leg inverter supplying a n-phase machine.

5.2.2.1. *Three-leg inverter seen from the star-coupled three-phase machine*

Here we consider a three-phase machine assumed to be star-coupled and to fulfill the general hypotheses expressed at the beginning of this chapter, and particularly the hypothesis in section 2.1.4.1 (magnetomotive and sinusoidal emfs).

In order to emphasize the problem of parasitic currents, we will assume that the neutral N' of the machine is physically linked to neutral point N of the inverter by an impedance of capacitor type. If we consider the capacitance to be 0, we will again find the ideal case and point N will not exist in reality. The capacitance will allow us to roughly model the impedance of the machine's earthing circuits and the inverter. We use this type of modeling when we are interested, for instance, in parasitic currents going through the bearings of a machine [DAH 08] where one of the origins is linked to the homopolar components of voltages imposed by the inverter [LEE 01].

Figure 5.5 shows the classical supply topology of the three-phase machine: star-coupled phases and supplied with a three-leg inverter.

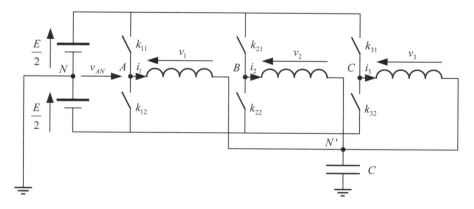

Figure 5.5. *Star-coupled three-phase machine supplied with a three-leg inverter*

The eight possible combinations of switches k_{ij} allow us to generate the eight inverter voltage vectors, $\vec{v}_k^{ond} = v_{AN}\vec{x}_1^n + v_{BN}\vec{x}_2^n + v_{CN}\vec{x}_3^n$:

184 Control of Non-conventional Synchronous Motors

$$\vec{v}_{N0} = -\frac{E}{2}\vec{x}_1^n - \frac{E}{2}\vec{x}_2^n - \frac{E}{2}\vec{x}_3^n \quad \vec{v}_{N1} = -\frac{E}{2}\vec{x}_1^n - \frac{E}{2}\vec{x}_2^n + \frac{E}{2}\vec{x}_3^n$$
$$\vec{v}_{N2} = -\frac{E}{2}\vec{x}_1^n + \frac{E}{2}\vec{x}_2^n + \frac{E}{2}\vec{x}_3^n \quad \vec{v}_{N3} = -\frac{E}{2}\vec{x}_1^n + \frac{E}{2}\vec{x}_2^n - \frac{E}{2}\vec{x}_3^n$$
$$\vec{v}_{N4} = +\frac{E}{2}\vec{x}_1^n - \frac{E}{2}\vec{x}_2^n - \frac{E}{2}\vec{x}_3^n \quad \vec{v}_{N5} = +\frac{E}{2}\vec{x}_1^n - \frac{E}{2}\vec{x}_2^n + \frac{E}{2}\vec{x}_3^n \quad [5.39]$$
$$\vec{v}_{N6} = +\frac{E}{2}\vec{x}_1^n + \frac{E}{2}\vec{x}_2^n - \frac{E}{2}\vec{x}_3^n \quad \vec{v}_{N7} = +\frac{E}{2}\vec{x}_1^n + \frac{E}{2}\vec{x}_2^n + \frac{E}{2}\vec{x}_3^n$$

Any one of these eight vectors can be broken down by applying the Concordia transformation in a unique sum of a voltage vector belonging to a principal vectorial subspace and another homopolar vectorial subspace:

$$\vec{v}_{Nk} = \vec{v}_{Nk}^{\alpha\beta} + \vec{v}_{Nk}^{z} = \left(v_{N\alpha}\vec{x}_\alpha + v_{N\beta}\vec{x}_\beta\right) + \left(v_{Nz}\vec{x}_z\right) \quad [5.40]$$

The detail of these vectors is given by:

$$\vec{v}_{N0}^{\alpha\beta} = 0.\vec{x}_\alpha + 0.\vec{x}_\beta \quad \vec{v}_{N1}^{\alpha\beta} = -\frac{E}{\sqrt{6}}.\vec{x}_\alpha - \frac{E}{\sqrt{2}}.\vec{x}_\beta$$
$$\vec{v}_{N2}^{\alpha\beta} = -\frac{E}{\sqrt{6}}.\vec{x}_\alpha + \frac{E}{\sqrt{2}}.\vec{x}_\beta \quad \vec{v}_{N3}^{\alpha\beta} = -\frac{2E}{\sqrt{6}}.\vec{x}_\alpha + 0.\vec{x}_\beta$$
$$\vec{v}_{N4}^{\alpha\beta} = +\frac{2E}{\sqrt{6}}.\vec{x}_\alpha + 0.\vec{x}_\beta \quad \vec{v}_{N5}^{\alpha\beta} = +\frac{E}{\sqrt{6}}.\vec{x}_\alpha - \frac{E}{\sqrt{2}}.\vec{x}_\beta \quad [5.41]$$
$$\vec{v}_{N6}^{\alpha\beta} = +\frac{E}{\sqrt{6}}.\vec{x}_\alpha + \frac{E}{\sqrt{2}}.\vec{x}_\beta \quad \vec{v}_{N7}^{\alpha\beta} = 0.\vec{x}_\alpha + 0.\vec{x}_\beta$$

and:

$$\vec{v}_{N0}^{z} = -\frac{\sqrt{3}E}{2}.\vec{x}_z \quad \vec{v}_{N1}^{z} = -\frac{E}{\sqrt{6}}.\vec{x}_z$$
$$\vec{v}_{N2}^{z} = -\frac{E}{\sqrt{6}}.\vec{x}_z \quad \vec{v}_{N3}^{z} = +\frac{E}{\sqrt{6}}.\vec{x}_z$$
$$\vec{v}_{N4}^{z} = -\frac{E}{\sqrt{6}}.\vec{x}_z \quad \vec{v}_{N5}^{z} = +\frac{E}{\sqrt{6}}.\vec{x}_z \quad [5.42]$$
$$\vec{v}_{N6}^{z} = +\frac{E}{\sqrt{6}}.\vec{x}_z \quad \vec{v}_{N7}^{z} = +\frac{\sqrt{3}E}{2}.\vec{x}_z$$

Voltages $\vec{v}_{\alpha\beta}$ and v_z applied to the principal and secondary fictitious machines are respectively expressed by:

$$\vec{v}_{\alpha\beta} = R_s \vec{i}_{\alpha\beta} + \Lambda_{\alpha\beta} \frac{d\vec{i}_{\alpha\beta}}{dt} = \vec{v}_{N\alpha\beta} \qquad [5.43]$$

$$v_z = R_s i_z + \Lambda_z \frac{di_z}{dt} = v_{Nz} - \frac{V_{N'N}}{\sqrt{3}} \qquad [5.44]$$

The capacitor being run through by a current equal to the sum of currents in the phases of the machine, we have:

$$V_{N'N} = \int \frac{\sqrt{3} i_z}{C} dt \qquad [5.45]$$

An ideal star coupling (zero capacitance, C) implies a current $i_z = \frac{i_1 + i_2 + i_3}{\sqrt{3}} = 0$ and hence a voltage applied to the homopolar machine of $v_z = 0$, whichever voltage v_{Nz} is imposed by the inverter. The homopolar machine is therefore *never* supplied when the real machine is star coupled and supplied by a three-leg inverter without a neutral point.

When capacitance C is not null, a non-zero current i_z circulates if voltage v_{Nz} is not zero. This is always the case when we consider the eight vectors $\vec{v}_{Nk}^z$ that can be imposed by the three-leg inverter. We will understand why some inverter control strategies aim to choose those voltage vectors presenting the weakest homopolar components (excluding inverter voltage vectors $\vec{v}_{N0}$ and $\vec{v}_{N7}$) from among the eight voltage vectors [LEE 01].

REMARK 5.1. these results still remain true if the emfs and magnetomotive forces have harmonics of rank 3 because this modifies the homopolar components of emf as well as the homopolar inductance. Only the homopolar current will be modified if the capacitance is non-zero.

REMARK 5.2. if we consider a triangle coupling, the impact of an emf harmonic of rank 3 is fundamental. In fact, in the case of triangle coupling it is the homopolar voltage v_z of the machine that we make 0. Current i_z is only 0 when the emf e_z is also 0.

By considering the ideal case of the star-coupling with no capacitance and the homopolar machine not being supplied, the real machine supplied by a real inverter is strictly equivalent to the principal fictitious machine supplied by a fictitious inverter. Instead of studying the supply of the real machine using one of the eight real vectors $\vec{v}_{Nk}$ given by [5.39], we prefer to study the supply of the fictitious machine by one of the eight principal vectors, $\vec{v}_k^{\alpha\beta} = \vec{v}_{Nk}^{\alpha\beta}$, obtained by projection of the real vectors into the subspace associated with the principal machine (see equation [5.41]). These two-dimensional vectors are classically represented in a plane and form the centered hexagon given in Figure 5.6. Vectors $\vec{v}_0^{\alpha\beta}$ and $\vec{v}_7^{\alpha\beta}$ appear in this figure only via the central point of the hexagon because they possess no projection in the principal plane.

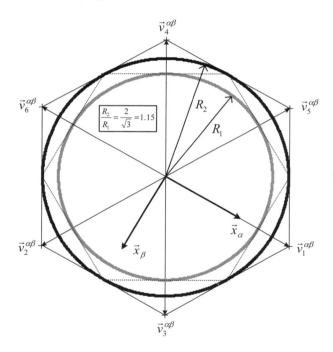

Figure 5.6. *The two-dimensional representation of voltage vectors supplying the principal machine*

In Figure 5.6 the gray circle of radius R_1 gives the limit of mean voltage vector $\langle \vec{v}_{\alpha\beta} \rangle(t)$ that can be imposed if we generate three mean sinusoidal voltages of reference at the level of the inverter (voltages v_{AN}, v_{BN} and v_{CN} in Figure 5.5). In this case, by control we ensure that the mean value on a PWM period of the homopolar voltage of the inverter is 0:

$$\langle v_{Nz} \rangle (t) = \frac{1}{T_{MLI}} \int_{T_{MLI}} v_{Nz} dt = 0 \qquad [5.46]$$

We then have a mean value of: $\langle v_{N'N} \rangle (t) = 0$.

The black concentric circle, of radius $R_2 > R_1$, indicates the limit of the mean voltage vector if we inject a homopolar component (often a harmonic of rank 3) in the reference voltages of the inverter. We are thus reminded that this overmodulation allows us to use at best, from the point of view of the DC voltage of the bus, the degree of freedom left by the star coupling (unsupplied homopolar with a harmonic multiple of 3 present in the inverter voltages).

In summary, the star coupling without a neutral point allows us to ensure that a fictitious machine will have 0 current and frees a degree of freedom for control, allowing us to better use the DC voltage of the bus (+15.5% of excursion).

The reader who would like to deepen his or her knowledge of the elements relative to the vectorial modeling of inverters is invited to read Chapter 8 "Multiphase voltage source inverters" of [MON 11].

5.2.2.2. Generalization to n-leg inverters: fictitious two-phase inverters. Example of machine-five-leg inverter association

The study of machines with more than three phases, supplied with inverters with a number of legs equal to the number of phases follows the same approach as that used in section 5.2.2.1.

The star coupling without a neutral point always allows the current in the homopolar fictitious machine to be 0 and frees a degree of freedom for control. In Table 5.3 [LEV 08], we find that the benefit introduced by the homopolar weakens as the number of phases increases. We could conclude that an increase in the number of phases is unfavorable for good use of the DC bus. We do, however, need to note that the calculations leading to Table 5.3 assume the machine is supplied in sinusoidal regime with an injection of homopolar voltage. In the "fictitious machine" approach, this means that we assume that we are only supplying a single fictitious machine. If this approach is enough in three-phase it is no longer the case if the number of phases n is greater than three because the optimal use of the DC bus depends on all the fictitious machines being supplied. Thus in [RYU 05], we find a modulation rate of 1.23 in the case of a five-phase machine that has been properly designed and whose two fictitious machines are supplied.

188   Control of Non-conventional Synchronous Motors

Number of phases n	Level of harmonics injection n	Index of maximum modulation Mi	Percentage of increase in the fundamental component
3	-1/6 of the fundamental	1.155	15.5
5	-0.062 of the fundamental	1.052	5.2
7	-0.032 of the fundamental	1.026	2.6
9	-0.02 of the fundamental	1.015	1.5

Table 5.3. *Benefit introduced in the modulation index in the case of homopolar injection*

If we omit this aspect of homopolar injection, which only intervenes in the cases where we work at the limits of the voltage possibilities of the inverter, supplying a multiphase machine has as a sole difference with respect to the three-phase case. It has to consider several projections of the 2^n characteristic vectors of the inverter (instead of 2^3) in several planes instead of a single plane. Thus, for a five-leg inverter we will find two planes with 30 non-zero vectors (2^5-2), see Figure 5.7.

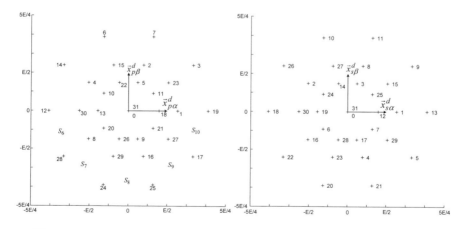

Figure 5.7. *Graphical representation of a five-phase inverter in the principal (left) and secondary (right) subspaces*

At the level of the inverter control, the different steps for the determination of duty cycles will be the following:

a) obtaining the voltage vectors for each fictitious machine (in three-phase we have only a single vector) as a function of torques required for each fictitious machine;

b) vectorial summation of the different voltage vectors of these machines;

c) possible addition of a homopolar voltage vector in order to optimize the use of the bus voltage;

d) calculation of the conduction durations by assuming we are in linear regime (non-saturation of the inverter) according to the expressions given in [KES 09a];

e) taking into account, if necessary, the saturation of the inverter by generating adapted references.

The analysis of the different steps emphasizes the fact that we apply a superposition theorem as long as there is no saturation of the inverter. This means that everything is as if each fictitious machine was supplied with a fictitious inverter. Thus, as long as we are not interested in the nonlinear aspects of the inverter control, we can use very classic intersective control techniques that implicitly fulfill step d, knowing that step c is less important as the number of phases increases.

5.3. Torque control of multiphase machines

Efficient position and speed controls require a torque control, i.e. stator currents. If several sets of n stator currents can generate the same torque, we will try and use the degrees of freedom given by the redundancy of phases to fulfill some criteria. Among the most frequently used is the criterion that minimizes Joule losses for a given torque. This will be used in the following sections.

5.3.1. *Control of currents in the natural basis*

5.3.1.1. *Statement of the method*

We speak of control of currents in the natural basis when the controlled currents are the real currents measured in the machine. Figure 5.8 gives the synoptic diagram of the torque control of a multiphase machine in natural basis. We will recognize:

– c^*: the reference torque;

– $\vec{k}$: the criterion used to elaborate the reference currents;

– $\vec{i}^*$ and $\vec{i}$: the current references and the measured currents;

– $\vec{v}^*$ and $\vec{v}$: the voltage references and the voltages applied to the machine; and finally

– V_{bus}: the voltage of the DC bus supplying the inverter.

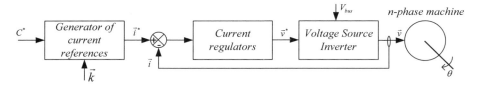

Figure 5.8. *Synoptic diagram of the torque control of a multiphase machine in the natural basis*

The first operation consists of elaborating the current references. Since we need to elaborate a vector (reference current vector) from a scalar (reference torque), it is necessary to introduce a criterion, and therefore a vectorial. Equation [5.47] clarifies the generation of current references from a reference torque:

$$\vec{i}^* = \vec{k}\ c^* \qquad [5.47]$$

We can, for instance, choose to work with a maximum torque for given statoric Joule losses. Given that these Joule losses are proportional to the square of the modulus of the current vector, this criterion is easily translated from equation [5.9] by the fact that the current vector has to be colinear to the emf vector. For a given current vector modulus, we then obtain the maximum torque. It therefore becomes:

$$\vec{i}^* = a\ \vec{\varepsilon},\ a \in \Re \qquad [5.48]$$

By reinjecting [5.48] in [5.9], we get $c^* = a\ \|\vec{\varepsilon}\|^2$, from where we get:

$$\vec{i}^* = \frac{c^*}{\|\vec{\varepsilon}\|^2}\ \vec{\varepsilon} \quad \Rightarrow \quad \vec{k} = \frac{\vec{\varepsilon}}{\|\vec{\varepsilon}\|^2} \qquad [5.49]$$

Equation [5.49] leads to several remarks:

– Vectors $\vec{i}$ and $\vec{\varepsilon}$ being colinear, the generated torque induces minimum global Joule losses. If there are effectively several strategies available, generating a given torque with minimum Joule losses allows optimal use (by neglecting the iron losses) from a thermal point of view of the machine.

– If the emfs only have harmonics of rank that are smaller than the number of phases n of the machine, term $\|\vec{\varepsilon}\|^2$ is constant and the currents possess the same harmonic ranks as the emfs. If we take the classic case of a machine with sinusoidal emf, we again find sinusoidal currents in phase with the emf.

– If the emfs possess a number of harmonics greater than the number of phases, n, term $\|\vec{\varepsilon}\|^2$ is no longer constant and the currents possess a number of harmonics greater than the emfs. Finally, if it's possible that term $\|\vec{\varepsilon}\|^2$ cancels out for a particular position θ of the rotor, there is no combination of currents that allows us to maintain a constant torque.

None of the cases shown previously have constant currents in steady-state regime and their control usually requires the use of correctors with large bandwidth (often with hysteresis). This is prohibited in the domain of high power because of the rich and poorly controlled spectral content. Let us note, however, that if the harmonic content of current is finite, the use of multifrequential resonating controllers can allow the perfect tracking of references in steady state [LIM 09].

5.3.1.2. *Example: five-phase machine with trapezoidal emf*

Here we show the results of controlling currents in the natural basis of the five-phase machine that were presented in section 5.2.1.4.1 for a machine that is star coupled and supplied with a five-leg inverter [KES 09b]. Figure 5.9 is a synoptic diagram of the experimental system put in place for these trials.

Torque reference c^* comes from the proportional integral (PI) controller ensuring the control of speed Ω. This speed is estimated by filtered numerical differentiation from the position measured by a synchro-resolver. The "optimal" current references are calculated from equation [5.49], the emf (i.e. vector $\vec{\varepsilon}$) being estimated in real time from mechanical angle θ.

The set is controlled by real-time set DSpace® 1005. The mechanical load is generated by a powder brake. The effective torque is measured by a torquemeter placed between the motor and the load.

Figure 5.10 shows the current references obtained for a control speed of 20 rad/s and a resistant torque of 2 N.m.

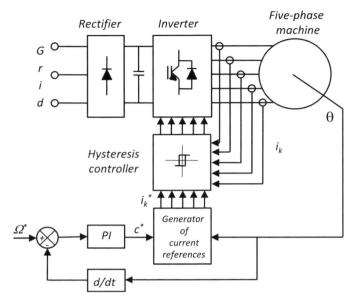

Figure 5.9. *Synoptic diagram of the experimental system of control of currents in the natural basis of a five-phase machine*

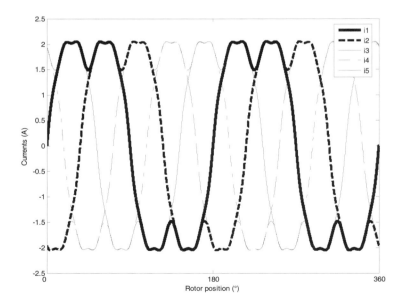

Figure 5.10. *"Optimal" current references of a five-phase machine with trapezoidal emf*

The currents, which are controlled by hysteresis controllers, allow us to obtain a constant torque with minimal Joule losses, as shown in Figure 5.11. The amplitude of torque oscillations results from the bandwidth of the hysteresis corrector, which is itself linked to the maximum authorized chopping frequency.

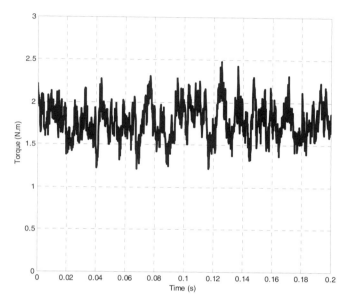

Figure 5.11. *Experimental torque of a five-phase machine obtained by control of currents in the natural basis*

5.3.2. *Control of currents in a decoupling basis*

5.3.2.1. *Statement of the method*

If we apply the decoupling transformation in section 5.2.1 to the characteristic equations of an n-phase machine, we are no longer studying the control of the real machine but a sum of N fictitious one- or two-phase machines. Figure 5.12 gives a synoptic diagram of the torque control of the machine in a decoupling basis. The variables carry indices d to indicate that it is about a decoupled variable and 1 to N to indicate the number of the fictitious machine of interest.

The torques reference distribution can be done in several ways according to the field of application. Either we use the principal machine (in which the fundamental component of the emf is projected) in normal running and the other fictitious machines are only used transiently in "transient overtorque" mode; or we make demands on all the fictitious machines. In this last mode, we can again take the case of control with minimum global Joule losses. By projecting the reference current vector expressed by [5.49] in each eigenspace associated with a fictitious machine, we obtain the reference current for each fictitious machine:

$$\vec{i}_d^* = c^* \frac{\vec{\varepsilon}_d}{\|\vec{\varepsilon}\|^2} \qquad [5.50]$$

$$c_d^* = \vec{\varepsilon}_d \cdot \vec{i}_d^* = c^* \frac{\|\vec{\varepsilon}_d\|^2}{\|\vec{\varepsilon}\|^2} \qquad [5.51]$$

According to the harmonic content of the real machine, the reference torque imposed on each of the fictitious machines can vary, although that of the real machine is constant. In the latter case, other strategies will be devised.

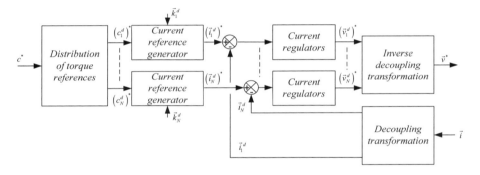

Figure 5.12. *Synoptic diagram of the torque control of a five-phase machine in a decoupling basis*

5.3.2.2. Case of machines whose fictitious machines have sinusoidal emf

In the case where each fictitious two-phase machine possesses sinusoidal emf, according to equation [5.51] and the remarks expressed in section 5.3.1.1, variables $\|\vec{\varepsilon}_d\|^2$ and $\|\vec{\varepsilon}\|^2$ are constant. The strategy that minimizes the global Joule losses for a given torque leads us to impose constant torque references in each fictitious machine if the torque reference imposed to the real machine is constant. We then end up controlling each two-phase fictitious machine as we would control the equivalent two-phase machine in the case of a star-coupled three-phase machine (where control is referred to as in the Park basis).

Each fictitious machine is assumed to possess an emf of the type:

$$\vec{e}_{\alpha\beta\_m} = \sqrt{\frac{n}{2}}\Omega\varepsilon_{max\_m}\left(\sin(h_m p\theta)\vec{x}_{\alpha\_m} - \sigma\cos(h_m p\theta)\vec{x}_{\beta\_m}\right) \quad [5.52]$$

Here, h_m corresponds to the emf harmonics of the real machine, which is projected in fictitious machine number m. This harmonic, assumed to be unique, generally corresponds to the first odd harmonics of each machine (see Table 5.1). We are reminded that σ allows us to take into account whether the vector rotates in the direct $(\sigma = +1)$ or the retrograde direction $(\sigma = -1)$.

If in each fictitious machine we have a new basis change, obtained by rotation of the basis $(\alpha\beta)$ as expressed by:

$$\begin{pmatrix}\vec{x}_{d\_m}\\\vec{x}_{q\_m}\end{pmatrix} = \begin{pmatrix}\cos(h_m p\theta) & \sigma\sin(h_m p\theta)\\-\sigma\sin(h_m p\theta) & \cos(h_m p\theta)\end{pmatrix}\begin{pmatrix}\vec{x}_{\alpha\_m}\\\vec{x}_{\beta\_m}\end{pmatrix} \quad [5.53]$$

We obtain emf, referred to as axes (dq):

$$\vec{e}_{dq\_i} = \sqrt{\frac{n}{2}}\varepsilon_{max\_i}\Omega\left(0.\vec{x}_{d\_i} - \sigma.\vec{x}_{q\_i}\right) \quad [5.54]$$

Insofar as some harmonics generate vectors rotating in the direct direction and others in the retrograde direction, the rotation can be in one direction or in the other.

The emfs of axes (dq), which are constant in steady-state regime, lead to "optimal" current references such as $i_{d\_i}^* = 0$ and $i_{q\_i}^* = k.c_d^*$. These justify the use of current PI-type controllers and a control strategy in PWM with constant frequency of the inverter. The parameters of current controllers can be determined using classic methods (compensation pole method, symmetrical optimum method, etc.) insofar as the fictitious machines are modeled by a first-order circuit made of the resistance of the stator phase, the inductance and the emf of the fictitious machine considered.

Figure 5.13 shows the synoptic diagram of current control of an n-phase machine in an extensive Park basis. If we wish to obtain better dynamic performances, we add the emf compensations (constant in a steady-state regime given the hypotheses taken in this section) to the PI controllers ensuring the control of currents.

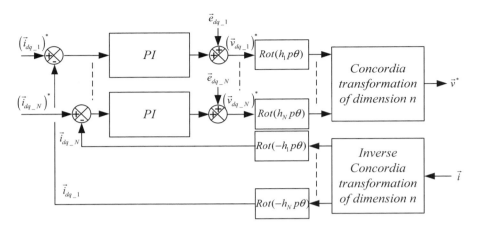

Figure 5.13. *Synoptic diagram of the current control of a multiphase machine in the generalized Park basis*

5.3.2.3. *Case of machines whose fictitious machines are not sinusoidal emf*

In the case of non-sinusoidal fictitious machines, there are two alternatives.

The first imposes minimum global Joule losses for a given torque. In this case, the torque references of fictitious machines are not constant when torque reference of the real machine is constant, and the choice of current correctors then arises. This approach, which has little advantage with respect to control in the natural basis detailed in section 5.3.1, will not be developed further.

The second alternative uses the approach developed in section 5.3.2.1. We then make a basis change in each fictitious machine by rotation of the basis associated with the being machine considered. We chose a rotation angle $h_m p\theta$ so that the mean value of the emf of axis q is as high as possible. In this case, imposing constant currents dq in a steady-state regime does not allow us to generate a constant electromagnetic torque but does however possess a mean value for minimum Joule losses.

Besides generating torque ondulations, the emf harmonics of ranks different from h_m generate harmonic currents that are responsible for additional losses and torque ripples. The solution here lies compensating for these harmonics by injecting them in the reference voltages. The synoptic diagram in Figure 5.13 remains valid, although the terms corresponding to the compensation of emf are not only necessary for obtaining good dynamic performance, but for canceling out harmonics current that are harmful in steady state.

5.3.2.4. *Example: five-phase machine with trapezoidal emf*

Here we give the results of current control in the extensive Park basis of the five-phase machine, described in section 5.2.1.4.1, that is star coupled and supplied with a five-leg inverter. Figure 5.14 shows a synoptic diagram of the current control system put in place for these trials.

We are reminded that the five-phase star-coupled machine is equivalent to the association of two two-phase fictitious machines: a principal machine, indexed p, in which we have harmonics 1 and 9 of the magnetomotive forces and emfs of the real machine; and a secondary machine, indexed s, in which we mainly find harmonics 3 and 7 of the same variables.

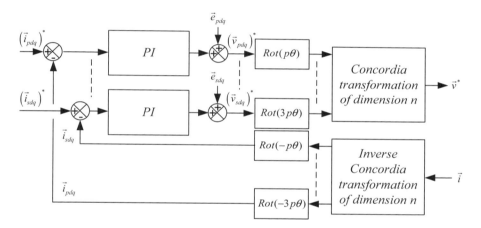

Figure 5.14. *Synoptic diagram of the current control of a five-phase machine in generalized Park basis*

The emf of fictitious machines indexed d or q in the new basis are obtained by rotation of variables of the principal machine of angle $p\theta$ and variables of the secondary machine of angle $3p\theta$, see Figure 5.15. These emfs are not constant because they are not made of a single harmonic, the real machine possessing quasi-trapezoidal emfs. In the emfs of fictitious machines we again find a constant value that is predominantly associated with a harmonic of rank 10, which comes from the rotation of harmonic 9 by θ in the principal machine and harmonic 7 by 3θ in the secondary machine.

For all the trials that have been carried out, the inverter is controlled in intersective PWM centered at a chopping frequency of 5 kHz. This frequency has been determined from the smallest time constant of the system corresponding to the electrical time constant of the secondary machine, $f_s = \dfrac{R_s}{2\pi \Lambda_s} = 114 Hz$.

The current references are constant and the currents are controlled via PI-type controllers. Inputs, allowing the emf to be compensated, are added to the output of current correctors.

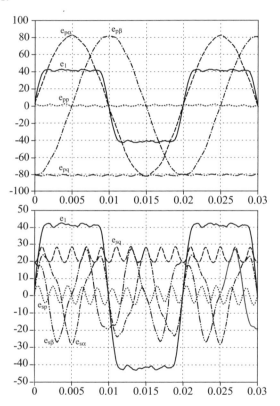

Figure 5.15. *The emfs of principal and secondary machines before ($e_{\alpha\beta}$) and after rotation (e_{dq}) compared to the emf of phase 1 (e_1)*

5.3.2.4.1. Current control without compensation of emfs

In the trials presented in this section, we impose the following current references:

– $i^*_{pd} = 0A$ and $i^*_{pq} = -7A$ for the principal machine;

– $i^*_{sd} = 0$ and $i^*_{sq} = 0A$ for the secondary machine.

The emfs are not compensated.

Figure 5.16 shows the current in one phase. We observe a strong harmonic 7 due to the corresponding component of the emf. The PI current controller as it is set is not capable of rejecting this perturbation at the speed considered. This current harmonic is responsible for additional losses as well as torque ripples. We realize these harmonic currents by checking in Figure 5.17 that, although the current references of the secondary machine are set to 0, strong harmonic currents (of rank 10) exist.

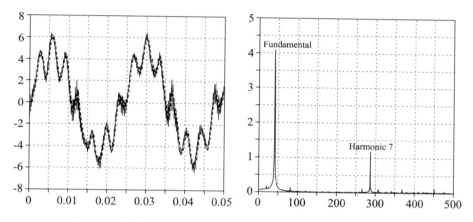

Figure 5.16. *Current in a phase and associated frequential spectrum when the emfs are not compensated*

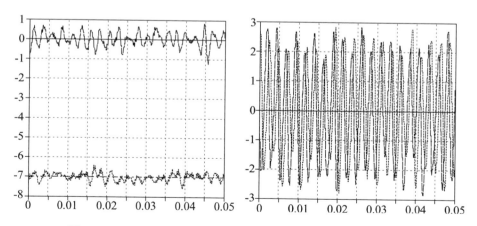

Figure 5.17. *The dq currents in the principal and secondary machines when the emfs are not compensated*

5.3.2.4.2. Current control with compensation of emfs

In the trials in Figures 5.18 and 5.19, the only compensation is for emf harmonic 7. The current in a phase is quasi-sinusoidal and the currents in the fictitious machines being properly controlled.

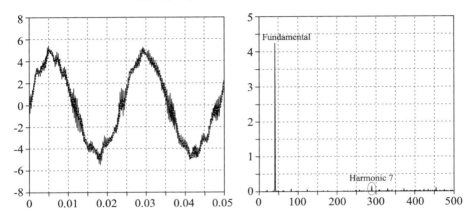

Figure 5.18. *Current in a phase and associated frequential spectrum when the emf harmonic 7 is compensated*

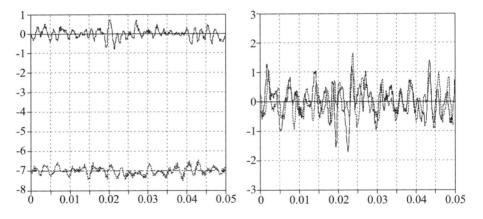

Figure 5.19. *The dq currents in the principal and secondary machines when emf harmonic 7 is compensated*

The fictitious machines now being considered properly controlled, we decide to generate a torque with each of the machines.

The secondary machine possesses a first emf harmonic (harmonic 3 of the real machine) equal to 30% of the first harmonic of the principal machine. The distribution of torque references in each of the fictitious machines, coming from equation [5.51], then fulfill ratio $c_s^* = \dfrac{0.3^2}{1^2} c_p^*$. Finally, if we assume that the electromagnetic torque generated by the machine is proportional to current of axis q, the references of currents of axis q then fulfill the ratio: $\left|i_{sq}^*\right| = 0.3 \left|i_{pq}^*\right|$.

We keep the same current references for the principal machine ($i_{pd}^* = 0A$ and $i_{pq}^* = -7A$), the references associated with the secondary machine becoming $i_{sd}^* = 0$ and $i_{sq}^* = 2.1A$. Figures 5.20 and 5.21 show the currents in one phase of the real machine and in the associated fictitious machines.

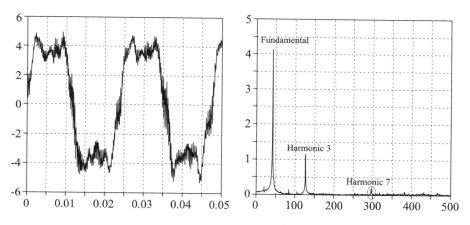

Figure 5.20. *Current in a phase and associated frequential spectrum when the two fictitious machines generate a torque*

The injection of current of harmonic 3 allows an increase in electromagnetic torque of almost 9% (30% of current of rank 3 and 30% of emf of rank 3) which leads to an increase in Joule losses of 9%. An increase of 9% in torque using only the principal machine would require an increase in Joule losses of almost 19% (1.09 times the current of the principal machine).

The use of fictitious machines to generate the total torque thus allows optimal use of the multiphase machine. Control in the extensive Park basis allows the use of PI controllers associated with an inverter controlled in PWM.

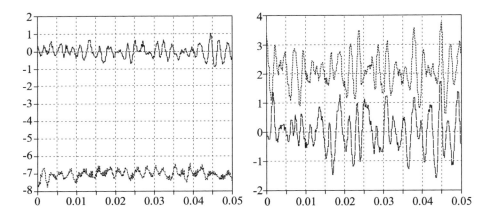

Figure 5.21. *The dq currents in the principal and secondary machines when emf harmonic 7 is compensated for*

Finally, Figure 5.22 shows the results of the speed control of this machine by using the torque distribution $c_s^* = 0.3^2\, c_p^*$. The speed references are bursts of plus or minus 10 rad/s.

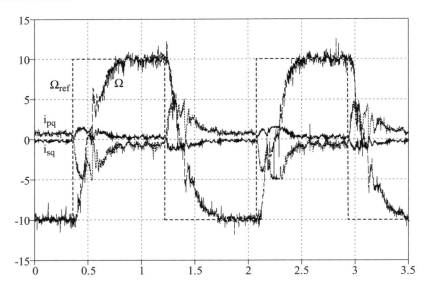

Figure 5.22. *The dq currents in the principal and secondary machines and speed response in the case of a controlled system*

5.4. Modeling and torque control of multiphase machines in degraded supply mode

The modeling and torque control of multiphase machines in degraded supply mode is a broad topic that cannot be developed in detail in this chapter. However, the techniques shown previously are, with some modifications, applicable to torque control in degraded supply mode.

5.4.1. *Modeling of a machine with a supply defect*

A first method models a machine with n symmetrical phases whose m phases would no longer be supplied as a machine with $n-m$ asymmetrical phases. By using a change of basis, such as the extensive Park transform, we control the system by controlling the constant current references [RYU 06]. This method has the major drawback of having to develop as many transforms (of models) as there are default cases. We can easily apply this method to machines with a small number of phases, but it is increasingly limited as the number of phases increase.

Another method models degradation and adds it to the initial model of the machine [CRE 10]. In this case the complete model is unique and therefore generally applicable to any number of phases. The use of particular controllers is necessary, however, to obtain satisfactory performance.

5.4.2. *Torque control of a faulty machine*

As for machines in normal functioning mode, control of the machine in degraded supply mode can be done in the natural basis and in a decoupling basis.

If we control the machine in the natural basis, we again have to calculate the current references allowing us to obtain a constant torque. Several authors propose calculations using optimization methods that cannot be used in real time (off-line method: [FU 94, PAR 07]). The method shown in section 5.3.1, however, can be applied in real time provided that we modify the emf vector according to the non-supplied phases [KES 09b].

If we control the machine in a decoupling basis, two cases arise. The first consists of designing a new model of the machine for each default case. In this case, the current references remain constant [RYU 06]. If we keep the same model of the machine and add a degradation model to it, we will either have to calculate new current references and check that the controllers are able to track them, or keep the same current references as in normal mode (constant) but adapt the number of degrees of freedom of control according to the number of non-supplied phases [KES 10, LOC 08] and adapt the bandwidth of the current controllers.

5.5. Bibliography

[CRE 10] CREVITS Y., KESTELYN X., LEMAIRE-SEMAIL B., SEMAIL E., "Modélisation causale pour la commande auto adaptée de machines alternatives triphasées en mode dégradé", *Revue Internationale de Génie Electrique (RIGE)*, February 2010.

[DAH 08] DAHL D., SOSNOWSKI D., SCHLEGEL D., KERKMAN R. PENNINGS M., "Gear up your bearings", *IEEE Industry Applications Magazine*, vol. 14, issue 4, 2008.

[FU 94] FU R., LIPO T.A., "Disturbance-free operation of a multiphase current-regulated motor drive with an opened phase", *IEEE Transactions on*, vol. 30, no. 5, pp. 1267-1274, 1994.

[GRE 94] GRENIER D., Modélisation et stratégies de commande de machines synchrones à aimants permanents à forces contre-électromotrices non sinusoïdales, thesis, Ecole Normale Supérieure de Cachan, September 1994.

[GRE 04] GRENIER D., STURTZER G., FLIELLER D., LOUIS J.-P., "Extension de la transformation de Park aux moteurs synchrones à distribution de champ non sinusoïdales", in LOUIS J.-P. (ed.), *Modélisation des Machines Électriques en Vue de Leur Commande, Concepts Généraux*, Hermes-Lavoisier, Paris, France, 2004.

[KES 03] KESTELYN X., Modélisation vectorielle multi-machines pour la commande des ensembles convertisseurs-machines polyphasés, Thesis, University of Lille 1, 4th December 2003. (Available at: http://www.univ-lille1.fr/bustl-grisemine/pdf/extheses/50376-2003-305-306.pdf, accessed 12.10.11.)

[KES 09] KESTELYN X., SEMAIL E., CREVITS Y., "Generation of on-line optimal current references for multi-phase permanent magnet machines with open-circuited phases", *International Electrical Machine and Drive Conference (IEMDC 09)*, May 2009.

[KES 10] KESTELYN X., SEMAIL E., "A vectorial approach for generation of optimal current references for multiphase permanent-magnet synchronous machines in real time", *IEEE Transactions on Industrial Electronics*, vol. 58, no. 11, pp. 5057-5065, November 2011, http://ieeexplore.ieee.org.docproxy.univ-lille1.fr/stamp/stamp.jsp?tp=&arnumber=5720307&isnumber=6011706.

[KES 11] KESTELYN X., SEMAIL E., "Multiphase voltage source inverters", in: MONMASSON E. (ed.), *Power Electronic Converters: PWM Strategies and Current Control Techniques*, ISTE Ltd, London and John Wiley and Sons, New York, 2011.

[KLI 83] KLINGSHIRN E.A., "High phase order induction motors - Part I - description and theoretical considerations", *IEEE Transactions on Power Apparatus and Systems*, vol. PAS-102, no. 1, pp. 47-53, 1983.

[LEE 01] LEE H-D., SUL S-K., "Common-mode voltage reduction method modifying the distribution of zero-voltage vector in PWM converter/inverter system", *IEEE Transactions on Industry Applications*, vol. 37, no. 6, pp. 1732-1738, 2001.

[LEV 08] LEVI E., DUJIC D., JONES M., GRANDI G., "Analytical determination of DC-bus utilization limits in multiphase VSI supplied AC drives", *IEEE Transactions on Energy Conversion*, vol. 23, no. 2, pp. 433-443, 2008.

[LIM 09] LIMONGI L.R., BOJOI R., GRIVA G. and TENCONI A., "Digital Current-control Scheme", *IEEE Industrial Electronics Magazine*, March 2009.

[LOC 06] LOCMENT F., SEMAIL E., PIRIOU F., "Design and study of a multi-phase axial-flux machine", *IEEE Transactions on Magnetics*, vol. 42, no. 4, pp. 1427-1430, 2006.

[LOC 08] LOCMENT F., SEMAIL E., KESTELYN X., "Vectorial approach based control of a seven-phase axial flux machine designed for fault operation", *IEEE Trans. Indust. Elect.*, vol. 55, no. 10, pp. 3682-3691, 2008.

[LOU 04a] LOUIS J-P., *Modélisation des Machines Électriques en Vue de Leur Commande, Concepts Généraux*, Traité EGEM, série Génie Électrique, Hermes-Lavoisier, Paris, France, 2004.

[LOU 04b] LOUIS J-P., *Modèles Pour la Commande des Actionneurs Électriques*, Traité EGEM, Série Génie Électrique, Hermes-Lavoisier, Paris, France, 2004.

[LOU 04c] LOUIS J-P., FELD G., MOREAU S., "Modélisation physique des machines à courant alternatif", in LOUIS J.-P. (ed.), *Modélisation des Machines Electriques en Vue de Leur Commande, Concepts Généraux*, Traité EGEM, série Génie Électrique, Hermes-Lavoisier, Paris, France, 2004.

[LOU 10] LOUIS J-P., FLIELLER D., NGUYEN N.K. and STURTZER G, "Synchronous motor controls, problems and modeling", in: LOUIS J-P.(ed.), *Control of Synchronous Motors*, ISTE Ltd, London and John Wiley and Sons, New York, 2010.

[MON 11] MONMASSON E., *Power Electronic Converters: PWM Strategies and Current Control Techniques*, ISTE Ltd, London and John Wiley and Sons, New York, 2011.

[PAR 07] PARSA L., TOLIYAT H.A., "Fault-tolerant interior-permanent-magnet machines for hybrid electric vehicle applications", *IEEE Transactions on Vehicular Technology*, vol. 56, no. 4, pp. 1546-1552, 2007.

[RYU 05] RYU H.M., KIM J.H., SUL S.K., "Analysis of multi-phase space vector pulse-width modulation based on multiple d-q spaces concept", *IEEE Transactions on Power Electronics*, vol. 20, no. 6, pp. 1364-1371, 2005.

[RYU 06] RYU H.M., KIM J.H., SUL S.K., "Synchronous-frame current control of multiphase synchronous motor under asymmetric fault condition due to open phases", *IEEE Transactions on Industry Applications*, vol. 42, no. 4, pp. 1062-1070, 2006.

[SCU 09] SCUILLER F., SEMAIL E., CHARPENTIER J.F., "Multi-criteria based design approach of multiphase permanent magnet low speed synchronous machines", *IET Trans. Electric Power Applications*, vol. 3, Issue 2, pp. 102-110, 2009.

[SEM 00] SEMAIL E., Outils et méthodologie d'étude des systèmes électriques polyphasés. généralisation de la méthode des vecteurs d'espace, thesis, USTL, June 30 2000. (available at: http://hal.archives-ouvertes.fr/.)

[SEM 03a] SEMAIL E., BOUSCAYROL A., HAUTIER J.P., "Vectorial formalism for analysis and design of polyphase synchronous machines", *European Physical Journal-Applied Physics*, vol. 22, no. 3, pp. 207-220, 2003.

[SEM 03b] SEMAIL E., KESTELYN X., CHARLIER A., "Sensibilité d'une machine polyphasée aux harmoniques spatiaux", *EF'2003*, Supelec, Gif-sur-Yvette, December 9-10, 2003.

[SEM 04a] SEMAIL E., KESTELYN X., BOUSCAYROL A., "Right harmonic spectrum for the back-electromotive force of an n-phase synchronous motor", *IEEE Industry Applications Conference*, vol. 1, pp. 78, October 3-7, 2004.

[SEM 04b] SEMAIL E., LOUIS J-P., FELD G., "Propriétés vectorielles des systèmes électriques triphasés", in: LOUIS J.-P. (ed.), *Modélisation des Machines Électriques en Vue de Leur Commande, Concepts Généraux*, Hermes-Lavoisier, Paris, France, 2004.

[WHI 59] WHITE D.C., WOODSON H.H., *Electromechanical Energy Conversion*, John Wiley & Sons, New York, 1959.

Chapter 6

Hybrid Excitation Synchronous Machines

6.1. Description

6.1.1. *Definition*

A hybrid excitation synchronous machine (HESM) is a machine associating the features of permanent magnet synchronous machines (PMSM) and coiled rotor synchronous machines (CRSM). It is a class of synchronous machine where the excitation flux is obtained by the contribution of two sources: one part of the excitation flux is created by permanent magnets; the other part by one or more excitation coils. This definition shows that the hybrid excitation concept is not linked to the synchronous machine. We could, for instance, design linear hybrid excitation actuators, or even hybrid excitation DC machines. We will exploit this to explain the simple functioning principle of hybrid excitation structures. The aim of such an association is obviously to associate the advantages of the two types of classical synchronous machines, i.e. a high mass and volume torque for PMSM and control of the excitation flux for CRSM.

In the first approach, excitation flux ψ_{exc} can be written as the sum of two fluxes: one generated by magnets ψ_a (which we will refer to as the magnet flux) and one generated by the excitation coils ψ_f (called the flux of excitation coils). This is summarized by the following equation:

$$\psi_{exc} = \psi_a + \psi_f \qquad [6.1]$$

Chapter written by Nicolas PATIN and Lionel VIDO.

This approach assumes the linearity of the ferromagnetic materials used to build the machine. This type of formula allows us, to some extent, to take into account the magnetic saturation, as shown in Figure 6.1.

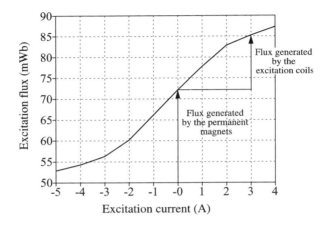

Figure 6.1. *Evolution of excitation flux as a function of excitation current*

This figure illustrates the evolution of maximum excitation flux as a function of excitation current typical of a HESM. The magnet flux is the excitation flux obtained for a zero excitation current. For a given current (for instance 3 A in Figure 6.1), the coil flux represents the difference between the total flux and the magnet flux.

The applications for which these machines can be competitive compared to existing machines are chiefly linked to transportation: electrical or hybrid vehicles and avionics. In the case of vehicles, these machines possess an additional degree of freedom (with respect to permanent magnet machines), which allows them to optimize the energy output on a running cycle [AMA 01]. In avionics, the HESM can be a good replacement for three-stage synchronous generators used to supply the onboard network, since they possess the same features (autonomous functioning and control of the excitation current to ensure a steady voltage) while having much less complex geometrical structures [PAT 06b]. From now on, we propose to classify HESMs.

6.1.2. *Classification*

Several classifications can discriminate between the various types of HESM (field lines, running as a motor or generator, structure). Here we use the most classic method, which separates the machines according to the field lines created by the two magnetic excitation sources [AMA 01, AMA 09, HLI 08, TAK 07]. We thus have two large classes of machines:

– HESM in series: the field lines generated by the excitation coils go through the magnets;

– HESM in parallel: the field lines generated by the excitation coils take a different path that allows the generation of flux without going through the magnets.

We will now present the features of machines falling into these two classes. We will associate one or more prototypes with each principle diagram.

6.1.2.1. *HESMs in series*

In this structure, the coil flux goes through the permanent magnets. This structure can be represented by a diagram of the basic electromagnetic actuator type (see Figure 6.2). The armature coil collects the excitation flux generated by the magnet and the excitation coil. The two excitation sources are assembled on the same branch of the magnetic circuit. Figure 6.3 shows a three-phase machine, with distributed winding and two pole pairs.

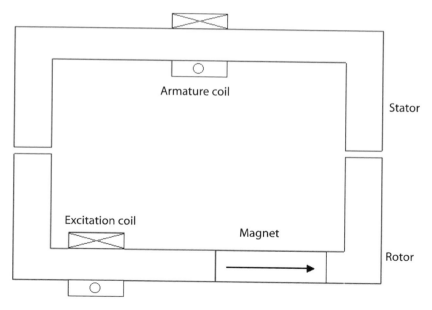

Figure 6.2. *Principle diagram of a hybrid excitation structure in series*

210   Control of Non-conventional Synchronous Motors

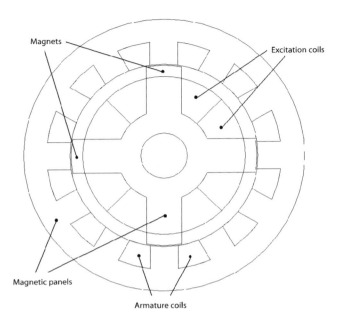

Figure 6.3. *Principle diagram of a HESM in series*

In these conditions, Figures 6.4 and 6.5 show that the excitation flux results from the contribution of flux generated by the permanent magnets and by the excitation coils. The flux of the excitation coil can be added to the magnet flux (overfluxing, see Figure 6.4) or be opposed to it (flux weakening, see Figure 6.5).

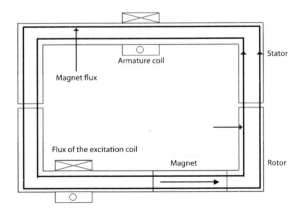

Figure 6.4. *Overfluxing*

Hybrid Excitation Synchronous Machines   211

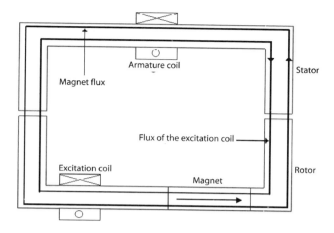

Figure 6.5. *Flux weakening*

As indicated by the name of the structure, we notice that the field line generated by the excitation coil goes through the permanent magnet.

Figures 6.6 and 6.7 show the design of HESMs in series. In Figure 6.6, the machine has a distributed winding. The rotor (six pairs of poles) has smooth poles, the magnets are assembled at the surface and the excitation coils are localized on the rotor. We can see this structure as a CRSM possessing an air-gap that is sufficiently large for magnets be put therein.

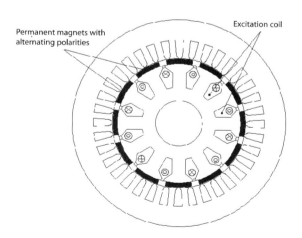

Figure 6.6. *A HESM in series [AMA 09]*

Figure 6.7 shows a prototype of a HESM with flux commutation (passive rotor). The excitation sources are in the stator, thus allowing us to discard the rings and sliding contacts.

The major advantage of HESMs in series lies in them running in flux weakening. The generation of a coil flux opposed to the magnet flux occurring at every point of the machine, the decrease in flux will lead to a decrease in iron losses (with Joule losses, however, due to the excitation coil). Hence, to some extent it will lead to an improvement in output in flux weakening zones.

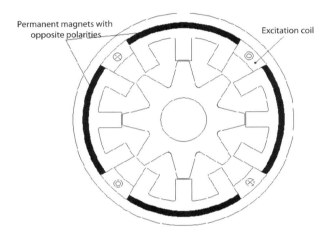

Figure 6.7. *A HESM in series [AMA 09]*

The major drawback of these structures is linked to their principle: the efficiency of the excitation coils is reduced by traveling through the magnet, the magnet being considered comparable with an important air-gap. In flux weakening mode there is a risk of demagnetization of the magnet by the excitation coils. It is necessary to take this into account during the dimensioning step of this type of machine. This problem makes it impossible to completely cancel the excitation flux.

The HESMs in parallel, which we present hereafter, allow us to partially solve this problem.

6.1.2.2. *HESMs in parallel*

When the flux of the excitation coils does not travel through the magnets, the HESMs are referred to as being in parallel. In this case, the HESMs' topologies can be reduced in number by the inventiveness of machine designers. Here, we restrict ourselves to presenting the functioning principle of these structures. For a more exhaustive analysis, we invite the reader to refer to [AMA 01, AMA 09] and [HLI 08]. From a global point of view, all these structures can be classified in two subcategories: HESMs in parallel, with or without short-circuit. We will describe these in the following sections.

6.1.2.2.1. HESMs in parallel with short circuit

When the excitation current is 0, the excitation flux is also 0[1]. For the magnet flux, there is a less reluctant path than that through the armature coil, as illustrated in Figure 6.8. In the absence of excitation current, the magnet flux is entirely picked up (short circuit, see Figure 6.7) by the excitation coil, and the flux collected by the armature coil is 0. HESMs using this principle therefore work rather like CRSMs whose excitation fluxes are improved thanks to the presence of permanent magnets.

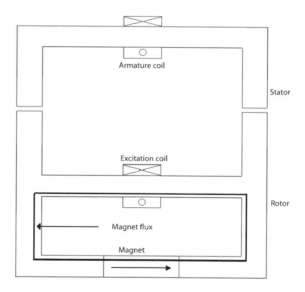

Figure 6.8. *Principle diagram of a hybrid excitation structure in parallel*

Figure 6.9 shows the principle of a rotating machine that works according to this principle.

---

1 The model proposed by equation [6.1] does not allow us to describe this class of machine.

214   Control of Non-conventional Synchronous Motors

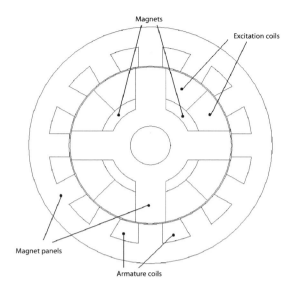

Figure 6.9. *Principle diagram of a HESM in parallel*

The functioning principle of these HESMs is shown in parallel in Figure 6.10 and with a short circuit in Figure 6.11.

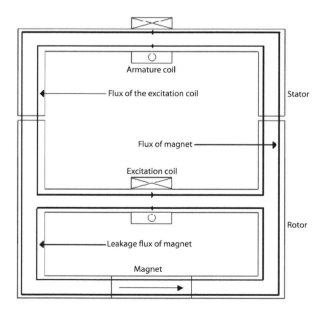

Figure 6.10. *Short-circuit of the magnet flux*

When the excitation coil is supplied, the magnet flux is progressively orientated towards the armature coil due to saturation of the magnetic branch having the excitation coil (see Figure 6.11). There does remain, however, the presence of a loss of flux.

Figure 6.11 is dedicated to the explanation of the overfluxing mode.

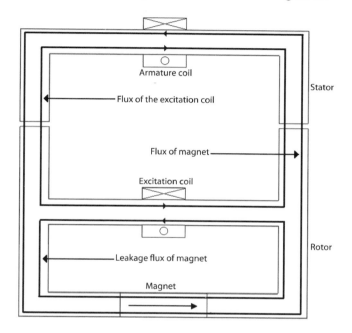

Figure 6.11. *Overfluxing*

Figure 6.12 presents the rotor of a real machine using this process (the stator is a classic stator with distributed winding). It is the claw-pole (or Lundell) machine, to which we have added magnets between the claws, referred to as inter-claw magnets. The excitation coil is global (not visible in Figure 6.12 because it is located under the claws).

The aim of these structures in parallel with short-circuit is to run safely. They allow us to fix a problem met by PMSMs supplied by an inverter during a defect in a short circuit on one of the bridge arms. In this case, the short-circuit current is only limited by the stator resistance and stator cyclical inductance of the PMSMs. With these HESMs in parallel and with short circuit, we can cancel out the excitation current in order to cancel out the short-circuit current. Safety is ensured even if the circuit supplying the coiled inductor shows a defect. Furthermore, if the magnet represents an important air-gap before the stator-rotor air-gap, the risk of demagnetizing the magnet due to the coiled excitation is considerably decreased.

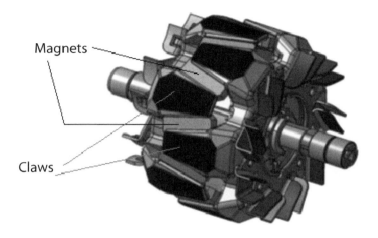

Figure 6.12. *Rotor of the hybrid excitation claw-pole machine [TAK 07]*

A few drawbacks of such structures need to be pointed out here:

– no matter what the function contemplated (overfluxing or flux weakening), part of the magnet flux is irremediably lost (see the losses flux in Figure 6.7);

– if HESMs are in series, the excitation coils do not work optimally;

– in HESMs in parallel, the magnet is not used completely;

– running is possible at the expense of Joule losses at the excitation circuit level.

HESMs without short circuit allows us to solve this problem to some extent. They are the topic of the following section.

6.1.2.2.2. *HESM in parallel without short circuit*

HESMs in parallel and without short circuit allow us to obtain a non-zero excitation flux, even if the excitation current is 0. Figure 6.13 shows an actuator that illustrates this principle. We will notice that a leakage flux exists that is characteristic of HESMs in parallel.

Hybrid Excitation Synchronous Machines 217

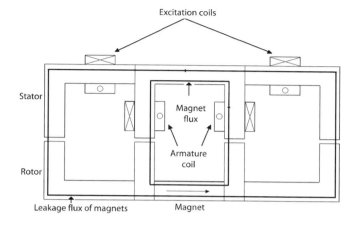

Figure 6.13. *Principle diagram of a hybrid excitation structure in parallel*

In this figure, we can notice that the excitation coils are represented in the stator. Figure 6.14 shows a rotating structure using the same principle.

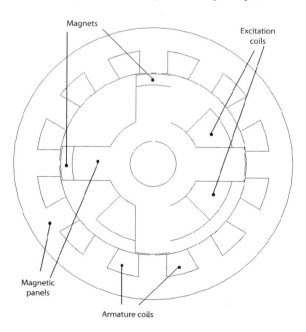

Figure 6.14. *Principle diagram of a HESM in parallel*

218   Control of Non-conventional Synchronous Motors

Figures 6.15 and 6.16 show the effect of excitation coils.

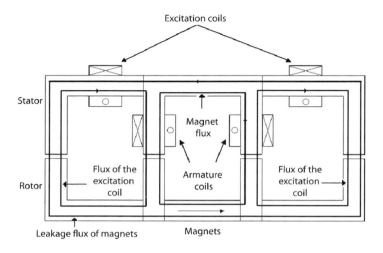

Figure 6.15. *Overfluxing*

During the operation in flux weakening mode, however, there are zones where the global flux increases. In this type of structure, flux weakening does not necessarily lead to a decrease in iron losses.

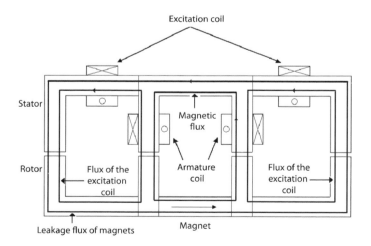

Figure 6.16. *Flux weakening*

Hybrid Excitation Synchronous Machines 219

Figures 6.17 and 6.18 show HESMs in parallel without short circuits. In the first structure (see Figure 6.17), the principle consists of taking a permanent magnet machine with flux concentration and replacing every other magnet with an excitation coil.

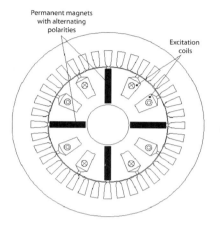

Figure 6.17. *HESM in parallel [AMA 09]*

This type of construction is applicable to practically every permanent magnet machine. Finally, Figure 6.18 shows a machine with hybrid excitation and flux commutation.

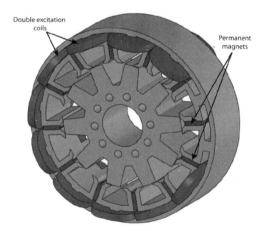

Figure 6.18. *HESM with hybrid excitation and flux commutation [HOA 07]*

220   Control of Non-conventional Synchronous Motors

Generally, running safety is ensured by these structures while the supply circuit of the inductor does not show defects at the same time as the armature. The excitation flux exists even in the absence of an excitation current, so these HESMs are good candidates for applications requiring the minimization of losses on a running cycle [AMA 01]. With these machines in series, the efficiency of the excitation coil is much better, since the coils do not have to overcome the reluctance produced by the permanent magnet.

6.2. Modeling with the aim of control

6.2.1. *Setting up equations*

Modeling the HESM involves the same principles as the classic synchronous machines. Assuming that the action of magnets is identical to that of the excitation coil, we can write the vector of stator fluxes as follows:

$$(\psi_{3s}) = (L_{ss}(\theta))(i_{3s}) + \Psi_{exc}(i_f, \theta) \qquad [6.2]$$

with:

$$\Psi_{exc}(i_f, \theta) = \Psi_0 \cdot (1 + k_H \cdot i_f) \begin{pmatrix} \cos(p\theta) \\ \cos\left(p\theta - \dfrac{2\pi}{3}\right) \\ \cos\left(p\theta + \dfrac{2\pi}{3}\right) \end{pmatrix} \qquad [6.3]$$

where p is the number of pole pairs of the machine, θ is the angular position of the rotor with respect to the stator (see Figure 6.19 describing the equivalent bipolar machine thus modeled) and k_H is a hybridization coefficient characterizing the contribution of the excitation current in the global excitation flux. We will note that for the boundary cases:

– $k_H = 0$ corresponds to a machine with permanent magnets only;

– $k_H \gg 1$ corresponds to a machine with coiled excitation alone.

Hybrid Excitation Synchronous Machines 221

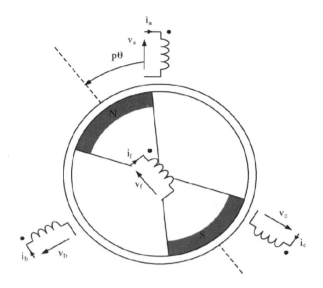

Figure 6.19. *Equivalent bipolar HESM*

In this model, we will consider a synchronous machine that potentially has salient poles, i.e. with a matrix of stator inductances (L_{ss}) function of the angular position of the rotor with respect to the stator. The classical formulation of the first harmonics of this matrix, written ($L_{ss}(\theta)$) is:

$$(L_{ss}(\theta)) = (L_{ss0}) + (L_{ss2}(\theta))$$ [6.4]

with:

$$(L_{ss0}) = \begin{pmatrix} L_{s0} & M_{s0} & M_{s0} \\ M_{s0} & L_{s0} & M_{s0} \\ M_{s0} & M_{s0} & L_{s0} \end{pmatrix}$$

and:

$$(L_{ss2}(\theta)) = L_{s2} \begin{pmatrix} \cos(2p\theta) & \cos\left(2p\theta - \frac{2\pi}{3}\right) & \cos\left(2p\theta + \frac{2\pi}{3}\right) \\ \cos\left(2p\theta - \frac{2\pi}{3}\right) & \cos\left(2p\theta + \frac{2\pi}{3}\right) & \cos(2p\theta) \\ \cos\left(2p\theta + \frac{2\pi}{3}\right) & \cos(2p\theta) & \cos\left(2p\theta - \frac{2\pi}{3}\right) \end{pmatrix}$$

The classical factorizations of these matrices call for Concordia matrices T_{32} and rotation matrices $P(.)$, defined as follows:

$$T_{32} = \sqrt{\frac{2}{3}} \begin{pmatrix} 1 & 0 \\ -1/2 & \sqrt{3}/2 \\ -1/2 & -\sqrt{3}/2 \end{pmatrix} \text{ and } P(\alpha) = \begin{pmatrix} \cos\alpha & -\sin\alpha \\ \sin\alpha & \cos\alpha \end{pmatrix}$$

We notice then that:

$$(L_{ss0}).T_{32} = (L_{s0} - M_{s0}).T_{32} \qquad [6.5]$$

and matrix $(L_{ss2}(\theta))$ can be written in the following way:

$$(L_{ss2}(\theta)) = \frac{3L_{s2}}{2} T_{32}.P(p\theta).K_2.P(-p\theta).T_{32}^t \qquad [6.6]$$

where matrix K_2 is a 2 × 2 matrix similar to the conjugation operation for complex numbers. It is written as:

$$K_2 = \begin{pmatrix} 1 & 0 \\ 0 & -1 \end{pmatrix}$$

The vectorial equation of stator fluxes must be completed by that of the flux generated in the excitation coil, i.e.

$$\psi_f = L_f i_f + \Psi'_0 + k_H \Psi_0 \left(\cos(p\theta) \quad \cos\left(p\theta - \frac{2\pi}{3}\right) \quad \cos\left(p\theta + \frac{2\pi}{3}\right) \right).(i_{3s})$$

$$[6.7]$$

In this equation, we have the reciprocity of the action between the stator armature coils and excitation coil. We also notice that the flux generated by the magnet in this excitation coil can be different from that generated in the armature coils by introducing a new parameter Ψ'_0 that, as we can see, has no influence on the voltage equations of the machine, since it is a constant. In fact, the voltage equations, both for the armature and the excitation coil, are obtained by application of Faraday's and Ohm's laws at the level of the coil:

$$(v_{3s}) = R_s.(i_{3s}) + \frac{d}{dt}[(\psi_{3s})] \qquad [6.8]$$

and:

$$v_f = R_f.i_f + \frac{d\psi_f}{dt} \qquad [6.9]$$

As a consequence, the application of *dq* transformation (i.e. *αβ* transformation plus rotation in the rotor reference frame with the excitation flux axis being *d*) leads us to the following *dq* equations (in reference frame *dq*):

$$v_f = R_f.i_f + \frac{d\psi_f}{dt} \qquad [6.10]$$

where J_2 is the matricial analogue of the pure imaginary *j* for complex numbers (i.e. $j^2 = -1$):

$$J_2 = P\left(\frac{\pi}{2}\right) = \begin{pmatrix} 0 & -1 \\ 1 & 0 \end{pmatrix}$$

and the armature flux equation gives:

$$(\psi_{dq}) = \left[(L_{s0} - M_{s0}).I_2 + \frac{3L_{s2}}{2}K_2\right](i_{dq}) + \sqrt{\frac{3}{2}}\Psi_0.(1 + k_H.i_f)\begin{pmatrix}1\\0\end{pmatrix} \qquad [6.11]$$

where I_2 is the second-order identity matrix.

Similarly, we can re-write:

$$\psi_f = L_f.i_f + \Psi'_0 + \sqrt{\frac{3}{2}}k_H\Psi_0.i_d \qquad [6.12]$$

Representation in the form of electrical diagrams is a natural goal for the electrical engineer and it is therefore interesting to address this formulation problem in the case of HESMs. The classical minimum solution is to describe the three-phase armature coils using a unique "single-phase" diagram. In the case of a dynamic model (that can be used to describe transient regimes), this single-phase representation requires the use of phasors that can be directly introduced from a real *dq* modeling. For this purpose, we can introduce the definition of a phasor $x_s = x_d + jx_q$.

We then obtain the phasor equation of the stator armature voltages:

$$\underline{v}_s = R_s \underline{i}_s + jp\Omega.\underline{\psi}_s + \frac{d\underline{\psi}_s}{dt} \qquad [6.13]$$

with:

$$\underline{\psi}_s = \left[(L_{s0} - M_{s0})\underline{i}_s + \frac{3L_{s2}}{2}\underline{i}_s^*\right] + \sqrt{\frac{3}{2}}\Psi_0.(1 + k_H.i_f) \qquad [6.14]$$

The major problem with this phasor modeling comes from the presence of a conjugation operation in the stator flux equation. This conjugation comes from the saliency term of the machine's inductance ($L_{ss}(\theta)$). It is therefore impossible to propose an equivalent electrical diagram without introducing a new component in the classic electrical library (consisting of voltage sources, resistances, capacitors, inductances and ideal transformers). Hence, it is preferable to give up a unified representation of equations on the two axes (d and q) of the machine in order to look for a representation using components. Physically, this is justified by the anisotropy of the salient pole machines, contrary to machines with smooth poles, such as asynchronous machines that allow a simple phasor representation.

REMARK 6.1. in the case of the synchronous machine with smooth poles, we again find the isotropic behavior of the machine so it is therefore possible to find the phasor representation but there is always an error source – the "purely real" excitation in this complex representation. This error in some way generates anisotropy by favoring axis d. This is all the more true in the case of a magnetically-saturated machine.

6.2.2. Formulation in components

To write equations in components of HESMs, we are going to introduce the following parameters:

– armature inductance of axis d: $L_d = L_{s0} - M_{s0} + L_{s2}$;

– armature inductance of axis q: $L_q = L_{s0} - M_{s0} - L_{s2}$;

– magnetic flux (constant) in $\alpha\beta$ reference frame: $\Phi_0 = \sqrt{\frac{3}{2}}\Psi_0$; and

– mutual inductance between the armature and excitation coil: $M = k_H.\Phi_0$.

The armature voltage equations (scalar) are then written:

$$v_d = R_s.i_d - p\Omega.\psi_q + \frac{d\psi_d}{dt} \quad [6.15]$$

$$v_q = R_s.i_q + p\Omega.\psi_d + \frac{d\psi_q}{dt} \quad [6.16]$$

with:

$$\psi_d = L_d.i_d + \Phi_0 + M.i_f \quad [6.17]$$

$$\psi_q = L_q.i_q \quad [6.18]$$

where from:

$$v_d = R_s.i_d - p\Omega.L_q i_q + L_d \frac{di_d}{dt} + M \frac{di_f}{dt} \quad [6.19]$$

$$v_q = R_s.i_q + p\Omega.(L_d.i_d + \Phi_0 + M.i_f) + L_q \frac{di_q}{dt} \quad [6.20]$$

we see that two back-electromotive forces (back-emfs) appear in the two equations, which we will refer to as e_d and e_q, respectively, and whose expressions are:

$$e_d = -p\Omega.L_q i_q \quad [6.21]$$

$$e_q = p\Omega.(L_d.i_d + \Phi_0 + M.i_f) \quad [6.22]$$

These two back-emfs are representative of the electromechanical energy conversion in the machine and as a consequence we have the expression of instantaneous mechanical power, p_m:

$$p_m = \begin{pmatrix} e_d & e_q \end{pmatrix} \begin{pmatrix} i_d \\ i_q \end{pmatrix} = -p\Omega L_q i_d i_q + p\Omega i_q.(L_d.i_d + \Phi_0 + M.i_f) = c_m.\Omega \quad [6.23]$$

where c_m is the instantaneous torque delivered by the HESM to the drive shaft (study in motor convention).

226   Control of Non-conventional Synchronous Motors

We then infer the electromechanical model of the machine with the expression of its torque as a function of electrical currents in the different coils:

$$c_m = p(L_d - L_q)i_d i_q + p(\Phi_0 + M.i_f)i_q \qquad [6.24]$$

Similarly to the introduction of $\Psi_{exc}(i_f, \theta)$ in the initial model (a,b,c) of the machine, in dq reference frame we can note an excitation flux $\Phi_{exc}(i_f)$ in the form:

$$\Phi_{exc}(i_f) = \Phi_0 + M.i_f \qquad [6.25]$$

There is a last step left for handling the equations for the electrical model of the machine. It requires the choice of localization of the magnetizing inductance that accounts for the machine's flux. We are going to proceed to modeling this in the form of a dynamic diagram in components (DDCs) of the HESMs by localizing the inductance at the level of the armature (on the equivalent circuit of axis d). In this case, we again take the equation of voltage, v_d, factorizing all the dynamic terms (i.e. in d/dt) by L_d:

$$v_d = R_s i_d - p\Omega L_q i_q + L_d . \frac{d}{dt}\left(i_d + \frac{M}{L_d} i_f\right) \qquad [6.26]$$

We then make a transformation ratio m appear, accounting for coupling between the component of axis d of the armature and the excitation coil: $m = M/L_d$. We have then emphasized a magnetizing current i_μ of the machine that accounts for its magnetic state:

$$i_\mu = i_d + m.i_f \qquad [6.27]$$

From these results we establish the equivalent diagram corresponding to axis d of the armature (see Figure 6.20). The diagram showing axis q of the armature, which completes the representation of the machine (see Figure 6.21), is immediately inferred from equation [6.20].

As indicated in the caption of Figure 6.20, this diagram is incomplete and more so for axis d because we see that current $m.i_f$ is injected in the circuit without being linked to a described supply. It now remains for us to describe the excitation coil that will have to be "naturally" connected to the diagram of axis d of the armature in the form of an equivalent diagram. For this purpose, we again take equation [6.9] by using expression [6.12] of the flux in the excitation coil. We thus get:

$$v_f = R_f . i_f + L_f \frac{di_f}{dt} + M \frac{di_d}{dt} \qquad [6.28]$$

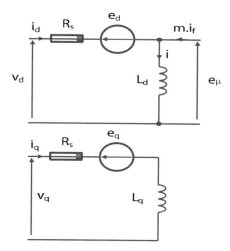

Figure 6.20. *Partial dynamic diagrams in components of the armature of the HESM (bottom) for axis d and (top) for axis q*

For a connection to be possible between the diagram shown in Figure 6.20 and the circuit representing the excitation, hawse have to introduce voltage $m.e_\mu$ into this equation because we see current $m.i_f$ appear in the armature circuit. We can then write:

$$m.e_\mu = \frac{M}{L_d}\left(L_d \frac{di_\mu}{dt}\right) = M\frac{d}{dt}(i_d + m.i_f) = M\frac{di_d}{dt} + \frac{M^2}{L_d}i_f \quad [6.29]$$

From equations [6.28] and [6.29], we can then write that:

$$v_f = R_f.i_f + \sigma L_f \frac{di_f}{dt} + m.e_\mu \quad [6.30]$$

In this equation, we make a dispersion coefficient σ appear that is expressed as:

$$\sigma = 1 - \frac{M^2}{L_d L_f} \quad [6.31]$$

The latter is similar to that existing in the asynchronous machine. It accounts for an imperfect coupling between axis *d* of the armature and the excitation coil. We will note that here the leakages have been added up at the level of the excitation coil.

228 Control of Non-conventional Synchronous Motors

This choice is arbitrary and we could have proceeded in the opposite way, with a magnetizing inductance placed at the level of the excitation coil and localizing the magnetic leakages on axis d of the armature. On the basis of equation [6.30] and of the partial diagrams in Figure 6.20, it is now possible to establish the complete DDCs of the HESMs, as indicated in Figure 6.21.

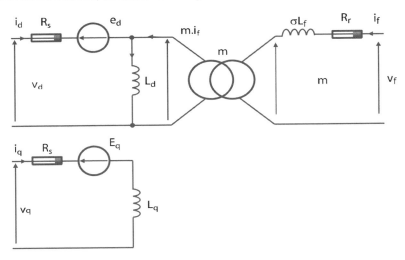

Figure 6.21. *Complete dynamic diagram in components of the HESM*

6.2.3. *Complete model*

The complete model of the machine associates the components of the dynamic model with the electromechanical conversion and the mechanical load of the motor. This set is represented diagrammatically in Figure 6.22.

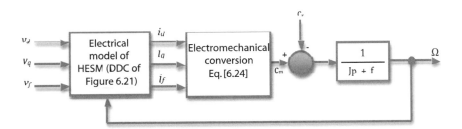

Figure 6.22. *Complete model of HESM (electrical/electromechanical) and its mechanical load (here an inertial load with viscous friction and an additional resistant torque c_r that is not defined)*

The model above is directly implantable in a "functional" simulator, such as Matlab/Simulink, due to its block diagram-type structure. However, the DDCs in Figure 6.21 are not directly useable, so it is necessary to convert the figure to block diagrams. This operation can be done systematically for any electrical diagram [PAT 06] starting from state variables carried by reactive components (inductances and capacitors). Here, we have three state variables, which are:

– current in the excitation coil (i_f);

– magnetizing current (i_μ); and

– the armature current of axis q, (i_q).

The interconnections between the three associated integrators are obtained by the application of Kirschhoff's laws and it eventually leads to the diagram in Figure 6.23.

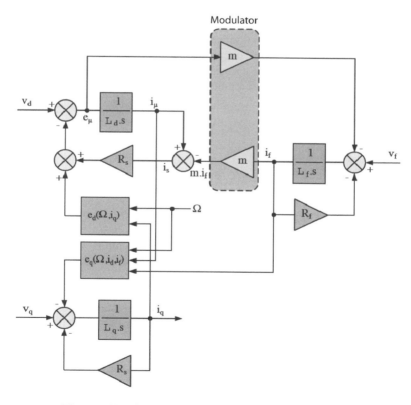

Figure 6.23. *Block diagrams corresponding to the DDCs in Figure 6.21 of an HESM*

6.3. Control by model inversion

6.3.1. *Aims of the torque control*

The torque control of HESM requires the optimization of the three currents injected in the equivalent coils of the *dq* model of the machine. In fact, the torque generated by the machine is a function of these three currents, as shown by equation [6.24]. There is thus an important redundancy in the degrees of freedom available for piloting. In the case of a machine with smooth poles, the search for a maximum torque output for a minimum injected torque requires cancellation of the current of axis *d* to the advantage of the sole current, i_q. Current i_d is eventually only used to overcome the limitations of the supply at high speed in order to ensure flux weakening of the machine.

This solution is no longer as easy in the case of machines with salient poles with the possibility of generating a torque of variable reluctance with the association of currents i_d and i_q when $L_d - L_q$ is not zero. The "balance" between these two components is a matter of a compromise between the benefit in torque brought by each component and the losses occurring in the machine. It is therefore an optimization problem of losses under torque constraint and possibly limitation of the voltage that can be provided by the uninterruptible power supply with a given DC bus voltage.

Finally, the situation is even more complicated with an HESM insofar as the magnetic flux of the machine is the superposition (in linear regime) of three sources:

– the permanent magnets;

– excitation current i_f; and

– armature current of axis *d*.

However, the process remains identical to that of machines with salient poles: it is about optimizing a criterion by gauging the contribution of each of these degrees of freedom.

As a consequence, we can see the torque control of HESMs as a structure organized into a hierarchy, such as that in Figure 6.6, in which the three input currents i_d, i_q and i_f (applied at the inputs of three dedicated regulation loops) are controlled by an optimizer, where one of the constraints is the desired motor torque on the drive shaft of the machine. This input is hence a torque input that can itself come from the speed or position regulation loop of the machine according to the targeted application.

Hybrid Excitation Synchronous Machines 231

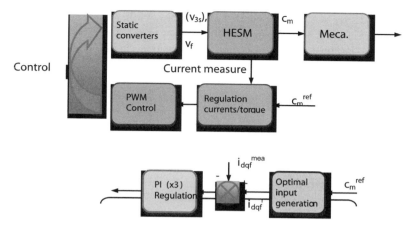

Figure 6.24. *Block-diagrams corresponding to the DDCs in Figure 6.3 of the HESM*

6.3.2. *Current control of the machine*

The torque piloting of HESMs physically requires current control in the coils of the machine. The regulation of these currents allows us to introduce limitations that have a safety role. Here, making a regulation loop can be based on an inversion control principle, in which control can be seen as the symmetrical process of the process thus piloted. We can illustrate this with the supply of the armature coil of axis q.

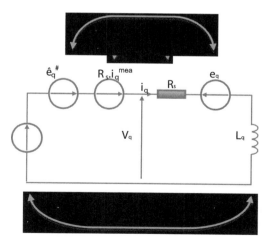

Figure 6.25. *DDCs of axis q of the armature and its control*

Figure 6.25 we see that the control must apply a voltage v_q equal to the sum of three voltages, $R_s.i_q$, e_q and $L_q.di_q/dt$. The first two voltages can be compensated for in open loop because it is about rigid relationships (in the sense of causality); whereas the third is piloted by a corrector acting as a regulator in the closed loop of current i_q. The control structure obtained is then that given in Figure 6.26.

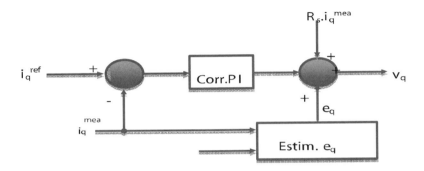

Figure 6.26. *Regulation loop of current i_q*

REMARK 6.2.   we notice that to compensate for voltage drop at the terminals of R_s, we use the measure (written with the exponent "mea") of current i_q here. A second possibility consists of using the input and not the measure. In practice, this type of solution prevents us from adding a measuring noise in the control and bears the loss of the sensor because it is about control that is partially in open loop.

By proceeding similarly, we can establish the regulation loops for i_d and i_f in Figure 6.27.

For the regulation of i_d, we notice that it is about the regulation of i_μ, which is hidden from the point of view of input by piloting of the input i_μ^{ref} with the help of inputs i_d and i_f (i.e. i_d^{ref} and i_f^{ref}).

When the three regulation loops are implanted, we can see the machine and its close control as a system relating to three reference inputs i_d^{ref}, i_q^{ref} and i_f^{ref}, and a motor torque c_m with controlled dynamics. In fact, the PI correctors used for each loop allow us to master the time constant of the system in closed loop (by pole placement) and the integrating effect of the corrector allows us to compensate for possible errors made in the compensation terms. The compensation terms are injected in the open loop in voltage inputs v_d, v_q and v_f applied to electronic power converters supplying the coils of the machine (inverter for the armature and chopper for the excitation).

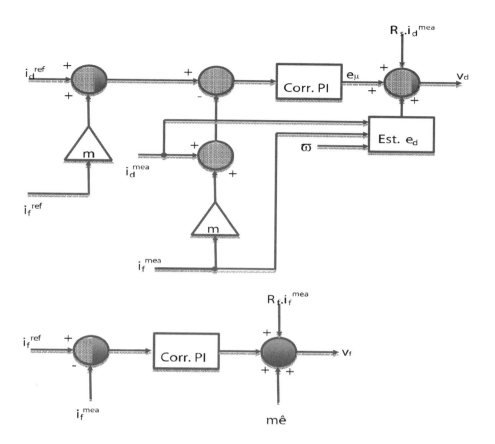

Figure 6.27. *Regulation loops of i_d and i_f*

6.3.3. *Optimization and current inputs*

As has already been mentioned, the three input currents are calculated by an optimizer in charge of obtaining the desired torque while minimizing a certain criterion: generally we try to minimize the losses in the machine (even in the converters). The function to be optimized is relatively complex and it is usually preferable to carry out the optimization work offline and to tabulate the inputs as a function of the different parameters, such as torque input and rotational speed of the machine.

The function to be optimized is relatively difficult to define, according to the model of losses used to describe the machine. In fact, we can include the iron losses and the mechanical losses, but here we are going to restrict ourselves to losses directly inferred from the model that was previously developed, i.e. copper losses p_{cu} in the different coils:

$$p_{cu} = R_s \cdot \left(i_d^2 + i_q^2 \right) + R_f \cdot i_f^2 \qquad [6.32]$$

Now criterion $J(.)$, which is the easiest to optimize within this framework, consists of the addition of p_{cu}, a constraint of the "equality on torque" type to this function by means of a Lagrange multiplier λ_1:

$$J(i_d, i_q, i_f, C_m^{ref}) = R_s \cdot \left(i_d^2 + i_q^2 \right) + R_f \cdot i_f^2 + \lambda_1 \cdot \left(c_m \left(i_d, i_q, i_f \right) - C_m^{ref} \right) \qquad [6.33]$$

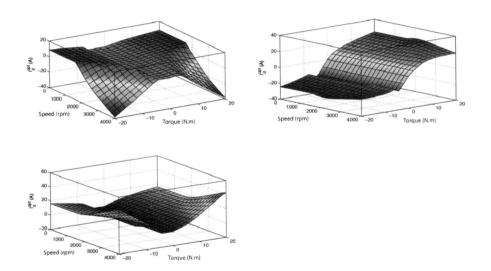

Figure 6.28. *Optimal currents to be injected in the machine as a function of the input torque (between -20 and +20 N.m) and the current rotational speed of the machine (between 0 and 4,000 rpm)*

We can easily integrate an additional constraint on vectorial the emf (e_d, e_q) of the machine so that it remains smaller or equal to a certain value, E_{smax}, set by the uninterruptible power supply (for instance 90% of the maximum value of voltage vector $(v_d, v_q)^t$ that it can generate for DC bus voltage U_0). In this boundary case, the optimization criterion is modified by the addition of a new Lagrange multiplier, λ_2:

– equality between estimated torque $c_m(i_d, i_q, i_f)$ and reference C_m^{ref} (associated with λ_1);

– equality between the back-emf of the armature and the maximum value allowed by the inverter, E_{smax} (associated with λ_2).

We will notice that this control is not optimal in the classical sense (i.e. of automatics) but rather a suboptimal control, because it is only truly optimal in steady state (constant torque/speed). In practice, it has a high performance as long as the dynamics of the input torque remain slow with respect to the response times of the current regulation loops. This is most often verified in the targeted applications (e.g. electrical or hybrid vehicles).

6.4. Overspeed and flux weakening of synchronous machines

6.4.1. *Generalities*

The torque control presented in the previous section minimized losses in the machine for a given torque input (the first constraint to fulfill) while limiting the back-emf of the machine. This last constraint (inequality) is that consisting of decreasing the global flux undergone by the armature (if necessary). We will notice that the armature is the site of a back-emf proportional to the flux and to the rotational speed of the machine. This is similar to the scalar case of the DC machine (DCM) for which back-emf, E, in the armature is ruled by the relationship:

$$E = k.\varphi.\Omega \qquad [6.33]$$

As a consequence, for a given supply (with a chopper, the supplied voltage lies between 0 and U_{dc}), the speed can not exceed a value Ω_{max} so that:

$$\Omega = \frac{U_{dc}}{k.\varphi} \qquad [6.34]$$

by assuming that the voltage drops in the armature (only in resistance in a DCM) are negligible.

In the case of a machine with adjustable excitation, we can control the value of φ as a function of an excitation current, i_f. In the case of a machine with linear functioning, we have a proportionality coefficient between these two variables, $\varphi = M_{af}.i_f$. We notice that when we want to increase the speed of the machine, we can decrease i_f because we then decrease $k.\varphi$. Even in the case of a saturated machine, we have a monotonously increasing relationship between flux φ and current i_f (which, at small i_f values, is almost linear), which leads to the same result.

6.4.2. *Flux weakening of synchronous machines with classical magnets*

This flux weakening function can be illustrated by a Behn-Eschenburg diagram for synchronous machines with smooth poles. In fact, for a given point of operation of the machine (in the torque-speed plane), we can define the single-phase vectorial diagram linking simple voltage and phase current by knowing the synchronous reactance X_s and by considering the negligible resistance R_s of the coil (see Figure 6.29). We will notice the voltage limit imposed by the inverter (amounting to 0.637 U_{dc} in the case of a space-vector modulation in linear regime – i.e. without overmodulation) in the two diagrams proposed. In Figure 6.29a we notice that the inverter can supply the required voltage, V, while in Figure 6.29b it is impossible. The solution is given in Figure 6.29c, where we see that the addition of a current component $I_d < 0$ on axis d decreases the amplitude of voltage vector V, which must be supplied by the inverter, and that the inverter enters the accessible zone. Obviously, the addition of component I_d decreases the value of I_q available because we can also define a limit common to the inverter and the machine, which is admissible current by phase I_{max} and leads us to write the following inequality:

$$\sqrt{I_d^2 + I_q^2} \leq I_{max} \qquad [6.35]$$

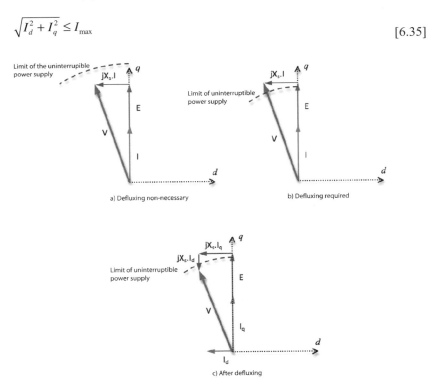

Figure 6.29. *Behn-Eshenburg diagrams*

6.4.3. *The unified approach to flux weakening using "optimal inputs"*

In the case of HESMs, we have seen that a current i_f was dedicated to the generation of the armature flux similarly to the magnets. It is also obvious in the DDCs established in Figure 6.21 that current i_f is not the only one to generate the armature flux. In fact, it is coupled with the armature current of axis *d*, written i_d. Now, current i_d is the only degree of freedom available for a possible flux weakening in synchronous machines with classical magnets. We can therefore distinguish two cases:

– The case of synchronous machines with smooth poles for which current i_d is only useful for the flux weakening insofar as it only intervenes in the expression of the motor torque.

– The case of machines with salient poles for which current i_d has an impact on the torque and the flux weakening of the machine.

The control technique previously proposed, however, tends to unify the optimal controls of the machine in the operating zones with constant available torque and power (i.e. flux weakening zone). It does this by the "tabulation" of input currents as a function of the operating point in the "torque/speed" plane.

REMARK 6.3.– This control principle is applicable not only to piloting of the machine by an inverter connected to a constant voltage source, but also in the case where the machine running as a generator, we wish to pilot voltage U_{dc}. In this case, the three-phase transistor bridge behaves as a PWM rectifier [PAT 06a].

6.4. Conclusion

To conclude, we suggest a basic comparison between the structures of HESM presented. This comparison is summarized in Figure 6.30, which shows the evolution of the image of the excitation flux as a function of the excitation current. The results come from finite element calculations carried out from Figures 6.3, 6.9 and 6.14. The CRSM curve represents the results obtained from Figure 6.9, where the magnets have been replaced with air. It allows us to measure the contribution of permanent magnets (HESM in parallel with short circuit). The structures in series are those that feature the best configuration for permanent magnets (HESM S). In fact, they thus feature the highest excitation flux when the excitation current is 0. On the other hand, because of the reluctance shown by the magnet, the impact of excitation coils is strongly reduced. The efficiency[2] of the hybrid excitation is therefore less important than for structures in parallel. The HESMs in parallel with

---

[2] One calls efficiency of the double excitation the derivative of the excitation flux with respect to the excitation current.

short circuit feature 0 excitation flux for a flux of 0. For the HESM in parallel without short circuit, the excitation flux can be cancelled out at the expense of Joule excitation losses. This type of machine requires a bidirectional supply circuit in current, similar to machines in series. The modeling of the machines in this chapter aims to translate the behavior of the machine in the form of a valid electrical diagram, both in transient regime and in sinusoidal steady-state regime. It allows us, moreover, to emphasize the "good state variables" of the system to then lead to their piloting by inversion of the model. Finally, because of the redundancy of degrees of freedom in controlling the machine, we can determine optimal inputs on the basis of the model developed. These inputs can then be generated on the basis of a measure of rotational speed by a "supervisor" delivering the torque input required by the application. This approach is all the more interesting from the point of view of its implantation. It allows us to generate unified control of the machine on the torque/speed plane (within the physical limits allowed). In other words, thanks to an adapted optimization criterion it allows us to ensure the machine runs in the operating zone with constant available torque and in the zone limited by constant power (flux-weakening zone).

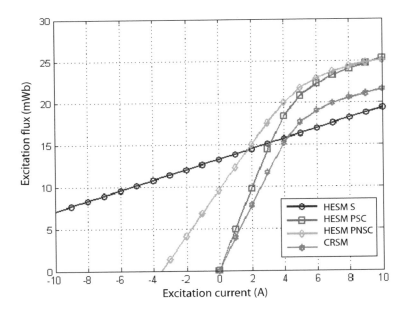

Figure 6.30. *Basic comparison of the different classes of HESM (Key: CRSM = coiled rotor synchronous machines; PNSC = in parallel without short circuit; PSC = in parallel with short circuit; S = ?*

6.5. Bibliography

[AMA 01] AMARA Y., Contribution à la commande et à la conception de machines synchrones à double excitation, PhD thesis, University of Paris XI, December 2001.

[AMA 09] AMARA Y., VIDO L., GABSI M., HOANG E., BEN AHMED A. H., LÉCRIVAIN M., "Hybrid excitation synchronous machines: Energy efficient solution for vehicle propulsion", *IEEE Transactions on Vehicular Technology*, vol. 58, no. 5, pp. 2137-2149, 2009.

[HLI 08] HLIOUI S., Étude d'une machine synchrone à double excitation. Contribution à la mise en place d'une plate-forme de logiciels en vue d'un dimensionnement optimal, PhD thesis, UTBM, December 2008.

[HOA 07] HOANG E., LECRIVAIN M., GABSI M., "A new structure of a switching flux synchronous polyphased machine with hybrid excitation", *European Conference on Power Electronics and Applications*, pp. 1-8, September 2007.

[PAT 06a] PATIN N., VIDO L., MONMASSON E., LOUIS J.-P., "Control of a DC generator based on a Hybrid Excitation Synchronous Machine connected to a PWM rectifier", *Proc. IEEE ISIE'06*, CD ROM, Montreal, Canada, July 2006.

[PAT 06b] PATIN N., Analyse d'architectures, modélisation et commande de réseaux autonomes, PhD thesis, ENS Cachan, December 2006.

[TAK 07] TAKORABET A., Dimensionnement d'une machine à double excitation de structure innovante pour une application alternateur automobile. Comparaison à des structures classiques, PhD thesis, ENS Cachan, January 2007.

Chapter 7

Advanced Control of the Linear Synchronous Motor

7.1. Introduction

This chapter presents some advanced controls of the permanent magnet linear synchronous motor (PMLSM). In section 7.1.1 we present a short historical and general introduction of this type of motor. We take a short look at state-of-the-art PMLSM technology in industry at present.

In section 7.2.1, we propose a brief overview of state-of-the-art of industrial controls available for linear synchronous motors. Then, advanced controls specific for linear motors are presented.

In particular in this chapter, control design is elaborated using the causal ordering graph (COG) principle that is applied to advanced analytical models of PMLSM, including cogging forces and end-effect forces. The COG principle facilitates the analysis of the physical model for controller tuning. Some advanced controls are presented, including estimator and feed-forward control, resonant controller and the nth derivative of the control variables in open-loop control.

7.1.1. *Historical review and applications in the field of linear motors*

One of the first transportation designs using linear motors was patented more than a century ago. It was designed by Alfred Zehden and patented in Switzerland in

---

Chapter written by Ghislain REMY and Pierre-Jean BARRE.

1902 [ZEH 02] and in the US in 1905. The patents focus on the use of linear induction motors (LIMs) to move passenger trains. In such an application, the primary was fixed by sections on top of a rail, along the railway. Each section successively received the power supply, allowing the train to move forward. The high fabrication cost of the rail, however, put an end to the development of this electric propulsion device.

The market in industrial applications using PMLSMs has only taken off since the end of the 1980s [MCL 88] with constant growth to date [EAS 02]. To understand such success, it is important to remember that industry has always tried to find solutions to reduce manufacturing time and to be very flexible. To achieve this, engineers have used agile machines, focusing on cheap and very simple concepts, recognized as the "lean" production practice that leads to lightening structures to adapt to the problem of size alone. In such a concept, LIMs have naturally become a vital component.

In 2004, the European and American markets for LIMs were worth €113 million ($153,000,000) and €95 million ($128,000,000) respectively [GIE 03]. It is a sector in full expansion, which is venturing into new fields of applications. Nevertheless, there is currently no mass market for linear motors. The business only survives due to innovations and the inventiveness of the applications. [CAS 03] presents a synthesis of the main applications of linear synchronous motors. Among them, we can find: the electromagnetic cannon, the magnetic levitation train (see Figure 7.1a), roller-coasters, ropeless elevators, conveyor belts, robotics, machine-tools, world-class high-accuracy positioning systems for the electronic and semiconductor industries (see Figure 7.1b).

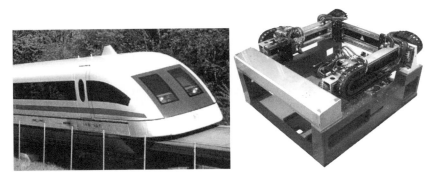

Figure 7.1. *Examples of linear motor applications:*
left) magnetic levitation train: Transrapid Maglev [THY 08]; and
right) Pick and place machine for electronic process application ETEL [COR 07]

7.1.2. Presentation of linear synchronous motors

A linear motor is used to move a system without any additional kinematic components in a linear translation[1]. Figure 7.2 presents a PMLSM. It can be classified as a flat, single-sided, slotted, short primary and iron-core LSM:

– a primary, lightweight moving part composed of a three-phase coil and a laminated stack used as a ferromagnetic circuit;

– a secondary, fixed part composed of a permanent magnet lying on a ferromagnetic yoke.

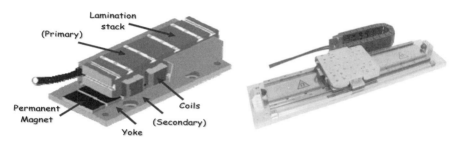

Figure 7.2. *Simplified scheme and photo of an iron core linear motor made by ETEL*

In a classical transmission device, the high number of mechanical linkages introduces flexibility, torsion and flexion of the screw, backlash, vibration, etc. For example, accuracy and repeatability suffer from the inherent limitations of the moving belt system. Therefore the limit on the positioning accuracy and high dynamic, especially for long-stroke application. One way to rigidify the system is to enlarge the diameter of the screw, leading to an increase in inertia, and to a decrease in the dynamics of the whole system.

Linear motors, however, provide smooth, high reliability, non-contact operation without backlash, longer lengths with no performance degradation, low maintenance and long life (with high mean time between failures) and lower costs of ownership. The mounting is composed of two integral linear bearings and an encoder. The drive train is simpler to install, as the linear motor replaces the ball screw, nut, end bearings, motor mount, couplings and rotating motor. Alignment with a linear motor is not critical (even for high-performance packages) and consists of mainly ensuring

---

1 Linear motors are mainly known as direct drive, as opposed to rotary motors which are coupled with a mechanical conversion device, such as a belt-pulley, conveyor/timing belt, rack and pinion, lead screw, etc., which are termed linear axis.

clearance is maintained for the moving coil during travel. More than one coil assembly can be used in conjunction with a single magnet assembly, as long as the coil assemblies do not physically interfere with each other. The stage features lightweight moving parts for higher acceleration of light loads. Resolution is of about 1 µm, and high accuracy and repeatability provide better quality control. In most applications, performance improvements can be expected with linear motors: repeatability and accuracy will be increased; and move times and settling times will be decreased.

Linear motors can achieve very high velocities (up to 10 m/s). However, there are several factors that limit the speed of the linear motor:

– Control must provide sufficient bus voltage to support the speed requirements.

– The encoder itself must be able to respond to that speed and its output frequency must be within the controller's capability. For example, with a 0.5 micron encoder and a speed of 5 m/s, the controller must handle 10 MHz.

– Finally the speed rating of the stage's bearing system must not be exceeded. For example, in a recirculating ball bearing, the balls start to skid (rather than roll) at about 5 m/s.

Furthermore, there is a trade-off in force for iron-core motors, as technology becomes limited by eddy current losses in the magnetic field. This is because linear motors use rare earth magnets that maintain their strength over time. When operating at high temperatures (>150°C), however, rare earth magnets can lose strength. Thus, consideration should also be given to motor cooling, which may be required to increase performance or improve thermal stability for the application. Both air and water-jacket cooling systems can be supplied. Furthermore, as linear motors are friction-free and the linear bearing system is normally a low-friction device, braking may be required for conditions of power-loss or power-off. For those vertical positioning applications, a linear motor will almost certainly need to consider a counterbalance and a braking system to avoid any back-driving.

Linear motors offer a very compact design, allowing motion systems manufacturers to revise the mechanical structure of the machines due to the exceptional possibilities of direct drive.

The recent success of linear motors, which were designed more than a century ago, is due to the progress of power electronics and its associated motion control components. In the near future, most linear translation will be designed with such linear actuators. However, one important constraint is that the permanent magnets of the secondary are unsheltered. This means that a protected environment is required to avoid destruction of the magnetic force or the influence of the magnetostatic field on the applications.

A number of books [BOL 85, BOL 97, BOL 01, NAS 76] present applications and structures of linear motors. [CAS 03] presents a synthesis of the different technologies of linear actuators. In this chapter, we have chosen to use only one type of linear motor, as shown in Figure 7.2. Many variants exist, however, even in the linear synchronous motor family.

7.1.3. *Technology of linear synchronous motors*

Currently, the linear motor market clearly focuses on two kinds of applications:

− Very high acceleration (>10g). The primary is lightweight, mainly with a short primary without ferromagnetic material, is classified as ironless. Indeed, the coils of the primary are assembled in a non-ferromagnetic coating, Bakelite for example, and surrounded on both sides by a magnet "sandwich", creating a U-shaped secondary. The ironless linear motor is used in pick and place machines for semiconductors and electronics manufacturing.

− Very high speeds with heavy loads to be moved, and thus significant high continuous and peak forces. In the primary, ferromagnetic circuits are added to the laminated stack, which is classified as an iron core. Such linear motors are used in applications where high force-to-volume ratios are desired.

The performance improvements of the linear motor in the past 10 years are strongly related to enhancements in the components used:

− Iron-cobalt sheets are now used and provide equivalent performances to iron-silicon ones, but with a lower mass. This allows the mover to achieve higher acceleration. Besides this, some materials are now able to reach an induction of 2.0 Tesla without saturation. Therefore, the peak forces have recently been increased by about 15%.

− Permanent magnets with a residual induction level superior to 1.2 Tesla are now available. The use of permanent magnets, such as the neodymium magnets (also known as NdFeB), however, is almost obligatory. Indeed, the permanent magnets of the secondary are surface-mounted, and thus the magnetic circuit is strongly modified on the primary's passing. The demagnetization curve is therefore important and eliminates the choice of the AlNiCo magnet family, which is very sensitive to demagnetization. One of the remaining limits is the effect of temperature on the magnets. The permanent magnets are demagnetized at high temperature so this limits the warming of the linear motors to around 150-200°C for neodymium magnets.

− Windings with a filling factor have improved with the use of press-mounted conductors weighing several tons. This requires the use of a constant cross-section (also called a straight) and open slots, but increases the winding coefficient to around 0.8, compared to the more common value of 0.6. Naturally, the level of peak force of the actuator is also increased.

– Position sensors, and more particularly the incremental linear encoders, are thermally compensated for to produce accurate measurement. Nowadays, accuracy is around several microns for the resolution of incremental encoders. Absolute linear encoders can be used for distances of < 1 m; incremental linear encoders > 1 m are more often used as they are cheaper. The measurement principle is established on the interferometer principle, combined with a digital resolution of 4,096 values for a grating period of the linear encoder. Thus, it is possible to achieve measurement accuracy at around the nanometer range. This measurement accuracy has to be separated from the accuracy of positioning (about 10 μm) or repeatability (about 1 μm). Indeed, the actuator (as depicted in Figure 7.2) is subject to mechanical and thermal stresses.

7.1.4. *Linear motor models using sinusoidal magneto-motive force assumption*

To define the model of a linear motor, the simplest assumption is to consider a linear motor such as an unrolled rotating motor of infinite length, as depicted in Figure 7.3, and to take a sinusoidal magneto-motive force assumption. The simplified model, also called the first harmonic model, is simply an adaptation of the classical model of a rotating synchronous motor [BOL 01, GIE 99, LOU 04].

Figure 7.3. *Rotating and linear motor analogy*

The linear motor model uses the following assumptions:

– the resistances and inductances of the three phases are identical;

– the ferromagnetic materials are ideal ($\mu r = \infty$);

– the slot effects are neglected;

– the primary has an infinite length;

– the permanent magnets are mounted on top of the yoke, so self-inductances then become constants equal to L;

– the eddy currents are neglected;

– the primary is considered to be a rigid mass;

– the bearings are considered as ideal (of infinite stiffness), so the attractive force of the magnets on the y-axis is then neglected.

Figure 7.4 presents the geometrical structure of a linear motor produced by ETEL.

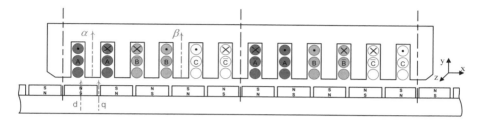

Figure 7.4. *Simplified representation of a linear synchronous motor (LMD10-050 produced by ETEL)*

The synchronous feature of a PMLSM[2] is directly visible in the equation of actuator speed. Indeed, the speed is dependent on the current frequency of the primary $i_{a,b,c}$, only:

$$f_{(Hz)} = \frac{1}{2.\tau_{p(m)}} \times v_{(m.s^{-1})} \qquad [7.1]$$

This equation is similar to the relation of the synchronous rotating motor:

$$\omega_{(rad.s^{-1})} = p_1 \times \Omega_{(rad.s^{-1})} \qquad [7.2]$$

If all the electrical state variables of the primary are constantly referenced in relation to the position of the secondary (self-control strategy), then it becomes interesting to control the actuator in the Park reference frame [LOU 04].

---

2 To make reading easier, the anagram PMLSM will replace the following expression: "Permanent magnet linear synchronous motor with short primary, iron core and that is single sided", as shown in Figure 7.4.

7.1.5. *Causal ordering graph representation*

Many papers and books (most in French) since the beginning of the 1990s have developed the causal ordering graph (COG) formalism [BAR 04, HAU 99a]. It has been a very efficient graphical formalism for establishing physical models by representing causal relations and for designing control structures by using a principle of systematic inversion. Figure 7.5 presents the two main processors of the COG formalism: the accumulative and the dissipative processors, respectively.

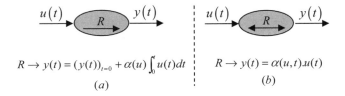

$$R \rightarrow y(t) = (y(t))_{t=0} + \alpha(u)\int_0^t u(t)dt$$
(a)

$$R \rightarrow y(t) = \alpha(u,t).u(t)$$
(b)

Figure 7.5. *Graphical COG symbols: (a) causal relation; and (b) rigid or non-causal relation*

Simple observation of the linear motor helps us to understand the causality of the energy transformation. Indeed, it appears that supplying the power cable with an adapted voltage will make the linear motor move to the desired position. This short analysis can be performed by applying the same principle using a graphical representation, here the COG, as depicted in Figure 7.6. We can see that the system inputs are the voltages of the primary (V_d, V_q) and the output is position x of the mover.

The PMLSM model in the first harmonic model can be represented in the Park reference frame in order to design vector control (including a self-control strategy).

This graphical representation offers the advantage of visually outlining the energy path, also called the "causal path". For example, the voltages induce the currents, which generate the electromagnetic force (emf), which moves the motor at a certain speed. This representation helps us to understand the couplings between the flux and thrust axes of the model in the Park reference frame. Indeed, the thrust generated by the q-axis T_q can be modulated by the d-axis current i_d, allowing us to control the field weakening of the motor.

Finally, by using a visual approach, COG formalism facilitates the understanding and definition of physical models. We will see in the next section that the main advantage of this formalism is to offer a systematical method for the control of structure design, using the COG inversion principle.

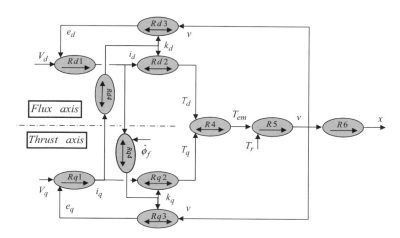

Flux axis (*d-axis*):	Thrust axis (*q-axis*):
$R_{d1} \rightarrow \left(R + L_d \cdot \dfrac{d}{dt}\right) \cdot i_d = V_d + e_d$	$R_{q1} \rightarrow \left(R + L_q \cdot \dfrac{d}{dt}\right) \cdot i_q = V_q - e_q$
$R_{d2} \rightarrow T_d = k_d \cdot i_d$	$R_{q2} \rightarrow T_q = k_q \cdot i_q$
$R_{d3} \rightarrow e_d = k_d \cdot v$	$R_{q3} \rightarrow e_q = k_q \cdot v$
$R_{d4} \rightarrow k_d = N_p L_q \cdot i_q$	$R_{q4} \rightarrow k_q = N_p (L_d \cdot i_d + \sqrt{3/2} \cdot \hat{\phi}_f)$
$R_4 \rightarrow T_{em} = T_q - T_d = k_q \cdot i_q - k_d \cdot i_d$,	$R_5 \rightarrow M \cdot \dfrac{dv}{dt} = T - T_r$ and $R_6 \rightarrow \dfrac{dx}{dt} = v$

Figure 7.6. *COG of the PMLSM*

7.1.6. *Advanced modeling of linear synchronous motors*

As with rotating motors, linear synchronous motors have some geometrical specificities that induce nonlinear thrust generation:

– emfs mainly have wave harmonics of ranks 5 and 7 that are related to concentrated or distributed windings. The emfs generate ripple forces on the thrust throughout all operating ranges. They are more prominent, however, in high-speed modes;

– fast transient currents enable high acceleration of the linear motor but induce local saturation of the magnetic circuit and cause inductances to behave in a nonlinear manner;

– the cogging and the end of primary effects, related to the geometric ends of the iron sheets (effects that only occur in the case of linear motors), generate detent forces that are particularly perceptible at low speeds. The detent effects are particularly disturbing for positioning devices because they occur even when there is no current supply.

In fine, these three phenomena result in the presence of ripple forces on the thrust generated by the PMLSM, with different areas of dominance.

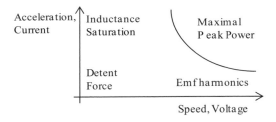

Figure 7.7. *Dominance zones of the ripple force amplitude*

With the exception of end effects, all these phenomena exist in synchronous rotating machines and have been extensively modeled in the literature [GIE 03, REM 07]. Figures 7.8 and 7.9 give some examples of linear actuator geometry with end-effect compensation specific to linear motors. They show that the secondary ends may be optimized. Indeed, the main idea is to reduce the magnitude of the detent forces due to interaction between the magnets and the ferromagnetic teeth of the primary.

Figure 7.8 is from a patent [STO 98] that aims to cancel the fundamental of the cogging force, knowing that the periodicity of the phenomenon can be directly defined by the least common multiple between the period of magnet disposal and the period of tooth disposal.

Advanced Control of the Linear Synchronous Motor   251

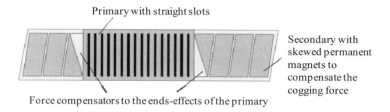

Figure 7.8. *Modification of the extremities of Siemens' LIMES400/120*

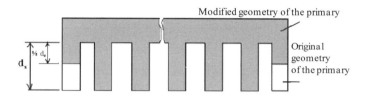

Figure 7.9. *Example of the primary geometry for limiting end-effects*

Figure 7.10 shows the primary ends of the LMD10-050 designed by ETEL [WAV 97]. Here, the principle is that the cogging phenomenon can be divided into two parts: the extremities and the central part, as shown in Figure 7.11. Here, the central part is defined by the extremities as having periodic disposal of the ferromagnetic teeth. The compensation for the detent forces generated in the central part of a PMLSM is therefore equivalent to the compensation of a rotating motor's detent forces, because it is a periodic area. The choice of the ratio between the tooth and the magnet width is then the way to reduce such cogging forces.

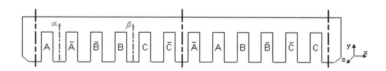

Figure 7.10. *Beveled ends of the primary, LMD10-050
produced by ETEL [WAV 97]*

The idea is to cancel the ripple forces generated by the end-effects, and therefore to cancel several harmonics. The result obtained is shown in Figure 7.11. The waveform of the detent force is thus greatly reduced in amplitude (less than 4 N for the peak value) for an engine with a peak value of about 580 N. If we compare this

252   Control of Non-conventional Synchronous Motors

with a more conventional geometry, this would have represented approximately 20 N in peak value.

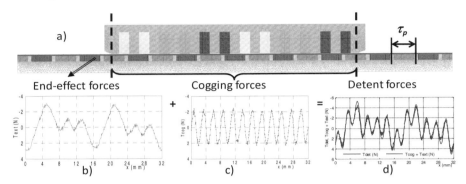

Figure 7.11. *Detent forces of the LMD10-050 by ETEL [GOM 08] using finite-element analyses in the linear case: (a) location of the phenomena; (b)forces of the end-effects; (c) cogging forces; and (d) detent forces according to position*

It is important to note that a rigorous scientific decomposition of the phenomena resulting from the central part and the end-effects is only valid in a linear regime, and not in a saturated case. In the saturated case, when operating at the maximum current of the actuator, we have found a relatively small difference to the linear case results of around a few percentage points.

These phenomena are dependent on the level of actuator saturation and on magnet position, which impact locally on the level of saturation. Some results can be obtained using only a finite element analysis [REM 07] or reluctance network modeling [GOM 08].

These phenomena are only noticeable at low speeds, however, because the moving mass of the primary acts as a first-order filter between the ripple force and the motor speed. The objective of this chapter is not to focus on the advanced modeling of linear actuators alone. Interested readers can find therefore, more details about PMLSM modeling in [GIE 99] and [BOL 01].

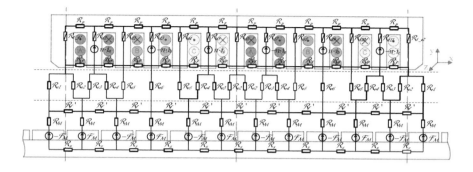

Figure 7.12. *Examples of equivalent circuits: network reluctance of a linear permanent magnet [GOM 08]*

7.2. Classical control of linear motors

7.2.1. *State-of-the-art in linear motor controls*

Nowadays, there are many strategies that can be used to control the thrust generation of a PMLSM [GIE 99]:

– Scalar controls regulating the amplitude of the signal to be controlled. This technique is mainly used in open-loop, as well as closed-loop, speed control for keeping the U/f ratio constant. This technique does not give satisfactory results for rotating motors in dynamic motion control. It is mainly used in applications such as roller coasters, where a very high positioning accuracy is not essential.

– Vector controls regulating the amplitude and phase vectors of the signal to be controlled. The control structure is usually placed in a reference frame linked to the secondary to perform decoupling of the flux and thrust of the motor. This technique is most widely used for controlling linear motors. To reduce the effects specific to linear motors (end effects, etc.), digital filters are inserted into the control structure in reality.

Both of these techniques currently use control structures with fixed and constant parameters. When the performances are no longer sufficient to control the system, however, the control structure becomes complicated:

– Auto-adaptive control or iterative learning control (ILC), which adjusts the variable parameters due to changes in the motor's parameters. In applications such as "pick and place", it is systematically used to increase the accuracy of repeatability.

– Sliding mode control, which switches between different control structures. The system is guided by a reference trajectory of successive commutations. The control is then insensitive to parameter and load variations. It is mainly used in applications such as magnetic levitation trains to differentiate between the control of starting and cruising phases.

– Control by neural network or fuzzy logic, where the mathematical model is not or is barely affordable by conventional identification. This has so far mainly been used in academic research.

The main difference between linear motor and rotating motor models is related to the fact that there are more harmonics with cogging forces (associated with the secondary ends) in a linear than a rotating motor. The conventional controls for rotating motors therefore cannot achieve maximum actuator performance. Indeed, the ripples of cogging force induce an imprecise thrust force control, which is insufficient in some cases (e.g. force feedback in the stick of a flight simulator) [REM 07].

The position sensor, however, is a much more sensitive component in linear actuators: it consists of a reading head on the primary and a graduated ruler that is deposited along the secondary (sometimes over several meters) and is therefore highly sensitive to its external environment. It is therefore able to detect mechanical deformations in torsion and/or bending, resulting from mechanical stress (high acceleration mode resonance, etc.) and thermal stresses. For applications calling for positioning accuracy of around the micrometer level, these distortions are particularly detrimental and lead to measurement errors, even measurement malfunction when the reading head moves away from the graduated ruler for more than 0.1 mm.

By expanding our research to industrial the control structures of synchronous machines, it appears that "nested loop" controls are the most widely used [GRO 01]. They mainly include controllers with proportional and integral (PI) or proportional, integral and derivative (PID) actions. They are easy to implement and tuning principles are now widely known in the industry. To satisfy the latest specifications that have become increasingly constraining, many applications now require more sophisticated control structures. So, two structures are currently competing:

– Control is obtained from a model of behavior (also called the black-box model), which is often characterized by controllers of a very high order (e.g. H∞ robust control). The difficulty with this kind of control is the order of model reduction. For the control structure to be able to operate in real time, the system transfer function has to be reduced from a high to a reasonable order. A high level of expertise is needed to accomplish these adjustments.

– Control is obtained from a model of knowledge that is deduced from the inversion of a model defined by the physical properties of the system's components. To be highly effective, this kind of control structure requires a physical model that is as close as possible to reality. The synthesis of the control structure is then facilitated by the use of graphical formalism, such as the Bond-Graph [DAU 99], the COG [HAU 98] or the energetic macroscopic representation [BOU 00].

In the next section, to optimize the thrust generation of a linear motor, different control architectures are studied using the inversion principles of the COG. First, from the first harmonic model of a linear motor, we compare different types of closed-loop controls. Then, using the models presented in the preceding sections, we establish some control structures using different control techniques, especially on the electrical and magnetic phenomena that generate disturbing ripples on the thrust. In particular, we specify the tuning techniques for each control strategy, and then present the advantages and limitations of these different commands.

Finally, we present several open-loop control strategies and focus on the problems associated with generation of the reference variable in the case of a command using the nth derivative of the reference variable.

7.2.2. Control structure design using the COG inversion principles

Since the 1990s, COG formalism has been developed to facilitate the modeling and control design of systems based on understanding energy exchanges, and particularly the principle of causality. This vision arises naturally in control design, as "since we know the cause, it is sufficient to apply the right cause to obtain the right effect" [HAU 99a]. Thus, regardless of the control structure type (continuous, discrete, sampled, etc.), we seek an implicit inverse model of the process that is to be driven.

The control structure design using the formalism of the COG is based on the following concepts:

The organization of the control structure is deduced systematically from an inversion principle of the graphical representation of the system model. We can usually find direct processors that represent dissipative phenomena and causal processors, which are time-dependent and represent the accumulation phenomena in the process. It is therefore natural for the inversions of both processor types to obey different laws, as shown in Figure 7.13a.

256   Control of Non-conventional Synchronous Motors

On one hand, it is possible to reverse a direct processor as:

$$R \rightarrow y = R(u), \quad R_c \rightarrow u_{reg} = C(y_{ref}),$$
$$\text{Si } u = u_{reg} \text{ and } C = R^{-1}, \text{ so } y = y_{ref}$$
[7.3]

where C is a time-independent function that represents the inverse of the surjective or bijective time-independent function R.

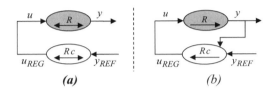

Figure 7.13. *Principles of indirect inversion of a CCG:*
(a) rigid relationship; and (b) causal relationship

On the other hand, the inversion of a causal processor requires an indirect inversion (feedback control), as shown in Figure 7.13b.

$$R \rightarrow y = R(u), \quad R_c \rightarrow u_{reg} = C(y_{ref} - \hat{y}),$$
$$\text{Si } u = u_{reg} \text{ and } C \rightarrow \infty, \text{ so } y \rightarrow y_{ref}$$
[7.4]

where C is the corrective function chosen according to the specifications of the desired system performance in closed loop. It can be demonstrated that for an indirect inversion, a proportional correction with high gain is sufficient to ensure convergence between the reference and the measurement of a linear time invariant system with non-minimum phase. Nevertheless, the controller choice is not defined by the COG inversion principle. Only control engineers can decide whether it is a proportional controller (P), an integral-proportional controller (IP) or a resonant controller, etc.

7.2.3. *Closed-loop control*

The indirect inversion principle of the COG is thus applied to the PMLSM model presented in Figure 7.6. Figure 7.14 shows a scheme of the control structure representation, also called maximum structure of control[3].

---

3 Note that it is only a maximum control structure for the model to be controlled. A more detailed model of the system would require us to redefine a new maximum control structure.

Advanced Control of the Linear Synchronous Motor

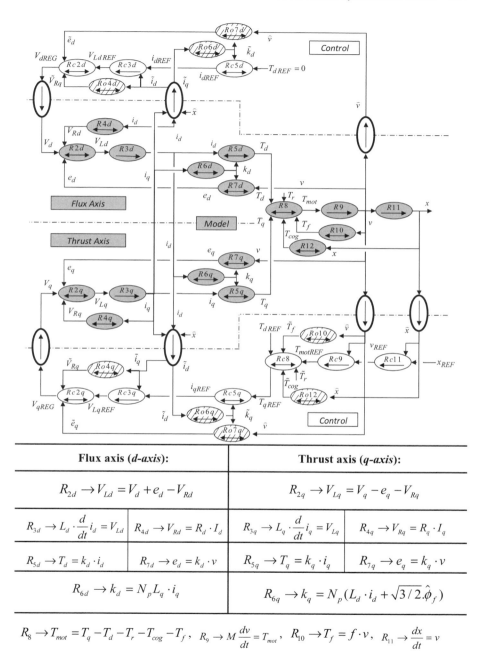

Flux axis (*d-axis*):		Thrust axis (*q-axis*):	
$R_{2d} \to V_{Ld} = V_d + e_d - V_{Rd}$		$R_{2q} \to V_{Lq} = V_q - e_q - V_{Rq}$	
$R_{3d} \to L_d \cdot \dfrac{d}{dt} i_d = V_{Ld}$	$R_{4d} \to V_{Rd} = R_d \cdot I_d$	$R_{3q} \to L_q \cdot \dfrac{d}{dt} i_q = V_{Lq}$	$R_{4q} \to V_{Rq} = R_q \cdot I_q$
$R_{5d} \to T_d = k_d \cdot i_d$	$R_{7d} \to e_d = k_d \cdot v$	$R_{5q} \to T_q = k_q \cdot i_q$	$R_{7q} \to e_q = k_q \cdot v$
$R_{6d} \to k_d = N_p L_q \cdot i_q$		$R_{6q} \to k_q = N_p (L_d \cdot i_d + \sqrt{3/2}.\hat{\phi}_f)$	

$R_8 \to T_{mot} = T_q - T_d - T_r - T_{cog} - T_f$, $\quad R_9 \to M \dfrac{dv}{dt} = T_{mot}$, $\quad R_{10} \to T_f = f \cdot v$, $\quad R_{11} \to \dfrac{dx}{dt} = v$

Figure 7.14. *Maximum control architecture using COG representation*

258   Control of Non-conventional Synchronous Motors

This type of control architecture is commonly called "cascaded closed-loop control". Indeed, Figure 7.14 shows the indirect inversion principle of the COG applied to causal processors of the (*R3d*, *R3q*, *R9*, *R11*) model. This structure requires currents, speed and position sensors to be available. Three concentric loops are thus formed. A maximum control structure is therefore established very quickly, "processor after processor".

Then, the *Rc3d* and *Rc3q* processors are inside the inner current loop, and both contain a PI controller. The *Rc9* processor is inside the speed loop and contains a P controller, and the *Rc11* processor is inside the outer position loop and contains a P controller. Figure 7.14 outlines the graphical simplicity of the COG representation that helps us to identify the energy links, successive energetic conversions and processors inside the cascaded closed-loop control.

Figure 7.15 shows us that the processors constituting the direct energy chain (*Rc2d*, *Rc2q*, *Rc3d*, *Rc3q*, *Rc5d*, *Rc5q*, *Rc8*, *Rc9*, *Rc11*) are involved in energy conversion. The presence of these processors is required to generate the manipulated variable from the reference variable. A perfect system in terms of energy means there is no dissipative loss; it could be controlled by a control structure without an estimation processor or the observers normally used to compensate for energy losses due to the effects of resistance or friction.

All hatched processors in the control structure contribute to the compensation of the imperfections of the process (for example, the resistive voltage drop of the resistance or friction force in the system). Indeed, *Ro4d* and *Ro4q* processors estimate the voltage drop of the resistance, such as:

$$Ro4q \rightarrow \tilde{V}_{Rq} = R_{est} \cdot \hat{i}_q \qquad [7.5]$$

When the processor input comes from the measurement of a variable, as for the estimation or observation processors, they are represented by a hatched oval in Figure 7.14. The other control processors are grey on a white background. Thus, at a glance, the COG representation directly shows the role and impact of different variables on the control strategy.

7.3.3.1. *Analysis of the current in a closed loop*

Using the COG's inversion principles, it is possible to define three types of current control architecture, as depicted in Figure 7.15. For ease of reading, we have chosen to show only the COG representation of the *q*-axis (also called the thrust axis).

Advanced Control of the Linear Synchronous Motor   259

The following assumptions are used:

– $V_q = V_{qreg}$, where the inverter is assumed to be ideal without delay;

– $e_q = e_{qest}$, and more broadly $V_{qreg} - e_{qest} = V_q - e_q$, the compensation for the emfs is assumed to be perfect in the speed loop;

– $i_{qmes} = i_q$, the initial angular value required to correctly achieve the Park transformation is assumed to be perfectly known, otherwise the value of i_q is decomposed on the two axes i_{dmes} and i_{qmes}. The measures are taken without delay and no measurement noise is considered.

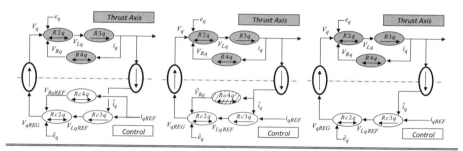

Model Relations

$R2q \rightarrow V_{Lq} = V_q - e_q - V_{Rq}$,	$R3q \rightarrow L_q \cdot \dfrac{d}{dt}(i_q) = V_{Lq}$,	$R4q \rightarrow V_{Rq} = R_q \cdot i_q$
Relations of case a	Relations of case b	Relations of case c
$Rc2q \rightarrow V_{qreg} = V_{Lqref} + \tilde{e}_q + V_{Rqref}$	$Rc2q \rightarrow V_{qreg} = V_{Lqref} + \tilde{e}_q + \tilde{V}_{Rq}$	$Rc2q \rightarrow V_{qreg} = V_{Lqref} + \tilde{e}_q$
$Rc3q \rightarrow V_{Lqref} = C_i(i_{qref} - \hat{i}_q)$	$Rc3q \rightarrow V_{Lqref} = C_i(i_{qref} - \hat{i}_q)$	$Rc3q \rightarrow V_{Lqref} = C_i'(i_{qref} - \hat{i}_q)$
$Rc4q \rightarrow V_{Rqref} = \tilde{R}_q \cdot i_{qref}$	$Ro4q \rightarrow \tilde{V}_{Rq} = \tilde{R}_q \cdot \hat{i}_q$	

Figure 7.15. *Three types of current control architecture*

The control architecture shown in Figure 7.15a comes directly from the COG inversion principle: a processor in the control structure to each processor of the model corresponds. All these processors are represented using a white background.

To compensate for the resistive voltage drop V_{rq}, the *R4q* processor is added to the control structure of Figure 7.15a and Figure 7.15b:

– in case *a*, the V_{Rq} value is deduced from the reference variable i_{qref}, hence the name is V_{Rqref}. It is equivalent to a feed-forward control. However, if there is an error between the reference and the actual currents, then this error is amplified by the resistance value;

– in case b, we can use the current measurement to estimate the ohmic voltage drop. Thus, the objective is for the model to be as close as possible to reality. It is therefore an estimation of the reference voltage, V_{Rqest}. It follows that the processor is shaded and is called $Ro4q$;

– in case c, if we focus on the case of a steady-state current that corresponds to $i_{qRef} = cst$, the ohmic voltage drop is constant and generates a constant error on current i_q. Thus, in steady-state current, the static error can be compensated for by an integral action associated within the proportional controller in the $Rc3q$ processor. In the end, if the estimation of the ohmic voltage drop is neglected ($Ro4q$ and $Ro4d$ processors) an integral controller must be added to the proportional one.

For the three cases studied, the relations ($R2q$, $R3q$, $R4q$) representing the process are equivalent:

$$i_q = (V_q - e_q) \cdot \frac{1}{R_q + L_q s} \qquad [7.6]$$

For the three structures depicted in Figure 7.14, the respective relations are written:

$$\begin{aligned}
&\left(i_{qref} - \hat{i}_q\right) C_i + \tilde{R}_q \cdot i_{qref} \\
&= \hat{i}_q \cdot (R_q + L_q s) \\
\\
&\left(i_{qref} - \hat{i}_q\right) C_i + \tilde{R}_q \cdot \hat{i}_q \\
&= \hat{i}_q \cdot (R_q + L_q s) \\
\\
&\left(i_{qref} - \hat{i}_q\right) C_i' \\
&= \hat{i}_q \cdot (R_q + L_q s)
\end{aligned} \qquad [7.7]$$

The transfer functions in the closed loop of the three current loops are then expressed as:

$$\frac{\hat{i}_q}{i_{qref}} = \frac{C_i + \tilde{R}_q}{C_i + R_q} \cdot \frac{1}{1 + \dfrac{L_q s}{C_i + R_q}}$$

Advanced Control of the Linear Synchronous Motor 261

$$\frac{\hat{i}_q}{i_{qref}} = \frac{1}{1 + \frac{L_q s}{C_i}} \qquad [7.8]$$

$$\frac{\hat{i}_q}{i_{qref}} = \frac{1}{1 + \frac{R_q + L_q s}{C'_i}}$$

The C_i and C'_i, controllers correspond to a P and a PI controller respectively:

$$\begin{cases} C_i = K_i \quad {}^{(en\ V/A)} \\ C'_i = K'_i \left(\frac{1 + \tau_i s}{\tau_i s} \right) = K'_i \left(\frac{\tilde{R}_q + \tilde{L}_q s}{\tilde{L}_q s} \right) \end{cases} \qquad [7.9]$$

The choice of the PI controller aims to compensate for the electrical time constant of the actuator:

$$\tau_i = \frac{\tilde{L}_q}{\tilde{R}_q}, \ (en\ s) \qquad [7.10]$$

ASSUMPTION 7.1.– the identification of R_q and L_q is perfect, and the expression in the case of PI controllers can be simplified:

$$\frac{\hat{i}_q}{i_{qref}} = \frac{1}{1 + \frac{L_q s}{K_i + R_q}}$$

$$\frac{\hat{i}_q}{i_{qref}} = \frac{1}{1 + \frac{L_q s}{K_i}} \qquad [7.11]$$

$$\frac{\hat{i}_q}{i_{qref}} = \frac{1}{1 + \frac{L_q s}{K'_i}}$$

Thus, with ideal identification of the resistance and inductance parameters, the three transfer functions have no static error. Nevertheless, the three cases show different time constants, in each case depending on controller coefficients. The proportional gains of cases *a* and *b* have almost the same role: in case *a*: $C_i = K_i - R_{qest}$; while in case *b*, $C_i = K_i$. To ensure stability, C_i is positive (which

262 Control of Non-conventional Synchronous Motors

means that $K_i > R_{qest}$). The main value of such a current controller is to improve system performance and especially the time response of the current loop. To reduce its time constant, it leads to $K_i = K'_i > R_q$ in equation [7.11] in cases b and c. In case a, if the ohmic voltage is known using a feed-forward control, the required gain value K_i for the controller will be lower than in other cases that are not anticipated. Here, the advantage of reducing the impact of the noise measured in the control is obvious.

In practice, determination of the controller coefficients has to be adapted to the physical limitations of the system. Here, the DC bus voltage U_s modulated through the inverter limits the voltage of the linear motor. Thus, the maximum voltage available is U_s. The dynamics and the bandwidth of the closed-loop current are thereby limited by the inverter's maximum voltage. Whatever the type of control (cases a, b or c), the controller cannot solicit more of the inverter output voltage that the U_s value. Consequently, using a proportional controller, for example, there are maximum gain values to prevent the inverter being saturated at $-U_s$ and $+U_s$ and so to keep a linear control of the linear actuator. High gain values will lead to saturation of the inverter and so to nonlinear control: a so-called "bang-bang" control strategy. Even if this control strategy gives the fastest possible acceleration of the linear motor, it generally produces a large error between the thrust generated and its reference, and produces a large range of harmonics on the thrust, which engenders unacceptable vibrations of the yoke. The determination of the gain, K_i, could be done by calculating the maximum current error and by fully scaling the inverter. Then, the maximum current error will directly correspond to the maximum response voltage on the inverter.

7.3.3.2. Application to the linear motor studied (see Appendix, section 7.8)

In case a in Figure 7.15, $K_i <= U_s / err_{max} \cdot i_q$, the maximum current error for a current step corresponds to the maximum current value. For the PMLSM studied, the DC bus voltage is limited to 300 V, so: $K_i = 300 \text{ V}/7.9 \text{ A} = 36.97 \text{ }\Omega$.

In case b, $K_i <= 36.97 + R_{qest} = 41.37 \text{ }\Omega$: the ohmic voltage drop is not anticipated.

In case c, $K_i <= 41.37 \text{ }\Omega$.

Figure 7.16 shows the three cases with the optimized gain to avoid inverter saturation. Response times are exactly the same and the limiting factor also remains the same in all three cases: i.e., DC bus voltage U_s.

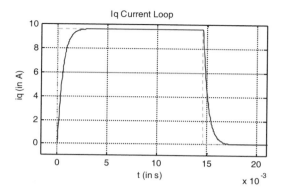

Figure 7.16. *Step response of current i_q (in simulation)*

The closed-loop bandwidth is identical in all three cases. For this linear motor, the constant L_q/R_q is about 4.9 ms. The time response in closed loop is about 0.52 ms and the current bandwidth is about 300 Hz.

7.3.3.3. *Influence of parameter variations*

In the *Ro4q* processor, for an estimation error of R_q resistance, such as $R_{qest} = \varepsilon \cdot R_q$, we get the following transfer functions:

$$\frac{\hat{i}_q}{i_{qref}} = \frac{K_i + \varepsilon \cdot R_q}{K_i + R_q} \cdot \frac{1}{1 + \dfrac{L_q s}{K_i + R_q}}$$

$$\frac{\hat{i}_q}{i_{qref}} = \frac{1}{1 + \dfrac{(1-\varepsilon) R_q + L_q s}{K_i}} \qquad [7.12]$$

$$\frac{\hat{i}_q}{i_{qref}} = \frac{1}{1 + \dfrac{R_q + L_q s}{(1-\varepsilon) R_q + L_q s} \cdot \dfrac{L_q s}{K_i'}}$$

For case *c*, the integral action of the PI controller compensates for the error of the estimated resistance value: the static error tends to 0. For cases *a* and *b* using P controllers, however, there is a static error dependent on the estimation error ε of the ohmic voltage drop:

$$\lim_{s \to 0} \frac{\widehat{i_q}}{i_{qref}} = \frac{K_i + \varepsilon \cdot R_q}{K_i + R_q}$$

$$\lim_{s \to 0} \frac{\widehat{i_q}}{i_{qref}} = \frac{1}{1 + \frac{(1-\varepsilon) R_q}{K_i}} \qquad [7.13]$$

$$\lim_{s \to 0} \frac{\widehat{i_q}}{i_{qref}} = 1$$

7.3.3.4. *Application to the linear motor studied (see Appendix, section 7.8)*

Figure 7.17 shows that cases *a* and *b* give a static error on current i_q, while the command with a PI controller converges to the correct value.

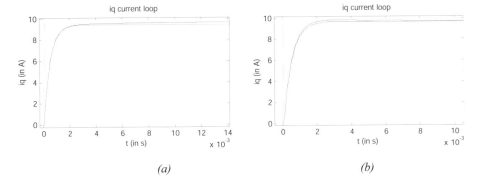

Figure 7.17. *Step response of current i_q (simulation): (a) with 20% R_q estimation error; and (b) with 20% L_q estimation error*

Figure 7.16b shows the response to a 20% estimation error of inductance L_q. The response in case c is affected by the estimation error: there is an overshoot of the reference but it finally converges with the reference value. Cases *a* and *b* are unaffected.

Unsurprisingly, the PI controller used makes the measured current converge to its reference value. The reasoning of the *d*-axis and that of the *q*-axis are the same because the two axes have similar behaviors. For linear motors with surface-mounted permanent magnets, however, the control architecture can be simplified by neglecting the *d*-axis component. Indeed, for such linear motors, and knowing that $L_q = L_d$, the expression of the thrust force was dependent on the current of the *q*-axis

i_q only. So, theoretically, current i_d does not influence the motion control of the linear motor. The feedback control on the *d*-axis current is therefore useful for performing a defluxing strategy or reducing the Joule losses. In this last case, current i_d is controlled to $i_{dref} = 0$.

7.3. Advanced control of linear motors

In the previous sections we have detailed how the controls of rotating motors can be applied to linear motors using the first harmonic model. Generally, the use of proportional and integral controllers is sufficient to assure speed control of linear motors. However, PI controllers appear to be insufficient due to the ripple force generated by the cogging force and end-effect force effects that only exists in linear motors.

Thus, using more sophisticated models, new control techniques are needed to take into account the specificities of linear motors:

– use of an auto-adaptive resonant controller to compensate for cogging forces;

– adding a feed-forward action in PI controllers in order to compensate for the detent forces. For example, it is possible to map the detent forces and to fill a lookup table. This could then be used in feed-forward anticipation as an added compensation of the detent forces;

– another solution would be to have a very detailed model and to deduce a more appropriate control strategy.

These are presented in the following sections.

7.3.1. *Multiple resonant controllers in a two-phase reference frame*

The principle of the resonant controller emerges from the need to eliminate the frequency error induced by the existing current and voltage harmonics [HAU 99b]. The equations of the resonant controller in the case of a sinusoid C_{sin} [SAT 98] and a co-sinusoid C_{cos} [FUK 01] are defined by:

$$C_{\sin}(s) = K_p + \frac{K_r}{s^2 + \omega_p^2}, \quad C_{\cos}(s) = K_p + \frac{K_r \cdot s}{s^2 + \omega_p^2} \qquad [7.14]$$

266   Control of Non-conventional Synchronous Motors

To simultaneously control the amplitude and phase of a sinusoidal signal, a solution is deduced from the previous two equations [WUL 00]:

$$C(s) = K_P + \frac{b_1 \cdot s + c_1}{s^2 + \omega_p^2} = \frac{a_2 \cdot s^2 + a_1 \cdot s + a_0}{s^2 + \omega_p^2} \qquad [7.15]$$

This resonant controller is now used in many areas:

– control of active filters to limit electrical pollution in grids [GUI 00];

– control of power converters for high-frequency variation of the grids [GUI 07];

– optimization of the torque control of a synchronous machine [ZEN 04].

For the variable speed drive of linear motors, the frequency of reference currents is greatly variable. Thus, the resonant controller is called self-adaptive or auto-adaptive. The reference pulse ω_p of the controller needs to be adapted to the desired or measured pulse ω at all times, as $\omega = \omega_p$. The numerator coefficients of equation [7.15] are also defined at all times to ensure stability and high-level controller gains.

Figure 7.18 shows the command structure of the thrust force in the steady two-phase reference frame (α, β) using resonant correctors.

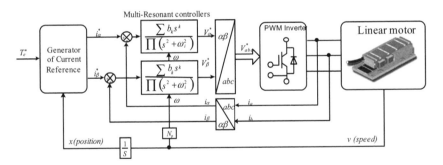

Figure 7.18. *Control structure with resonant correctors in the reference frame (α, β)*

Figure 7.19 shows the Bode diagrams of two-frequency series resonant controllers in open loop (BO) and closed loop (BF), in amplitude and phase. The first controller is set to operate at pulse ω_p and the second at pulse ω_{p5}.

Advanced Control of the Linear Synchronous Motor 267

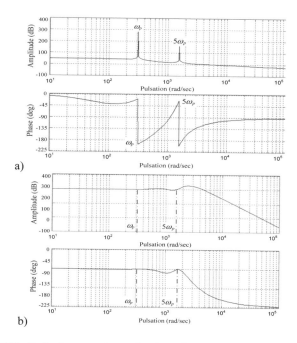

Figure 7.19. Bode diagrams in amplitude and phase of a resonant controller:
(a) in open loop; and (b) in closed-loop

Some of the properties of resonant correction are:

– at the pulse ω_p, this transfer function has a module tending to an infinitely large value (about 300 dB, see Figure 7.19a). The gain is higher at the desired pulse and not across the frequency bandwidth. Thus, for a given stability criterion, the gain in value can be higher for the resonant controller than it can be for a PI controller. The convergence between the reference and the measurement is then faster than a conventional solution;

– the value of the gain being high at the pulse ω_p, if the conditions of stability are guaranteed the frequency error will remain low, even for parameter estimation errors. This allows the resonant controller to be efficient;

– reducing the frequency error at the reference frequency causes the output variable to be in phase with the input reference variable, as shown in the closed-loop phase of the Bode diagram, in Figure 7.19b.

The two-phase reference frame is determined using the Concordia and inverse Concordia transformations, which are defined by the equations [LOU 04]:

268   Control of Non-conventional Synchronous Motors

$$T_3 = \sqrt{\frac{2}{3}} \begin{bmatrix} 1 & -\frac{1}{2} & -\frac{1}{2} \\ 0 & \frac{\sqrt{3}}{2} & -\frac{\sqrt{3}}{2} \end{bmatrix}, \text{ and } T_3^{-1} = \sqrt{\frac{2}{3}} \begin{bmatrix} 1 & 0 \\ -\frac{1}{2} & \frac{\sqrt{3}}{2} \\ -\frac{1}{2} & -\frac{\sqrt{3}}{2} \end{bmatrix} \quad [7.16]$$

The transformation is only useful because the linear motor studied has a star connection and is balanced. This means that out of the three phases to be supplied, there are only two independent variables. Hence, a mathematical representation, or rather a reference frame, is needed where two parameters, called i_α and i_β, will be sufficient to independently control the currents.

Finally, two current references are generated to compensate for the influence of the emf on the thrust force:

$$\begin{cases} i_{\alpha ref} = \dfrac{T_{eref} \cdot \left[-\sin(\theta) + \lambda_5 \cdot \sin(5\theta) \right]}{\sqrt{\dfrac{3}{2}} \cdot N_p \cdot \hat{\phi}_f \cdot \left(1 - \lambda_5^2\right)} \\ i_{\beta ref} = \dfrac{T_{eref} \cdot \left[\cos(\theta) + \lambda_5 \cdot \cos(5\theta) \right]}{\sqrt{\dfrac{3}{2}} \cdot N_p \cdot \hat{\phi}_f \cdot \left(1 - \lambda_5^2\right)} \end{cases} \quad [7.17]$$

A resonant controller is then applied to two frequencies:

– on the fundamental of the emf to control the thrust force;

– on the rank 5 harmonic of the emf, using coefficient λ_5 to eliminate the ripple force of the thrust generated.

There are many techniques to generate these reference variables. For one reference force, two reference currents must be generated, with as many frequency references as the number of frequencies to be controlled in the resonant controllers. There are therefore several degrees of freedom in the generation of reference currents.

Furthermore, the resonant controller is not directly applied to an error between the reference and the measurement of the emf. It is very difficult to determine the emf during motion without unplugging the power supply. Thus, the resonant controller is used on the current error only, and the harmonics of emfs are defined with a reference current generator, using the equations described in equation [7.17].

Advanced Control of the Linear Synchronous Motor 269

The work of [ZEN 05] specifies the tuning principle of a multiple resonant controller, and applies it to a multiple auto-adaptive resonant controller.

The tuning technique for each step of calculation is defined as follows:

– to evaluate the fundamental value of the reference pulse ω_p, from the measured speed: $\omega_p = \pi \cdot v / \tau_p$;

– to control the harmonics of ranks 1 and 5. Then, the denominator coefficients are known:

$$(s^2 + \omega_p^2)(s^2 + \omega_{5p}^2);$$

so the characteristic polynomial of the closed-loop current is then expressed:

$$P(s) = \left(s - \frac{R}{L}\right) \cdot \prod_{i=1}^{\beta}\left(s^2 + \omega_i^2\right) + \frac{1}{L} \cdot \sum_{\alpha=0}^{2\beta} b_\alpha s^\alpha \qquad [7.18]$$

– the numerator coefficients are calculated using a pole placement technique known as the "symmetrical optimum", which guarantees the same time response for each harmonic [LOR 97, SHI 78].

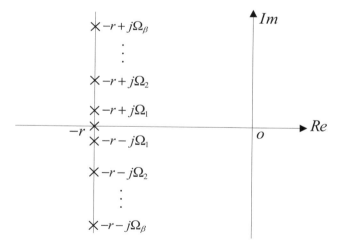

Figure 7.20. *Pole placement technique*

so, the characteristic polynomial is simplified as follows:

$$P_{GSM}(s) = (s+r)\prod_{i=1}^{\beta}\left[(s+r+j\Omega_i)(s+r-j\Omega_i)\right] \quad \{r, \Omega_i \in \mathbf{R}; \; i \in \mathbf{N}\} \quad [7.19]$$

– limits for the resonant controller are fixed in terms of bandwidth and maximum phase;

– finally, the characteristic polynomial is solved using Gaussian elimination on the coefficient matrix for coefficients b_n [ZEN 05].

In the case of the motor studied here, the resonant controller must compensate for 1.6% of the rank 5 harmonic of the emf [REM 07].

Experimental tests have been conducted in order to apply this resonant controller strategy to LSP120C, a linear motor created by Indramat, which has a two-phase reference frame (α, β). Figure 7.21 shows that the results are satisfactory and that the estimated force completely follows the reference force [ZEN 04]. Furthermore, the zoom of Figure 7.21 reveals that the time response of the resonant controller is about 5 ms during fast acceleration. Furthermore, the real-time implementation of a multiple resonant controller with variable frequency can be achieved in a computation time less than the time step of feedback control: 26 μs of time calculation[4] on the 50 μs admissible for the current loop, as can generally be found in industrial drives [GRO 01].

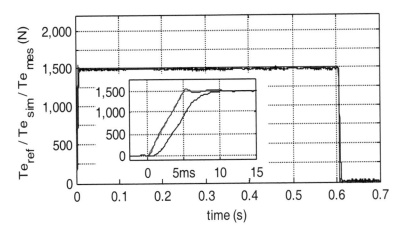

Figure 7.21. *Results of force control with multiple resonant controllers for a trapezoidal force reference and a zoom during acceleration*

---

4 These results are obtained by taking into account the possibilities of the technology used: DS1005 dSPACE ® 400 MHz [ZEN 05].

Advanced Control of the Linear Synchronous Motor 271

The results in Figure 7.22 show that the resonant controller allows the motor current to follow the reference current perfectly:

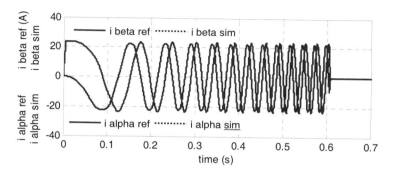

Figure 7.22. *Results of currents when using a multiple resonant controller*

We have proved the use of the resonant controller technique in controlling linear motors.

An auto-adaptive resonant corrector is nothing more than a highly selective filter, with self-adaptive properties at given frequencies (see Figure 7.19). Equation [7.15] shows that the expression of the resonant controller is very close to that of the lead controller. However, it is the peak width in Figure 7.19 that guarantees the immunity of control from parasitic frequencies. So, if the value of the estimated frequency reference is not perfectly identified, a peak that is too fine will induce a greater risk of stalling the frequency tracking of the resonant controller.

Thus, the resonant controller can be a "miraculous" solution only if the resolution of the speed sensor is sufficient. Similarly, to control high-order harmonics, the speed error will be proportional to the harmonic rank and the risk of divergence will increase proportionately. To our current knowledge, there is no study showing the limits of the resonant controller, and this is beyond the scope of this book.

7.3.2. *Feed-forward control for the compensation of detent forces*

The principle of feed-forward control is to replace the estimation and observation processors using measurements of quantities (e.g. position, velocity and currents) with reference values given by reference variables and derivatives. Figure 7.23 shows a feed-forward control structure.

The feed-forward control is widely used in fields such as machine tools with linear motors [GRO 01]. In practice, this control structure allows us to keep a cascaded closed-loop control and add feed-forward compensation to each loop entry. The main idea is to reduce delays from one loop to the next. Yet, such compensation can only be effective if the system runs exactly as anticipated by the reference. Thus, although faster and more precise, this control structure is less resilient to unforeseen external forces.

It is rarely necessary to anticipate the current loop using PI controllers, however, as it is sufficiently fast and accurate, as shown in Figure 7.23. Indeed, although there are some disturbances in the motor current (e.g. from the pulse-width modulation – PWM) and because the current error remains low in amplitude, it generally leads to the improvements of this kind of control not being discerned on the current loop performances. Nevertheless, on the closed loop speed control for linear motors, the improvement can be very significant [GRO 01].

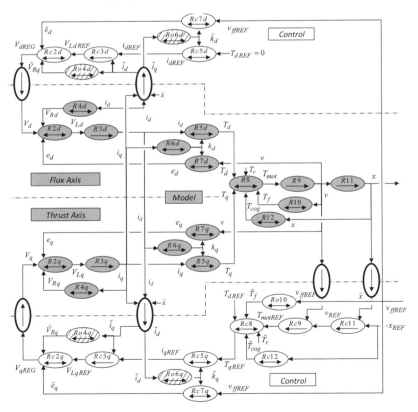

Figure 7.23. *Feed-forward control structure*

7.3.3. Commands by the nth derivative for sensorless control

The inversion principle of the control structure by the nth derivative is defined in the COG formalism as a direct inversion [BAR 04]. Figure 7.24 shows the direct inversion of rigid and causal processors:

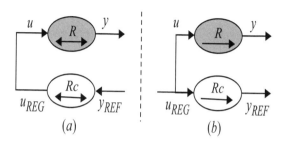

Figure 7.24. *Direct inversion principle*

Figure 7.24b shows that the inversion of a causal processor does not require measurement of the controlled variable, so no sensor is required here. Control by the nth derivative is based on the following principle: if a linear time invariant with minimum phase is a n-order system between an input control u and output variable y, then the evolution of this system can be established from the nth derivative of y:

$$\sum_{j=0}^{n} a_j \frac{d^j y(t)}{dt^j} = \sum_{i=0}^{m} b_i \frac{d^i u(t)}{dt^i} \qquad [7.20]$$

where a_i and b_i are constant coefficients. In all cases $m \leq n$, which ensures compliance with natural causality. The relationship expressed in the Laplace domain takes the form of the following transfer function:

$$P(s) = \frac{Y(s)}{U(s)} = \frac{b_0 + b_1 s + \dots + b_m s^m}{a_0 + a_1 s + \dots + a_n s^n} \quad ; \quad m \leq n \qquad [7.21]$$

Control by the nth derivative derived in open loop is written:

$$\sum_{i=0}^{m} b_i' \frac{d^i u_{REG}(t)}{dt^i} = \sum_{j=0}^{n} a_j' \frac{d^j y_{REF}(t)}{dt^j} \quad , \quad \text{with} \quad \begin{cases} b_i' = b_i \\ a_j' = a_j \end{cases} \qquad [7.22]$$

The assumption hidden in equation [7.22] is an important concept of control by the nth derivative: in fact, the performance of this control is based solely on the quality of the identification of the system's parameter. This control structure is also known as early control [BEA 04]. It is important to note that three variables are essential to the PMLSM control in open loop:

– the input variables (here, the voltage supply);

– the controlled variables (here, the position of the linear motor); and

– the nth derivative of the controlled variable. This means the variable that defines all other variables in the system from a defined initial state (here, the third derivative of the position, i.e. the jerk).

Thus, this technique is applied to the PMLSM, with the position of the linear motor x as the output variable, and V_q as an input voltage of q-axis. First, the canonical representation of the first harmonic model of a linear motor is used in the Park reference frame:

$$\frac{X(s)}{V_q(s)} = \cdot \frac{k_q}{k_q^2 + R_q \cdot f_{vis}} \cdot \frac{1}{s + \left(\frac{L_q + M}{k_q^2 + R_q \cdot f_{vis}}\right) \cdot s^2 + \left(\frac{L_q \cdot M}{k_q^2 + R_q \cdot f_{vis}}\right) \cdot s^3} \quad [7.23]$$

Figure 7.25 shows a COG representation of an open-loop control structure using the first harmonic model in the Park reference frame.

Here, in order to control the linear motor, it is necessary to define the inductance voltage V_{Lq} for a given displacement. This can be obtained by adapting equation [7.23] to a canonical equation that links acceleration and jerk:

$$V_{Lq}{}^{(0)}_{(s)} = \frac{L_q \cdot f_{vis}}{k_q} \cdot X^{(2)}_{(s)} + \frac{L_q \cdot M}{k_q} \cdot X^{(3)}_{(s)} \quad [7.24]$$

To control the position x of the linear motor, first we need to have defined the third derivative of the position, i.e. the jerk. Then, from this variable and all the initial values of the state variables, all of the states of the system can be calculated, which means that we can find the acceleration, velocity and position at any time. Finally, using the inductance voltage is enough to rebuild all the system states starting from an initial state.

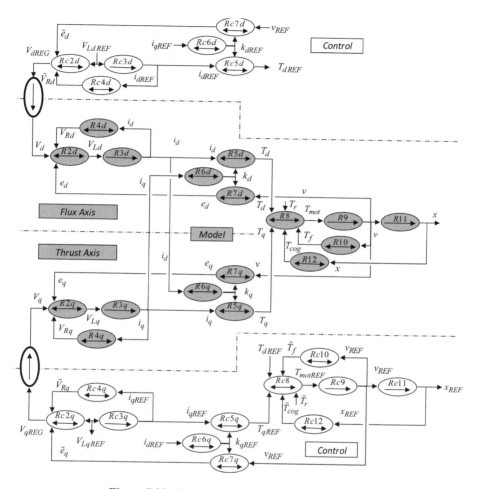

Figure 7.25. *Maximum control structure in open loop*

Although the control in Figure 7.25 involves a large number of processors, reading is very intuitive as it allows us to recreate the set of states of the system.

This control architecture is simulated in the case of the motor studied, for a displacement of 0.2 m at 2 m/s and 20 m/s², using a jerk bang-bang. The following results are obtained:

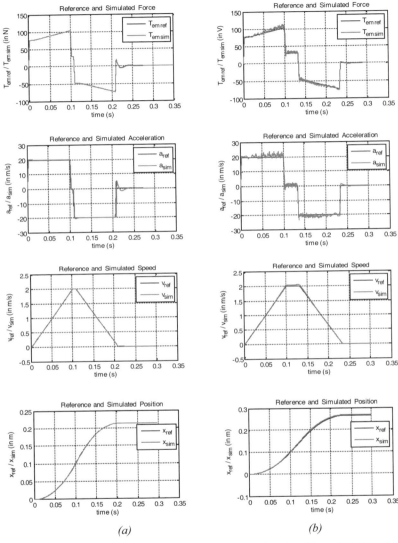

Figure 7.26. Results of the force, acceleration, speed and position of LMD10-050: (a) Simulation using the first harmonic model; and (b) simulation using the full model

Figure 7.26a shows the simulation of the nth derivative control technique in the case of PMLSM using the first harmonic model. The results show that such a technique is efficient. Figure 7.26b shows the application of this technique in a more realistic way, because it includes the emf harmonics, the inductance saturation effects and the detent force. Furthermore, it shows that the results deviate slightly from those in Figure 7.26a, where an error of 4 mm in position can be seen at the end of a 260 mm displacement (with 1.5% as the maximum position error). Moreover, the speed has an error of about 0.05 m/s, regarding the 2 m/s desired (with 2.5% as the maximum speed error). Finally, for the thrust generated, the maximum error is 4 N, regarding the 20 N desired, for a 20% error. Although there is no external disturbance, the force control is not precise enough.

Thus, it is necessary to extend the force control to the q-axis using the full first harmonic model of the linear motor. This means no longer neglecting the influence of the d-axis, including the emf e_d.

If we analyze the control structure in Figure 7.25 more precisely, between the input and the output variables, the direct chain has three causal processors ($Rc11$, $Rc9$ and $Rc3q$), and the nth derivative of this system is a third-order system. In the representation in Figure 7.25, the variables corresponding to the third-order derivative are the inductance voltages V_{Ld} and V_{Lq} of axes d and q, respectively. This means that the inductance voltages must mathematically be of at least class C^3 to respect the continuity of the whole system.

Thus, with a very accurate model of the system and being able to generate the nth derivative of position x, it is possible to obtain absolute control of the linear motor. Figure 7.25 shows that this kind of control does not use any current sensor.

When using this control technique, it is important to highlight the weak points.

If the friction and detent forces are not neglected, generation of the reference variable is much more complicated. In fact, it becomes necessary to know the derivative of the friction and the detent forces, which are highly nonlinear, in order to generate the derivative of the reference acceleration:

$$V_{LqREF} = \frac{L_q}{k_q} \cdot \frac{d}{dt}\left(M\frac{d^2x}{dt^2} + T_{det} + T_s\right) = \frac{L_q \cdot M}{k_q}\frac{d^3x}{dt^3} + \frac{d}{dt}(T_{det} + T_s) \qquad [7.25]$$

So even if the third derivative of the reference position is known, it is difficult to directly define the inductance voltage reference. This reference is not causal and therefore generates significant discontinuities in the force reference, mainly at each speed inversion on the friction force.

Figure 7.27 compares the reference, measurement and simulation results to a bang-bang jerk profile with 20 m/s², at 2 m/s for a 20 cm displacement.

278 Control of Non-conventional Synchronous Motors

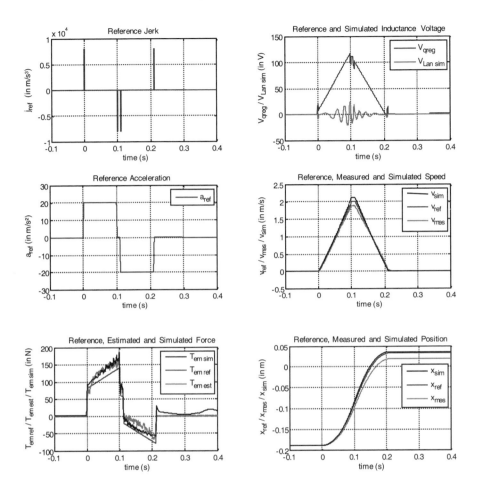

Figure 7.27. *Results of force, jerk, acceleration, speed and position in the open loop of the LMD10-050 motor*

The open-loop control uses the strict inversion of the model. It is therefore the ideal control structure, since it allows us to entirely anticipate the behavior of the system. Nevertheless, the generation of the reference variable (n^{th} derivative) in the open-loop control is the major problem with this kind of control. The physical limits of the linear motor must be taken into account in the generation of reference variables, since there are no sensors to protect the engine.

Obviously, the generation of the reference variable is the main problem in the case of our nonlinear model of a linear motor. The flatness control appears to be a possible alternative for controlling the linear motor in open loop [FLI 92]. This approach may be coupled to learning methods by successive iteration (known as iterative learning).

7.4. Conclusion

It is clear that linear motors have undeniable advantages for rapid implementation in industrial applications: increasingly, systems are using direct drive (they are cheaper, have fewer components in the transmission, etc.). Initially, the formalism of the COG allows us to find the same control structures as industrial structures using the first harmonic model of the actuator. Obtaining very high performances from such actuators requires the definition of a highly detailed physical model of the linear motor concerned.

This chapter has shown the inherent specificities of linear motors and their impact on thrust generation. In particular, the generation of force ripple can be controlled by the control design using resonant controllers, which can dynamically compensate for multiple parasitic harmonics or frequencies. Another approach that has been discussed in this chapter is the open-loop control of linear motors using a direct inversion of the COG representation. Such a sensorless control strategy could be very efficient in friction-free linear motors equipped with very high-speed inspection machines and used in the semiconductor industry. However, the drawback with this model is that the open-loop control is very sensitive to modeling errors and to variations in parameter.

Finally, knowing how to graphically design the control structure of an actuator enables us to obtain the adapted performance of the system. The next step is to redesign the geometry of the motor itself: Even if the efficiency of the motor might be reduced, the main idea is to improve the whole system performance. An example of this mechatronic approach could be to geometrically design a linear motor with a perfect sinusoidal cogging force instead of trying to reduce the force to low amplitude. Even if the cogging forces are greater in amplitude, it may be easier to have a sinusoidal waveform to compensate for cogging forces in the control design. And so, better performances for system positioning will be achieved and the vibration of the mechanical yoke could be reduced.

7.5. Nomenclature

$\tilde{X}$	Estimated or calculated X variable,
$\widehat{X}$	Measured X variable
$\hat{X}$	Maximum value of the X variable
V_d, V_q	Voltage of the d-axis and the q-axis, in Park reference frame [V]
V_{dref}, V_{qref}	Voltage references of V_d, V_q, in Park reference frame [V]
e_d, e_q	Electromotive forces e_d of the d-axis and e_q of the q-axis, in Park reference frame [V]
$\tilde{e}_d, \tilde{e}_q$	Estimations of e_d, e_q, in Park reference frame [V]
i_d, i_q	Currents of the d-axis and q-axis, in Park reference frame [A]
i_{dref}, i_{qref}	Reference current i_d and i_q, in Park reference frame [A]
k_d, k_q	Electrical factors of d-axis and q-axis, in Park reference frame [N/A]
$\tilde{k}_d, \tilde{k}_q$	Estimations of k_d and k_q, in Park reference frame [N/A]
L_d, L_q	Inductances L_d of d-axis and L_q of q-axis, in Park reference frame [H]
$\tilde{L}_d, \tilde{L}_q$	Estimations of L_d, L_q, in Park reference frame [H]
L	Inductance of one phase of the primary of the PMLSM, in the abc reference frame [H]
R	Resistance of one phase of the primary of the PMLSM, in the abc reference frame [Ω]
M	Mass of the primary of the PMLSM [kg]
T_{em}, T_{emref}	Thrust force and reference of thrust force [N]
T_r	Load force [N]
τ_p	Pole pitch – the distance between two consecutive magnetic poles – for the PMLSM [m]
N_p	Electrical constant of position: $N_p = \pi / \tau_p$ [rad/m], for a rotating motor: $N_p = p$, with p being the number of pole pairs
x	Linear displacement [m]

ϑ	Mechanical angle of the rotor for a rotating motor [rad]
ϑ_e	Electrical angle [rad]
$\hat{\phi}_f$	Maximum value of the magnetic flux of induction for each phase in the *abc* reference frame [Wb]
v	Linear speed [m/s]
Ω	Mechanical angular speed for a rotating motor [rad/s]
ω	Electrical angular speed or electrical pulsation [rad/s]

7.6. Acknowledgment

This work has been supported by ETEL's Research Team, including Ralph Coleman, Motion Control Research Manager, and his team working on the control of linear actuators and collaborating with the L2EP (Department of Electrical Engineering and Power Electronics of Lille, France).

7.7. Bibliography

[ANO 04] ANORAD, "Precision gantry systems, (HPG) high precision gantries, (SPG) super precision gantries, (UPG) ultra precision gantries", *Catalogue Anorad*, 2004. (Available at: http://www.rockwellautomation.com/anorad/index.html, accessed December 2008.)

[BAR 98] BARRE P.J., CARON J.P., HAUTIER J.P., LEGRAND M., *Systèmes Automatiques, Tome 1 Analyses et modèles*, Ellipses Marketing, 1998.

[BAR 04] BARRE P.J., Commande et entraînement des machines-outils à dynamique élevée – formalismes et applications, thesis, USTL de Lille, France, December 2004.

[BOL 85] BOLDEA I., NASAR S.A., *Linear Motion Electromagnetic Systems*, Wiley Publishing, New York, 1985.

[BOL 97] BOLDEA I., NASAR S.A., *Linear Electric Actuators and Generators*, Cambridge University Press, Cambridge, 1997.

[BOL 01] BOLDEA I., NASAR S.A., *Linear Motion Electromagnetic Devices*, Taylor & Francis, London, 2001.

[BOU 00] BOUSCAYROL A., DAVAT B., DE FORNEL B., FRANÇOIS B., HAUTIER J.P., MEIBODY-TABAR F., PIETRZAK-DAVID M., "Multimachine multiconverter system: application for electromechanical drives", *European Physics Journal - Applied Physics*, vol. 10, no. 2, pp. 131-147, 2000.

[BRA 00] BRANDENBURG G., BRUCKL S., DORMANN J., HEINZL J., SSCHMIDT C., "Comparative investigation of rotary and linear motor feed drive systems for high precision machine tools", *Proceedings of the 6th International Workshop on Advanced Motion Control*, pp. 384-389, April 2000.

[CAS 03] CASSAT A., CORSI N., WAVRE N., MOSER R., "Direct linear drives: Market and performance status", LDIA2003, *Proceedings of the 4th International Symposium on Linear Drives for Industry Applications*, Birmingham, UK, September 8-10, 2003.

[CON 89] CONSTANT O., "L'Aérotrain, 15 ans après", *Voies Ferrées*, no. 52, September-October 1989.

[COR 07] CORSI N., COLEMAN R., PIAGET D., "Status and new development of linear drives and subsystems", LDIA2007, *Proceedings of the 6th International Symposium on Linear Drives for Industrial Applications*, Lille, France, 2007.

[DAU 99] DAUPHIN-TANGUY G., RAHMANI A., SUEUR C., "Bond graph aided design of controlled systems", *Simulation Practice and Theory*, vol. 7, pp. 493-513, 1999.

[EAS 02] EASTHAM J.F., TENCONI A., PROFUMO F., GIANOLIO G., "Linear drive in industrial application: State of the art and open problems", *Proceedings of the International Conference on Electrical Machines (ICEM'02)*, CD-ROM, Bruges, Belgium, August 2002.

[ETE 07] ETEL, *Linear Motors Handbook*, ETEL, October 2007, (Available at: http://www.etel.ch, accessed December 2008.

[FAV 00] FAVRE E., BRUNNER C., PIAGET D., "Principes et applications des moteurs linéaires", *J'Automatise*, no. 9, March-April 2000.

[FLI 92] FLIESS M., LEVINE J., MARTIN P., ROUCHON P., "Sur les systèmes non linéaires différentiellement plats", *Comptes Rendus des Séances de l'Académie des Sciences*, vol. I-315, pp. 619-624, 1992.

[FUK 01] FUKUDA S., YODA T., "A novel current-tracking method for active filters based on a sinusoidal internal model", *IEEE Transactions on Industry Applications*, vol. 37, no. 3, pp. 888-895, 2001.

[GIE 99] GIERAS J.F., PIECH Z.J., *Linear Synchronous Motors: Transportation and Automation Systems*, CRC Press, New York, 1999.

[GIE 03] GIERAS J.F., "Status of linear motors in the United States", *Proceedings of the 4th International Symposium on Linear Drives for Industry Applications*, LDIA2003, Birmingham, UK, September 8-10, 2003.

[GOM 05] GOMILA G., *Le Moteur Linéaire, sans Rival en Vitesse et Précision*, Mesures, Paris, France, no. 774, 2005.

[GOM 08] GOMAND J., "Analyse de systèmes multi-actionneurs parallèles par une approche graphique causale - application à un processus électromécanique de positionnement rapide", Electrical Engineering Thesis, Arts et Métiers Paristech (ENSAM) de Lille, France, 4 December 2008.

[GRO 01] GROß H., HAMANN J., WIEGARINER G., *Electrical Feed Drives for Automation Technology - Basics, Computation, Dimensioning*, VCH Publishers, Germany, 2001.

[GUI 00] GUILLAUD X., HAUTIER J.P., WULVERICK M., CRESPI F., "Multiresonant corrector for active filter", *Industry Applications Conference 2000. Conference Record of the 2000 IEEE*, vol. 4, pp. 2151-2155, October 8-12, 2000.

[GUI 07] GUILLAUD X., DEGOBERT P., TEODORESCU R., "Use of resonant controller for grid-connected converters in case of large frequency fluctuations", *EPE2007, 12th European Conference on Power Electronics and Applications*, Aalborg, Denmark, September 2007.

[HAU 98] HAUTIER J.P., CARON J.P., *Systèmes Automatiques, Tome 2 Commande des Processus*, Ellipses Marketing, Paris, France, 1998.

[HAU 99a] HAUTIER J.P., CARON J.P., *Convertisseurs Statiques: Méthodologie Causale de Modélisation et de Commande*, Editions Technip, Paris, France, 1999.

[HAU 99b] HAUTIER J.P., GUILLAUD X., VANDECASTEELE F., WULVERYCK M., "Contrôle de grandeurs alternatives par correcteur résonant", *Revue Internationale de Génie Électrique*, vol. 2, pp. 163-183, 1999.

[LOR 97] LORON L., "Tuning of PID controllers by the non-symmetrical optimum method", *Journal Automatica (Journal of IFAC)*, vol. 33, no. 1, 1997.

[LOU 04] LOUIS J.P., *Modèles pour la Commande des Actionneurs Électriques* (Traité EGEM, Série Génie Électrique), Hermès-Lavoisier, Paris, 2004.

[MCL 88] MCLEAN G.W., "Review of recent progress in linear motors", *Electric Power Applications, IEE Proceedings B*, vol. 6, pp. 380, 1988.

[NAS 76] NASAR S.A., BOLDEA I., *Linear Motion Electric Machines*, Wiley Publishing, New York, 1976.

[REM 07] REMY G., Commande optimisée d'un actionneur linéaire synchrone pour un axe de positionnement rapide, Electrical Engineering Thesis, Arts et Métiers Paristech (ENSAM) de Lille, France, December 12, 2007.

[SAT 98] SATO Y., ISHIZUKA T., NEZU K., KATAOKA T., "A new control strategy for voltage-type PWM rectifiers to realize zero steady-state control error in input current", *IEEE Transactions on Industry Applications*, vol. 34, no. 3, pp. 480-486, 1998.

[SHI 78] SHINNERS S.M., *Modern Control System Theory and Application*, Addison-Wesley, Boston, USA, 1978.

[STO 98] STOIBER D., Synchronous Linear Motor, Krauss-Maffei AG, US Patent US005744879A, April 28, 1998.

[THY 08] THYSSENGRUP AG – TRANSRAPID, High-Tech for "Flying on the Ground", Maglev System Transrapid, 2008 (Available at: http://www.thyssenkrupp.com/documents/transrapid/TRI_Flug_Hoehe_e_5_021.pdf, accessed December 2008.)

[WAV 97] WAVRE N., Permanent-Magnet Synchronous Motor, U.S. Patent 05642013A, June 24, 1997.

[WUL 00] WULVERYCK M., Contrôle de courants alternatifs par correcteur résonant multifréquentiel, Thesis, Université des Sciences et Technologies de Lille, France, June 2000.

[ZEH 02] ZEHDEN A., Elektrische Beförderungsanlage, Patent no. 26847, Charlottenburg, Germany, June 1902.

[ZEN 04] ZENG J., REMY G., DEGOBERT P., BARRE P.J., "Thrust control of the permanent magnet linear synchronous motor with multi-frequency resonant controllers", *Maglev 2004, 18th International Conference on Magnetically Levitated Systems and Linear Drives*, vol. 2, pp. 886-896, Shanghai, China, December 2004.

[ZEN 05] ZENG J., High-Performance Control of the Permanent Magnet Synchronous Motor using Self-Tuning Multiple-Frequency Resonant Controllers, PhD Thesis, Lille University of Science and Technology, France, 21 September 2005.

7.8. Appendix: LMD10-050 Datasheet of ETEL

IRONCORE LINEAR MOTOR   LMD 10-050-XXX-
                                              Reference   Winding Code

Direct Drives & Systems

DESIGN CONSTANT

		UNIT	VALUE
Fp	Peak force	N	554
Fu	Ultimate force	N	643
Fc130	Continuous force (coil @ 130°C)	N	172
Fc80	Continuous force (coil @ 80°C)	N	130
Fs	Stall force (coil @ 130°C)	N	122
Pp	Peak Power dissipation (@ 20°C)	W	822
Pc130	Continuous power dissipation (coil @ 130°C)	W	77.3
Pc80	Continuous power dissipation (coil @ 80°C)	W	36.4
Km	Motor constant (@ 20°C)	N / √W	24.6
Ko	Damping coefficient zero impedance	N / (m/s)	603
Te	Electrical time constant	ms	3.29
Rth130	Thermal resistance (coil @ 130°C)	K / W	1.42
Rth80	Thermal resistance (coil @ 80°C)	K / W	1.65
Mm	Motor weight (magnetic way exclude)	kg	1.6
Mw	Magnetic way mass	kg / m	6.2
Fa	Attraction force	N	1770
Fh	Hysteresis force	N	2
Fd	Detent force	N	5
Td	Maximum continuous recommended speed	m / s	10

WINDING CONSTANT

	Winding Code	UNIT	2 phases	3 phases
				3QA
Kt	Force constant	Nm / Arms		88.8
Ku	Back EMF Constant [peak value](*)	V / (m/s)		72.5
R130	Electrical resistance at 130°(*)	Ω		12.5
R80	Electrical resistance at 80°(*)	Ω		10.8
R20	Electrical resistance at 20°(*)	Ω		8.8
L1	Electrical inductance (*)	mH		28.8
Ip	Peak current	Arms		7.9
Iu	Ultimate current	Arms		9.9
Ic 130	Continuous current @ 130°c	Arms		2
Ic 80	Continuous current @ 80°c	Arms		1.5

Note :
All data ± 10%   (*) : Terminal to terminal   Version : 1.0   Date : 16.09.99

© ETEL SA - SUBJECT TO MODIFICATION WITHOUT PREVIOUS NOTICE

ETEL SA   2112 Môtiers (NE) - Switzerland
Phone : +41 32 862 01 23 -   Fax : +41 32 862 01 01
E-mail : etel@etel.ch   http://www.etel.ch

Chapter 8

Variable Reluctance Machines: Modeling and Control

8.1. Introduction

The term "variable reluctance machine" (VRM) covers a very large range of devices directly or indirectly using the variation of the air-gap permeance to convert electromechanical energy. Different topologies, associated with specific supply modes, have been designed, studied and tested to fulfill the specifications of various applications with very variable performances. Although the majority of structures have not (yet) moved beyond the prototype, two families − synchronous reluctance or synchro-reluctant machines (Synchrel) and the switched reluctance machines (SRM) − have very interesting performances and potentials. The latter already being well established in industry [MUL 99].

The rotors of these two families of machines are devoid of all sources of magnetomotive force (permanent magnets and supplied winding). Only the stator supports polyphase winding (concentrated for SRM and generally distributed for Synchrel). This gives them an undeniable robustness and genuine advantages for high-speed applications. In both cases, synchronous operation can be obtained when their supplies are supervised by an appropriate control. The performances obtained are then comparable, or even greater in some cases, to that of induction machines of the same volume [STA 93].

Chapter written by Mickael HILAIRET, Thierry LUBIN and Abdelmounaïm TOUNZI.

These machines are not devoid of defects. Thus, the principal drawback of Synchrel machines lies in their power factor, which is relatively limited in the case of basic structures. Its improvement involves the adoption of specific topologies that increase the complexity of manufacture and the price [MEI 86, SAR 81]. On the other hand, SRMs generate a significant pulsating torque inherent to their operating principle. This torque can be decreased by the control but to the detriment of the efficiency of the machine-converter set [HAN 10]. These structures are also known to have vibrations and acoustic noise of magnetic origin. These are due to the action of attractive radial magnetic forces between the stator and rotor teeth, particularly when in conjunction [MIN 08, OJE 09a, OJE 09b]. The advantages of Synchrel and SRM machines, particularly in conjunction position, do however, compensate for their drawbacks, which make them attractive for diverse applications, particularly those at high speeds.

Whichever electromechanical converter is used, a control combining robustness and good performances is dependent, among other things, on an accurate modeling of the machine. The design of the control requires an analytical model of the structure that is developed under a certain number of simplifying hypotheses, which limit its accuracy. A compromise is hence to combine model simplicity and control efficiency. Generally, the saturation of magnetic materials is not insignificant in the case of Synchrel and SRM. This is due to an air-gap, often very weak, feature inherent to these machines. This aspect can possibly be taken into account in the model of the machine.

The current chapter deals with the description, modeling and control of variable reluctance machines belonging to two families: Synchrel and SRMs. It is divided into two parts. Each of them introduces, in the first section, the description and presentation of the operating principle of the structure while introducing the static converter most often used to ensure its supply. The second section of the chapter is devoted to the analytical model of the machine. It uses notions introduced in other chapters about the modeling and control of synchronous machines. The effects more specific to variable reluctance machines (VRM) will be dealt with. Finally, the third section of this chapter will be devoted to different control strategies. In the absence of an excitation circuit, these strategies can significantly differ from those used for classical electromechanical converters. The advantages and limitations of each of them will be emphasized.

8.2. Synchronous reluctance machines

8.2.1. *Description and operating principle*

The Synchrel machine (SRM) is a structure whose stator, made of laminated steel sheets, is identical to that of a classic synchronous or induction machine with an isthmus of weak opening slots. It is equipped with a polyphase winding with p_1 pole pairs that is generally distributed and supplied with a sinusoidal polyphase source to generate a rotating field. The rotor of this machine is salient, with N_r teeth with slots that are usually larger than the stator slots. In its basic version, it can also be made of laminated steel sheets. The transverse cut of an example of a Synchrel machine with $N_r = 4$ is shown in Figure 8.1.

Figure 8.1. *Synchronous reluctance machine*

Due to the disparity in sizes, the effect of the air-gap variation due to the stator slots is ignored and the stator is assumed to be smooth. This hypothesis is widely used in the case of classic machines[1].

The operating principle of the synchronous reluctance machine is based on minimizing the reluctance seen by the armature magnetic field passing by. We can easily show that a synchronous operation, devoid of strong torque ripples, is ensured if N_r and p_1 fulfill the following condition [SAR 81]:

$$N_r = 2 p_1 \qquad [8.1]$$

Thus the rotor, having as many teeth as stator poles, will rotate at the same speed as the stator rotating field so that the latter tends to meet a minimum reluctance. From then on, operation is synchronous, with a speed Ω whose expression is given by:

$$\Omega = \omega / p_1 = 2\omega/N_r \qquad [8.2]$$

---

[1] Actually, this effect introduces torque ripples, which can be attenuated by acting on the topology or through controlling the machine [HAM 09].

where ω represents the supply pulsation of stator variables.

As in the case of a synchronous machine, the Synchrel machine must be supplied according to the position of the rotor. Consequently, a rotor position sensor is necessary. As a matter of fact, this structure is a synchronous machine with salient poles whose excitation circuit has been suppressed [LOU 10]. It therefore only works thanks to the reluctance torque.

As we will see later, the reluctance torque is proportional to the difference between the inductances of direct axis L_d and quadrature L_q under the hypothesis of linearity of the characteristic $B(H)$ of magnetic materials. As a consequence, the increase in the capabilities of this structure (mass torque and power factor) needs to maximize L_d and minimize L_q.

Besides the classical structure in which a minimum air-gap as weak as possible is needed, different other topologies, more or less complex, have been proposed. They sometimes have performances superior to those of squirrel cage induction machines for a same volume. Figure 8.2 shows performances higher to the transverse cut of two of the most interesting topologies in terms of mass torque and power factor – the Synchrel with flux guides [STA 93] and the axially-laminated Synchrel [VAG 00].

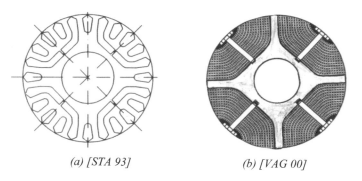

(a) [STA 93]   (b) [VAG 00]

Figure 8.2. *Synchrel: (a) with flux guides; and (b) that is axially laminated*

Whichever configuration is adopted, the search for optimal Synchrel performance involves a significant level of saturation, which can have a considerable influence on the behavior of these machines.

Synchrel machines are generally supplied by classical pulse width modulation (PWM) inverters. Figure 8.3 shows the converter used in the three-phase case.

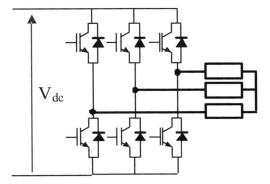

Figure 8.3. *Two-level inverter for the supply of a Synchrel machine*

Equipped with their converter, the Synchrel machine can, *a priori*, fulfill the same functions as the squirrel-cage induction machines. Devoid of a rotor coil, they are interesting in terms of manufacturing cost and restricted heating at the rotor. In their classic topology, they suffer from a relatively limited power factor, which is prejudicial to the dimensioning of the static converter. The use of specific rotors (with flux guides or ones that are axially laminated) allows us to considerably improve the power factor, bringing it closer to that of classic induction machines, but this is done at the expense of manufacturing price. As a rule of thumb, these structures are limited to low- or medium-power applications.

Due to the absence of a rotor circuit, these machines are well adapted to high-speed operations. In this case, they are often equipped with a massive rotor where currents are induced during the transient operations. This phenomenon needs to be taken into account in the modeling of the machine.

Whichever application is targeted, it is necessary to control the structure through the control of the static supply converter. The design of the control of the system machine-converter needs a model of the set. The following section is devoted to the mathematical model of the Synchrel machine.

8.2.2. *Hypotheses and model of a Synchrel machine*

As previously mentioned, the Synchrel machine can be considered as a synchronous machine with salient poles without an excitation circuit. As a consequence, its model can be directly inferred from that of this structure. For the sake of simplification, we only show the classic model of this machine. We will be able to refer to other works to take into account the saturation phenomenon and/or currents induced in a massive rotor [LUB 03, TOU 93].

By adopting the simplifying hypotheses (see section 1.3.2 and the notations in 1.3.3 of [LOU 10]) devoted to modeling in view of the control of synchronous actuators, the matricial equation ruling the electrical variables of the stator of a three-phase Synchrel machine is written in the following form:

$$(v_3) = R_s.(i_3) + \frac{d(\psi_3)}{dt} \qquad [8.3]$$

where R_s is the phase resistance and (v_3), (i_3) and (ψ_3) represent the voltage, current and flux vectors, respectively, relative to the stator phases of the machine.

The flux vector can be written in the following form:

$$(\psi_3) = (L_{ss}(\theta)).(i_3) \qquad [8.4]$$

where $(L_{ss}(\theta))$, the inductance matrix, is expressed as:

$$(L_{ss}(\theta)) = (L_{ss0}) + (L_{ss2}(\theta)) \qquad [8.5]$$

with θ being the mechanical position. The classical development of L_{ss0} and $(L_{ss2}(\theta))$ [LES 81] is given in Chapter 1 of [LOU 11].

The expression of electromagnetic torque developed by the structure is obtained directly from the derivative of the coenergy, which leads to:

$$C_{em} = \left.\frac{\partial W_{cm}}{\partial \theta}\right|_{(i_3)=cste} = (i_3)^t.\left(\frac{\partial (L_{ss}(\theta))}{\partial \theta}\right)^t.(i_3) \qquad [8.6]$$

Since there is no excitation circuit, the stator currents are used for magnetization of the machine and generation of torque at the same time. This differs from the case of the synchronous machine in which the excitation current (or magnets in the case of a PMSM) plays a dominant role. However it is possible to supply the machine with a strategy similar to that of self-controlled synchronous machine.

The study and control of Synchrel machines are usually done using the diphase model. To develop this model, we use the Park transform of angle $p_1\theta$ (see Chapter 1 of [LOU 11]). We then obtain the expressions of the different electrical variables, given in the referential linked to the rotor. Thus, by applying this transform to equation [8.3], the latter is re-written in the following form:

$$\begin{cases} v_d = R_s.i_d - p_1.\Omega.\psi_q + \dfrac{d\psi_d}{dt} \\ v_q = R_s.i_q + p_1.\Omega.\psi_d + \dfrac{d\psi_q}{dt} \end{cases} \qquad [8.7]$$

where v_d, v_q, i_d and i_q, respectively, represent the stator voltages and currents according to direct d and in quadrature q axes. ψ_d and ψ_q, the fluxes according to these same axes, are expressed:

$$\psi_d = L_d.i_d$$

and:

$$\psi_q = L_q.i_q$$

where L_d and L_q are the direct inductance and the inductance in quadrature, that are written in a manner similar to those of the synchronous machine with salient poles, i.e.:

$$L_d = L_{s0} - M_{s0} + \frac{3}{2}L_{S2}$$

and:

$$L_q = L_{s0} - M_{s0} - \frac{3}{2}L_{S2},$$

The same transform, applied to relationship [8.6], leads to the following expression of the electromagnetic torque:

$$C_{em} = p_1.(L_d - L_q).i_d.i_q = p_1.(\psi_d.i_q - \psi_q.i_d) \qquad [8.8]$$

8.2.3. *Control of the Synchrel machine*

Control of the Synchrel machine is based on the vector control approach through the variables expressed in referential $d - q$. The electromagnetic torque is proportional to the product of stator currents i_d and i_q, as shown by relationship [8.8]. We then have a degree of freedom in the choice of currents since, only the product is imposed for a given load torque. Several control strategies exist [BET 93]. They are distinguished by the criterion to be optimized during the generation of current references. In fact, we can control the Synchrel machine in order to maximize a criterion, such as the efficiency output [LUB 03], power factor [BOL 96] or electromagnetic torque [BET 93]. The global diagram of speed control is given in Figure 8.4.

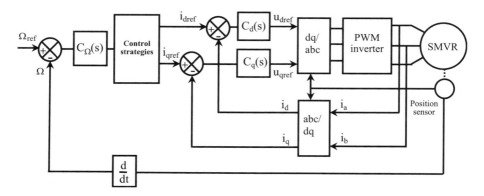

Figure 8.4. *Block-diagram of the vector control of the Synchrel machine*

From the speed reference and a corrector ($C_\Omega(s)$), we can determine the reference torque of the machine. According to the control strategy adopted, we infer the reference values of currents i_d and i_q. The regulation loops of currents i_d and i_q (correctors $C_d(s)$ and $C_q(s)$) and the inverse Park transform allows us to end up with the three-phase reference voltages that the PWM inverter must generate.

In the following, we will present the most common control strategies. To illustrate them, we will show the simulation results for a machine with cut sections, represented in Figure 8.1, whose electrical and mechanical parameters are given in Table 8.1. J represents the inertia moment of the system and f the viscous friction coefficient. The inductances L_d and L_q can be measured using the voltage scale interval method (current response) by positioning the rotor according to axis d or q.

$R_s(\Omega)$	L_d(H)	L_q(H)	J(kg.m²)	f(Nm/(rd/s))
7.8	0.54	0.21	0.038	0.0029

Table 8.1. *Parameters of a Synchrel machine (600 W, 1,500 rpm)*

8.2.3.1. *Vector control with* i_d *constant*

In the applications that need good dynamics at low speed (quick torque response), we often prefer to control the Synchrel machine using a constant current

i_d [BET 93]. This allows us to impose flux in the machine, since the inductance of axis d (small air-gap) is large compared to that of axis q (large air-gap). The principle of this control is similar to that of a DC machine with separate excitation. The component the of stator current along axis d, playing the role of excitation, allows us to set the value of the flux in the machine (nominal flux Ψ_{dn}). The component along axis q plays the role of the armature current and allows us to control the torque. The torque can then be written:

$$C_{em} = K \cdot i_q \qquad [8.9]$$

with:

$$K = p_1 \cdot (L_d - L_q) \cdot i_{dref} \qquad [8.10]$$

and:

$$i_{dref} = \frac{\Psi_{dn}}{L_d} \qquad [8.11]$$

This strategy allows performant drive systems to perform and enables us to impose the nominal torque, from stand-still to nominal speed.

The control strategy block shown in Figure 8.4 can be completed by that of Figure 8.5. Current i_{dref} has a constant value corresponding to the rated flux in the machine (equation [8.11]). The reference value of the current i_{qref} is determined according to the type of control used in the application (speed or torque). In the case of speed regulation, i_{qref} is obtained through a speed corrector. It is limited to ensure the protection of the machine in terms of current magnitude.

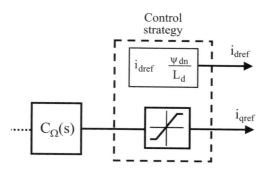

Figure 8.5. *Control strategy with constant i_d*

296   Control of Non-conventional Synchronous Motors

The block diagram of the stator current's regulation loop along axis d is given in Figure 8.6 (the regulation loop according to axis q is identical to within the indices). To synthesize the correctors (generally of porportional integral (PI) controller type), we re-write equations [8.7] in the following form:

$$\frac{L_d}{R_s}\frac{di_d}{dt}+i_d=\frac{1}{R_s}\left(p_1\Omega L_q i_q+v_d\right) \qquad [8.12]$$

$$\frac{L_q}{R_s}\frac{di_q}{dt}+i_q=\frac{1}{R_s}\left(-p_1\Omega L_d i_d+v_q\right) \qquad [8.13]$$

These relationships make the coupling terms $p_1\Omega L_q i_q$ and $p_1\Omega L_d i_d$ appear between the two axes. They are considered perturbations that need to be compensated for by the control (estimated term, $\tilde{e}_q$). The corrector gains are calculated using the classic automatic methods.

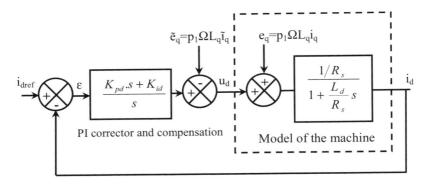

Figure 8.6. *Regulation loop of current i_d*

Figure 8.7 shows the block diagram of speed regulation. The transfer function of the system is obtained from the fundamental relationship of the dynamics of rotating systems:

$$\frac{J}{f}\frac{d\Omega}{dt}+\Omega=\frac{1}{f}\left(C_{em}-C_{ch}\right) \qquad [8.14]$$

where C_{ch} represents the load torque. Coefficient K (torque coefficient), which appears in Figure 8.7, is defined by equation [8.10].

The structure of the speed corrector is of PI controller type. To calculate the value of the speed corrector gains (K_p and K_i), we assume that current i_q instantaneously follows its reference (this hypothesis is often fulfilled due to the difference in the time constants of the speed loop and the current loop).

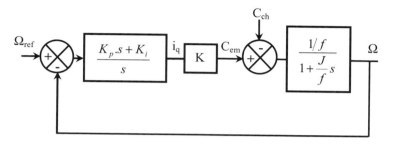

Figure 8.7. *Speed regulation loop*

The results of applying this control strategy to the case of simulation will be presented in section 8.2.3.4.

8.2.3.2. *Maximum torque control*

Stator current vector i_s can be represented in referential $d - q$ linked to the rotor (Figure 8.8). In steady state, it is fixed and has a constant norm in this referential.

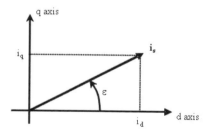

Figure 8.8. *Definition of control angle ε*

We can then define angle, ε, referred to as the control angle, as well as variable m, so that:

$$m = \tan \varepsilon = \frac{i_q}{i_d} \qquad [8.15]$$

The control strategy consists of determining the value of m (or ε) in order to obtain the maximum torque for a given stator current value.

The expression of electromagnetic torque is:

$$C_{em} = p_1(L_d - L_q) \cdot i_d i_q \qquad [8.16]$$

The root mean square (RMS) value of the current in one phase of the stator can be expressed as:

$$i_s = \sqrt{\frac{i_d^2 + i_q^2}{3}} \qquad [8.17]$$

By using this relationship in expression [8.16], we can express the electromagnetic torque as a function of i_s and coefficient m:

$$C_{em} = 3p_1 \cdot (L_d - L_q) \cdot i_s^2 \frac{m}{1+m^2} \qquad [8.18]$$

We can then show that the machine develops the maximum torque for a given current if the following condition is fulfilled:

$$m = 1, \text{ namely } \varepsilon = \pi/4 \implies i_d = i_q = \sqrt{\frac{3}{2}} i_s \qquad [8.19]$$

In this case, the expression of the maximum torque is:

$$C_{em(max)} = \frac{3}{2} \cdot p_1 (L_d - L_q) \cdot i_s^2 \qquad [8.20]$$

The optimum control angle therefore has as a value of $\pi/4$, which corresponds to $i_d = i_q$. It is then sufficient to use this equality in the control for the machine to operate at its maximum torque. The current absorbed is then at a minimum for a given torque, which minimizes Joule losses in the machine.

In the case of this strategy, the control block of the diagram in Figure 8.4 can be completed by that of Figure 8.9.

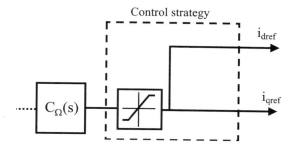

Figure 8.9. *Control strategy at maximum torque*

8.2.3.3. *Maximum power factor control*

By disregarding the different losses in the machine, the power factor can be defined as the ratio of the converted mechanical power to the apparent power:

$$FP = \frac{C_{em}\Omega}{3V_s I_s} \qquad [8.21]$$

where V_s, the RMS voltage of a stator phase, can be written in the following form by disregarding the voltage drops:

$$V_s = p_1 \Omega \Psi_s \qquad [8.22]$$

By expressing flux Ψ_s as a function of its components according to axes d and q:

$$\Psi_s = \sqrt{\frac{\psi_d^2 + \psi_q^2}{3}} \qquad [8.23]$$

we end up with the following expression of the power factor:

$$FP = \frac{(L_d - L_q) \cdot m}{\sqrt{L_d^2 + L_q^2 m^2} \cdot \sqrt{1 + m^2}} \qquad [8.24]$$

With regard to m, the power factor is at its maximum for:

300   Control of Non-conventional Synchronous Motors

$$m = \frac{i_q}{i_d} = \sqrt{\frac{L_d}{L_q}} \qquad [8.25]$$

This only depends on the ratio of L_d/L_q, as can be seen below:

$$FP_{(max)} = \frac{\frac{L_d}{L_q} - 1}{\frac{L_d}{L_q} + 1} \qquad [8.26]$$

The control strategy block of the diagram of Figure 8.4 can then be completed by that of Figure 8.10:

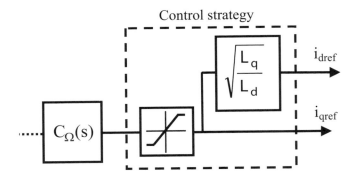

Figure 8.10. *Control strategy at maximum power ratio*

8.2.3.4. *Simulation results*

In Figure 8.11, we present the simulation results obtained for the control strategy with i_d constant. The calculations have been done by using the electrical and mechanical parameters given in Table 8.1. The value of current i_{dref} is set at 2.5 A. At $t = 0.5$ s, a speed level of 400 rpm is applied. It induces a current draw i_q of 7.0 A (we notice that current i_q enters into the limitation). An anti-windup device is put in place to avoid non-negligible draw when current i_q is limited. At $t = 2.0$ s, a load torque level of 4 N.m is applied. The rotational speed drops slightly and stabilizes again around 400 rpm. We notice that current i_d is barely disrupted by the variations in i_q.

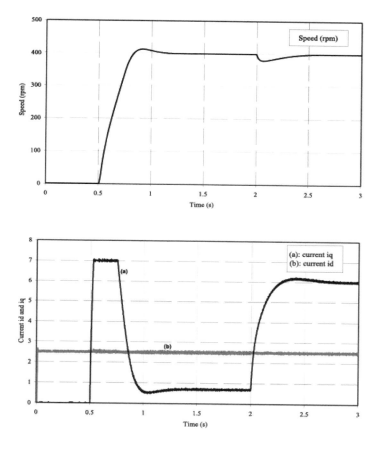

Figure 8.11. *Control strategy at i_d constant (i_{dref} = 2.5 A); Simulation results*

Figure 8.12 shows a comparison between the different control strategies in terms of Joule losses and power factor by considering the complete model of the machine. The operating point is the same as that in Figure 8.11 (400 rpm, 4 N.m). At $t = 4.0$ s, we go from the control strategy at constant i_d to the control strategy at maximum torque. At $t = 5.0$s, we use the control strategy at maximum power factor. We notice that the Joule losses are at a minimum for the second control strategy and that the maximum power factor obtained is $F_p = 0.44$. For the machine studied here, with the saliency ratio L_d/L_q being weak ($L_d/L_q = 2.6$), the maximum power factor is relatively limited.

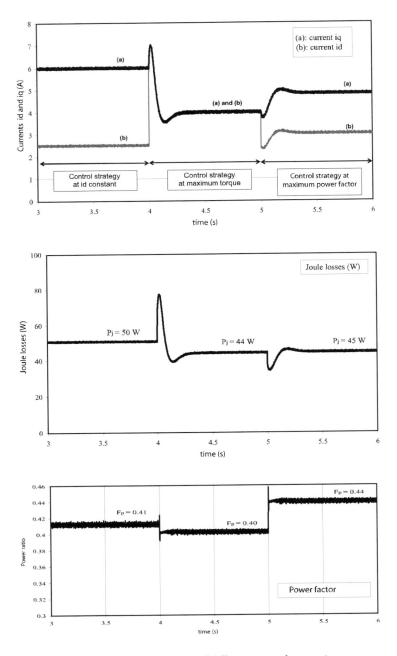

Figure 8.12. *Comparison of different control strategies*

Finally, Figure 8.13 represents the variations of the power factor as a function of saliency ratio L_d/L_q (this ratio depends on the structure of the machine) for a control strategy at maximum power factor ($m = \sqrt{L_d/L_q}$) and a control strategy at maximum torque ($m = 1$). In the case of the control strategy at maximum power factor, we observe that the latter becomes interesting for supply dimensioning (static converter) from a saliency ratio > 6.

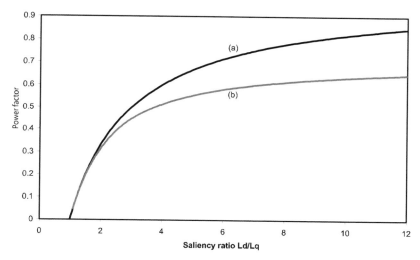

Figure 8.13. *Power factor as a function of saliency ratio L_d/L_q: (a) control at maximum power factor; and (b) control at maximum torque*

8.2.4. *Applications*

The controlled Synchrel machine has a very interesting performance. Except for the limited power factor in classic structures that can largely be improved by adopting specific rotors, the machine's performance can be comparable to that of induction machines. References mentioning its use in industrial applications are rare, although it is present in some catalogs produced by manufacturers of electrical machines for powers up to a few kW [ISG 11, LUC 04, SAV 99].

8.3. Switched reluctance machines

8.3.1. *Description and principle of operation*

The switched reluctance machine (SRM) consists of a stator and a rotor with laminated steel sheets equipped, respectively, with N_s and N_r teeth that are regularly distributed. The magnetic circuit of the rotor is totally passive (there are no conductors or magnets); whereas the stator supports the armature winding. Thus, a concentrated coil is wound around each tooth. The coils surrounding diametrically opposed teeth, in the electrical sense, are connected to create a winding of q phases with p_1 pole pairs. Figure 8.14 shows an example of a SRM structure. It possesses eight stator teeth and six rotor teeth. The armature winding, with four phases, has one pole pair.

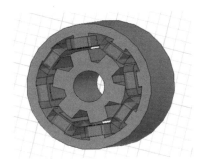

Figure 8.14. *Example of an 8/6 SRM with four phases*

SRM operation can easily be made explicit. Similarly to the current in an electrical circuit trying to follow the path with the lowest impedance, the flux in a SRM tries to follow the easiest path between two points. As a consequence, when the winding of one phase is supplied, the electromagnetic field generated is going to attract the closest rotor tooth, so that it tends to align along the axis of the phase in order to minimize the flux path in the air and apply the maximum flux rule in the magnetic circuit: this is the reluctant effect. To ensure "continuous" running, it is enough to sequentially supply the phases (one phase after the other) to attract the closest rotor tooth as it goes along. It is easy to understand the need for a position sensor to ensure the supply. SRM operation does not require a rotating field, so the coils are ideally supplied with current pulses. The current pulses can be unipolar or bipolar square-wave current. In fact, reluctance minimization is independent of the field polarities.

In both cases, in order for the operation described above to be possible, the following relationship, between N_s, N_r and p_1, has to be fulfilled:

$$\pm (N_s \pm N_r) = 2p_1 \qquad [8.27]$$

When a unipolar (or bipolar) supply of phases occurs periodically, at pulsation ω, the rotor moves at a speed Ω:

$$\Omega = \frac{\omega}{N_r} \quad (\text{or } \Omega = \frac{2\omega}{N_r}) \tag{8.28}$$

During rotation of the rotor, two particular positions are distinguished, which are very important in the modeling and control of the SRM (see Figure 8.15):

– The opposition position: for which the stator and rotor teeth are furthest away from each other. As a consequence, the magnetic circuit has a minimum inductance (maximum air-gap).

– The conjunction position: for which the stator and rotor teeth are facing each other. The magnetic circuit shows a maximum inductance.

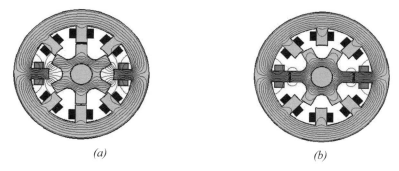

Figure 8.15. *Opposition position (a) and conjunction position (b) of a SRM*

These two positions confine the energy cycle for a given current magnitude, as shown in Figure 8.16.

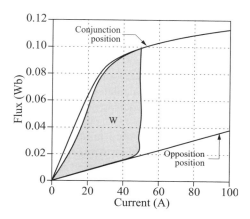

Figure 8.16. *Energy cycle during an electrical period*

It is possible to show that the average electromagnetic torque $<C_{em,k}>$ generated by a phase is proportional to the surface W of this cycle as well as the number of rotor teeth N_r [KRI 01, MIL 01]:

$$< C_{em,k} > = \frac{N_r}{2\pi} W \qquad [8.29]$$

As a consequence, for a given number of teeth and level of current, the torque is a function of the difference between the flux in conjunction position and in opposition position. The difference between these two fluxes must be as high as possible which implies the weakest minimum air-gap possible.

As mentioned previously, the SRM is generally supplied with unipolar square-wave current. The most frequently used converter then consists of an asymmetrical half-bridge per phase (see Figure 8.17).

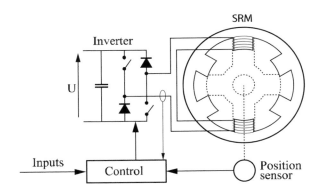

Figure 8.17. *Representation of the asymmetrical half-bridge*

The arms being controlled independently of each other, in the case of a defect on a component, the functioning of the other bridges is not interrupted [GAM 08]. In this case, the control angles must be recalculated in order to maximally decrease the torque ripples and optimize the global efficiency of the set machine/converter.

The SRM has for a long time been used in applications requiring step-by-step operation. Due to its low cost and high robustness, as well as the advances in power and control electronics, it is increasingly being used in applications similar to those of classic structures; it is particularly appreciated in industrial applications requiring high-speed running (alterno-starter, high-speed machining, etc.) [VIS 04, VRI 01]. However, its composition and principle of operation require very specific modeling and control with respect to other synchronous machines.

8.3.2. Hypotheses and direct model of the SRM

The SRM is a structure whose phase windings are, in the majority of cases, magnetically independent. We can therefore write the following electrical equation for each of the phases of the machine:

$$u = R_s i + \frac{d\psi}{dt} \qquad [8.30]$$

where R_s is the phase resistance. u, i and ψ represent the voltage at the terminals, the current and the total linkage flux of the phase winding, respectively. The latter is function of current i and the position of the rotor θ, as shown in the Figure 8.18, drawn for a 8/6 SRM.

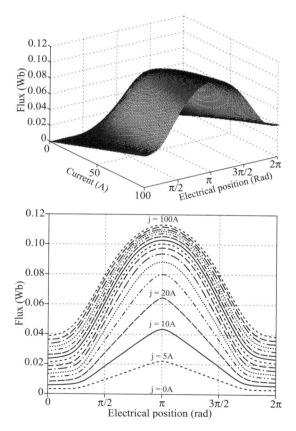

Figure 8.18. *Linkage flux networks as a function of current and position: (top) 3D view; and (bottom) 2D view*

308   Control of Non-conventional Synchronous Motors

By adopting the hypothesis of a linear magnetic material behavior, it is possible, at first, to express the linkage flux as a function of inductance in the form:

$$\psi(\theta,i) = L(\theta)\, i \qquad [8.31]$$

Electrical equation [8.30] then becomes:

$$u = R_s i + \frac{d\psi}{di}\frac{di}{dt} + \frac{d\psi}{d\theta}\frac{d\theta}{dt} = R_s i + L(\theta)\frac{di}{dt} + i\Omega\frac{dL}{d\theta} \qquad [8.32]$$

In the case of unsaturated operation of the magnetic circuit, phase inductance $L(\theta,i)$ varies only according to position. The coenergy [LOU 04, MAT 04] is then written as:

$$W_{cem,k}(\theta,i) = \int_0^i \psi(\theta,i)\, di = \int_0^i L(\theta) i\, di = \frac{1}{2}L(\theta)i^2 \qquad [8.33]$$

We then infer the expression of the electromagnetic torque generated by a phase supplied with a constant current i:

$$C_{em,k} = \frac{1}{2}i^2\frac{dL(\theta)}{d\theta} = \frac{1}{2}N_r i^2 \frac{dL(\theta_e)}{d\theta_e} \qquad [8.34]$$

The equation of torque allows us to make a few fundamental observations:

– as the sign of the torque is independent of the direction of the current, unipolar square-wave currents are often used;

– the machine can run as a motor ($dL/d\theta > 0$) or as a generator ($dL/d\theta < 0$).

The torque generated by the different phases is equal to the algebraic sum of torques generated by each of the q phases:

$$C_{em} = \sum_{k=1}^{q} C_{em,k} \qquad [8.35]$$

In practice, the non-saturation hypothesis of the magnetic circuit is invalid; the phase inductance is a function of the position of the rotor and the phase current. In this case, the electrical expression becomes:

$$u = R_s i + \frac{d\psi}{di}\frac{di}{dt} + \frac{d\psi}{d\theta}\frac{d\theta}{dt} = R_s i + \left(L(\theta,i) + \frac{dL(\theta,i)}{di}i\right)\frac{di}{dt} + i\Omega\frac{dL}{d\theta} \qquad [8.36]$$

The expression of the electromagnetic torque is obtained by a derivative of the coenergy with respect to the position, namely:

$$C_{em,k} = \frac{dW_{cem,k}(\theta,i)}{d\theta}\bigg|_{i=cte}$$

$$W_{cem,k} = \int_0^i L(\theta,i) i \, di$$

[8.37]

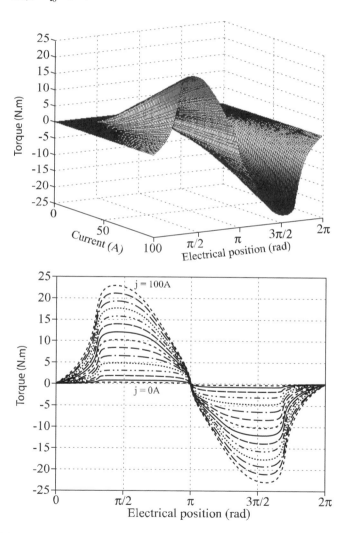

Figure 8.19. *3D torque networks as a function of current and position: (top) 3D view; and (bottom) 2D view*

310   Control of Non-conventional Synchronous Motors

The evolution of the inductance and/or torque can be expressed by an analytical expression or via a two-dimension table. Then, these analytical expressions or tables can be integrated in the machine's control laws (see section 8.3.3.1).

Figure 8.19 represents the behavior of the torque generated by a phase as a function of the position and magnitude of the supply current of a SRM. The nonlinear feature of these characteristics gives us an insight into the difficulties of controlling this structure.

8.3.3. *Control*

Contrary to other machines, the SRM only operates when it is self-controlled. The global structure of the control is generally made of q fast internal current regulation loops and one external speed regulation loop, as shown in Figure 8.20.

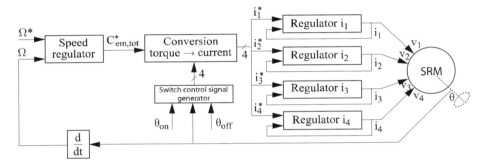

Figure 8.20. *Global control structure*

The (linear or nonlinear) speed corrector has as output the total torque reference. The latter can be distributed over the q phases in two different ways, which will be detailed in the following sections. The conversion of the torque reference C^*_{tot} into a reference current is the most important and specific part of control in the SRM.

As a rule of thumb, the phases of the SRM are supplied, according to the rotor position, in order to generate the required torque: it is self-controlled. We then distinguish two setting parameters (see Figure 8.21) established with respect to the inductance variation of one phase:

– start angle θ_{on}: this is the electrical angle at the beginning of magnetization with respect to the opposition position. It corresponds to the switch-on time of the switches;

– the electrical angle at the end of magnetization, written as θ_{off}.

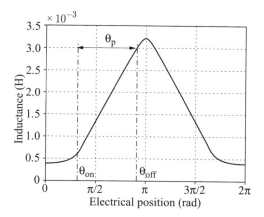

Figure 8.21. *Representation of a phase inductance*

The period during which the phase is supplied is the "magnetization period". It is written as θ_p and is equal to $\theta_{off} - \theta_{on}$.

Figure 8.22 shows an example of the supply sequence of the windings of a SRM with four phases ($q = 4$) during an electrical period. The impulses are shifted by $2\pi/q$ radians. In this example, the start angle θ_{on} is $\pi/6$ and the end of magnetization angle θ_{off} is $5\pi/6$, namely θ_p equal to $2\pi/3$.

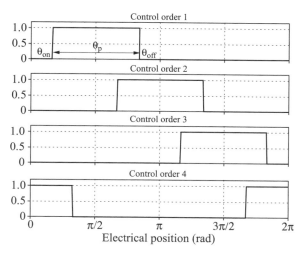

Figure 8.22. *Supply sequence of phases ($q=4$; $\theta_{on} = \pi/6$; $\theta_{off} = 5\pi/6$)*

312   Control of Non-conventional Synchronous Motors

The torque being indirectly controlled through current regulation, the transformation of the total reference torque, obtained from the output of the speed regulator, into q reference currents can be done in two ways:

– one, referred to as "instantaneous torque control" (ITC), aims for the generated torque to follow instantaneously the reference torque. In this case, the reference current of the conduction phase is calculated at each sampling period in order to ensure this equality;

– the other is referred to as "average torque control". In this case, the average value of the generated torque must be equal to the reference torque. This leads to reference currents in the form of pulses.

The choice of control type depends on the goals of the application being considered. These aims can be: the minimization of torque ripples; the maximization of efficiency; the maximization of torque at high speed; etc.

8.3.3.1. *Instantaneous torque control*

In this approach, total reference torque $C^*_{em,tot}$ is at first distributed on the phases in conduction according to the self-control (θ_{on}, θ_{off} and θ) and according to a distribution law. From the one-phase reference torques thus obtained ($C^*_{em,k}$) and measure of the position, the reference currents (i^*_k) are calculated by using a torque table $C_{em}(\theta,i)$ (see section 8.3.2 and Figure 8.19 (bottom)) generally obtained by finite element calculations or experimental tests. Linear interpolations allow us to infer the reference current of each phase as a function of the electrical position of the rotor and the reference torque per phase.

By assuming current regulators with a sufficiently large bandwidth, this method has the advantage of considerably minimizing the torque ripples because the torque is instantaneously controlled. The goal thus amounts to finding the best reference current profile (i.e. to how best to distribute the total reference torque over the q phases) that allows us to reach the required torque and to optimize one or more goals, such as Joule losses and/or torque ripples [KJA 97].

A first solution consists of sequentially imposing the reference torque on each of the phases. Calculation of the reference current of the active phase can then be obtained in two different ways:

– In the hypothesis of a phase inductance, which can be modeled using a trigonometric function $L(\theta,i) \approx L_0 + L_1\cos(\theta)$ with L_0 and L_1 two positive constants, the expression of the electromagnetic torque can be written as

$C_{em,k} = \frac{1}{2} N_r L_1 i_k^2 \sin(\theta)$ (see section 8.34). In order to ensure the equality between generated torque $C_{em,k}$ and reference torque $C^*_{em,tot}$, the reference current is calculated as follows:

$$i_k = \begin{cases} \sqrt{\frac{2}{N_r L_1} \frac{C^*_{em,tot}}{\sin(\theta)}} & ; \quad \theta_{on} < \theta < \theta_{off} \\ 0 & ; \quad \text{otherwise} \end{cases} \quad [8.38]$$

– Generally, the phase inductance cannot be modeled by such a trivial trigonometric function, given the strong saturation feature of the machine. In this case, the reference current is obtained from a two-dimensional torque table, as shown in Figure 8.23 below, and the use of linear interpolations.

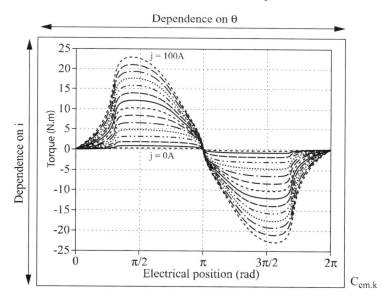

Figure 8.23. *Electromagnetic torque of one phase as a function of rotor position and current*

However, these solutions are not advocated because of strong torque ripples as well as the increase in Joule losses.

An alternative consists of sharing the total requested torque between two successive phases over a large range in order to decrease the quite significant torque ripples at commutation times [KRI 01]. Generally, the torque is distributed over the q phases in the following way:

314   Control of Non-conventional Synchronous Motors

$$C^*_{em,tot} = \sum_{k=1}^{q} C^*_{em,k} = \sum_{k=1}^{q} C^*_{em,k} \, f_k(\theta_e) \quad \text{with} \quad \sum_{k=1}^{q} f_k(\theta_e) = 1 \qquad [8.39]$$

where f_k and $C^*_{em,k}$ are the distribution function and the torque reference of the kth phase, respectively. Several functions can play the role of $f_k(\theta_e)$ [KRI 01, MIL 01]. A possible solution, represented in Figure 8.24, consists of manipulating the trigonometric functions so that:

$$f_1(\theta_e) = \begin{cases} 0.5 - 0.5\cos(k(\theta_e - \theta_0)) & ; \theta_0 < \theta_e < \theta_1 \\ 1 & ; \theta_1 < \theta_e < \theta_2 \\ 0.5 + 0.5\cos(k(\theta_e - \theta_2)) & ; \theta_2 < \theta_e < \theta_3 \\ 0 & ; \text{otherwise} \end{cases} \qquad [8.40]$$

with:

$$\Delta = \theta_p - 360°/p \qquad \theta_0 = \theta_{on} \qquad k = 180°/\Delta$$
$$\theta = \theta_{on} + \Delta \qquad \theta_2 = \theta_{on} \, 360°/q \qquad \theta_3 = \theta_{on} + \theta_p = \theta_{off}$$

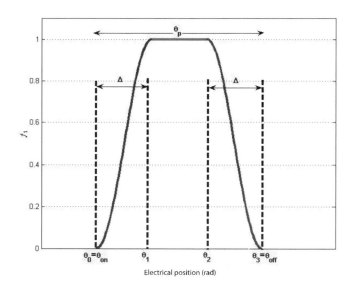

Figure 8.24. *Distribution of torque sharing function*

The choice of angles θ_{on} and θ_{off} of the self-control directly affecting the profile of the reference current, the commutation from one phase to the other must be done over a sufficiently long duration Δ in order to ensure a progressive transfer of the total torque. Furthermore, in motor operation with the adopted logic of torque distribution, it is necessary that the conduction takes place during the phase inductance growing (generation of a positive torque). We then have an additional condition:

$$0° < [\theta_0, \theta_3] < 180° \qquad [8.41]$$

The global block diagram of control, in the case of the instantaneous torque control, is given in Figure 8.25.

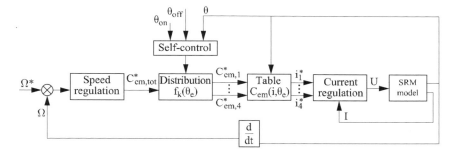

Figure 8.25. *Block diagram of the instantaneous torque control*

The regulation of currents is generally ensured by a hysteresis regulator for simplicity of design, although it introduces strong current ripples and hence torque ripples in steady state. An alternative consists of adding a linear or nonlinear regulator associated with a PWM.

8.3.3.2. *Average torque control*

The ATC is the method most often used to control torque. The phases are supplied by current pulses of constant value that generate a strongly rippled instantaneous torque whose average value is equal to the reference torque.

The three setting variables are: θ_{on}, θ_{off} and the reference current I_{ref}. Several combinations of these variables can generate the same average torque at a given speed. The choice of their values can be done through the optimization of various objectives, such as the efficiency or ripple rate according to the type of application desired. The setting variables are then stored in two-dimensional tables [HAN 08, HAN 10, REK 07].

316   Control of Non-conventional Synchronous Motors

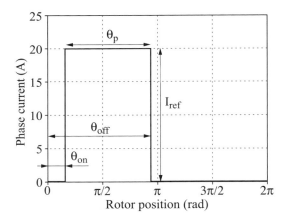

Figure 8.26. *Setting parameters of the square wave control*

Once reference current I_{ref}, as well as angles θ_{on} and θ_{off} are determined, the regulation of currents is ensured in the same way as for ITC.

The global block diagram of ATC is shown in Figure 8.27.

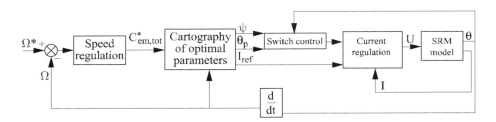

Figure 8.27. *Block diagram of the average torque control*

8.3.3.3. *Comparison of the two controls*

In order to illustrate the assertions of the last two sections, Figure 8.28 shows the comparison of the two types of control in steady state with a 5.4 N.m. load. In the case of ATC, the reference current is constant; whereas it is modulated in the case of ITC, which allows us to decrease the torque ripples.

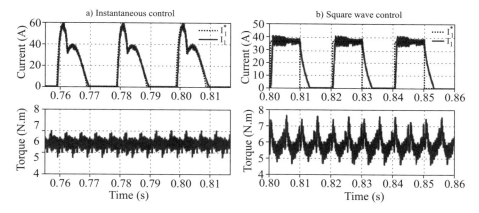

Figure 8.28. *Comparison of: (a) instantaneous; and (b) average torque controls*

The performances of the ITCs depend on the exact measure of the position and dynamics of current regulation chosen. The current in the phase must be able to perfectly follow its reference; otherwise it generates significant torque ripples, mainly at the commutation time.

These limitations are essentially due to the supply source. In fact, when the speed increases, the electromotive force (emf) increases, which decreases the *di/dt* ratio for a given supply voltage. In the end, the current can no longer reach I_{ref} and the voltage applied to the phase is saturated at its maximum value, which corresponds to the operating mode referred to as "full wave". Torque control is then only achieved by variation in angles θ_{on} and θ_{off}; the ITC then becomes the ATC.

The condition imposed by equation [8.41] is another source of limitation for the ITC. An anticipated commutation where angle θ_{on} is negative is not possible in instantaneous control. The advantage of such a commutation lies in the fact that it allows the current to increase quickly while the inductance is weak. When the current is already well-established in the phase, it generates a high torque that compensates for the negative torque generated at the beginning of the magnetization phase.

From the point of view of practical implantation, the memory space occupied in the case of ATC is larger because it involves manipulating three tables of optimal parameters versus one table in the case of the ITC. On the other hand, the number of arithmetic operations is larger for the ITC, where there are one or (at most) two functions to perform, besides the linear interpolations. The principal elements of this comparison are summarized in Table 8.2.

318   Control of Non-conventional Synchronous Motors

	Instantaneous torque control	**Average torque control**
Memory space	1 table $C(\theta,i)$	3 tables θ_{on}, θ_{off}, I_{ref}
Algorithmic cost	2 function cosine at most (generally 2 phases are supplied simultaneously) + 4 interpolations (1 per phase)	3 linear interpolations
Torque ripple	Small	Large
Dependence of the position	Calculation of current references at each sampling period according to the torque and position reference	Reading of the current reference at each sampling period according to the torque reference and the speed
Torque/speed plane	Decreased surface	Increased surface
Peak phase current	Greater than with the ATC, for an equal RMS current	-

Table 8.2. *Comparison algorithmic cost – the memory site for the two methods*

The ITC is then better for applications that do not tolerate torque ripples (machine tools [VIS 04] or robotics) with high performance restricted to a certain range of running. As opposed to the ITC, the ATC is able to cover the whole torque/speed plane of the machine, but with limited performance. The ATC is essential in order to exploit the performance of the machine (maximize the torque, efficiency, etc.).

8.3.3.4. *Continuous conduction*

Figure 8.29 shows the energy cycle in plane (ψ,i) for two different speeds: 500 and 5,000 rpm. The quantity of potentially useable energy is delimited by two flux curves at the opposition and conjunction positions. The energy conversion is better at 500 rpm, where almost all the surface is used. This is not the case at 5,000 rpm where the energy cycle is greatly reduced with respect to the quantity of potentially useable energy. In fact, at high speed and with a constant voltage source, the emf induces an unavoidable current drop in the phase.

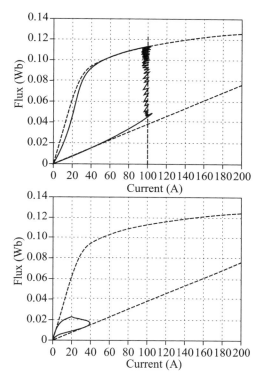

Figure 8.29. *Representation of the energy cycle at low speed (500 rpm) (top) and at high speed (5,000 rpm) (bottom)*

The decrease in energy converted per phase causes a torque and power drop from the nominal speed of the SRM according to equation [8.29]. In order to increase the torque at high speeds, we can decrease the number of coils per phase. However, this solution leads to a decrease in torque at low speeds [HEN 06, REK 07, REK 08, SCH 09]. A second solution consists of supplying the SRM without demagnetizing the phases at each cycle: it is what is referred to as "continuous conduction" (see Figure 8.30). Work carried out on this operating mode has been very recent, although the advantages of an "excitation" (a continuous component) have been known for a long time [MUL 94].

Continuous conduction is done by increasing conduction angle θ_p at values greater than 180°: the fact that angle θ_p is greater than half an electrical period (180°) can lead to an incomplete demagnetization of phases and hence a progressive growing of the flux and current. The risk in this operating mode is that we can lose control of the current which carries on increasing until the protection system is set

320   Control of Non-conventional Synchronous Motors

off. High-performance control of the current is necessary to control the divergence risks [HAN 08, IND 05, LOU 06, REK 07]. A possible solution consists of adjusting the magnetization duration θ_p through an efficient control of the current. See Figure 8.31 [HAN 08].

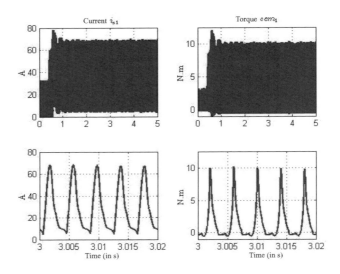

Figure 8.30. *Representation of phase current (top) and electromagnetic torque in continuous conduction (bottom)*

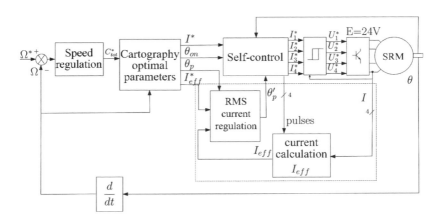

Figure 8.31. *Representation of the control law of the SRM in continuous conduction*

However, this type of supply has drawbacks, namely [HIL 09]:

– a decrease in the global efficiency of the set machine/converter, essentially due to Joule losses;

– an increase in torque ripples due to sensitivity of the control to an update of the mechanical position (self-control of the machine). As a consequence, a relatively short sampling period is necessary (in the range of 1 to 20 µs).

To conclude, continuous conduction is an excellent way to increase the power peaks without increasing the nominal power of the machine.

8.3.4. *Applications*

Here we consider applications using industrial drives at variable speeds, from large-scale economy drives (household appliances) to high-performance drives (aeronautics, the motor industry, etc.).

In the naval propulsion field, compactness and greater maneuverability are the improvements looked for. Thus, the SRM has been used to create naval propulsion units of 7.5 kW or 2.2 kW [MAR 00, RIC 96a].

The SRM fulfills the demands of the aeronautical world [NAA 05, RAD 92]. In planes, or even helicopters, work has been carried out to design alterno-starters that are directly coupled to the turbine. General Electric [RIC 96b] has worked on a device represented in Figure 8.32 whose features are the following: 177 N.m from 0 to 13,500 rpm and 250 kW at steady state from 13,500 to 22,200 rpm (330 kW peaks during 5.0 s). It has been chosen for its operating qualities in extreme environments (high temperatures) and for its tolerance to defects (it can run in degraded modes).

Figure 8.32. *Alterno-starter with variable reluctance (structure12/8) of a reactor (General Electric - Sundstrand aerospace)*

322   Control of Non-conventional Synchronous Motors

The SRM, being robust by design, has therefore naturally attracted the attention of motor industry manufacturers, both for use as a motor and as a generator. The studies and publications on this topic are numerous [FAH 04, FAI 05, KAL 02, KRI 06, RAH 00, RAH 02, WU 02, ZHU 07]. The typical application is the alterno-starter [FAH 04, FAI 05]. It is coupled to the thermal motor and can also be used as a traction motor in hybrid vehicles, thus allowing the recovery of braking energy.

Prototypes of hybrid vehicles having the SRM as a traction motor have been built in Australia. The Commonwealth Scientific and Industrial Research Organisation has worked with the Australian motor industry to develop two hybrid prototypes, one with parallel structure (ECOmmodore [HOL]) and the other with a structure in series (aXcessaustralia LEV [AXE]). SR Drives in partnership with Green Propulsion, a Belgian company specialized in the development of clean vehicles, has designed two SRMs (50 kW and 160 kW, see Figure 8.33) for a project involving hybrid propulsion of a bus [SRD].

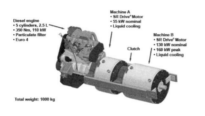 

Figure 8.33. *Hybrid vehicle application for an urban bus [SRD]*

The SRM has also been used in certain domestic applications, due to its cost and ergonomy [MUL 99]. We can cite, for example, the motors of Neptune of Maytag washing machines or those marketed by SR Drives.

Figure 8.34. *VRM for a washing machine [SRD]*

The SRM is also used in many other types of applications or products, such as pumps, compressors, sliding doors, in robotics, etc. [SRD].

To finish, let us cite a few manufacturers who market this machine, such as Allenwest, Brighton Ltd in the UK, Sicme-Motori in Italy [MUL 94] and SRD Ltd in the UK, which is a subsidiary of Emerson Electric Company in St. Louis, Missouri.

8.4. Conclusion

This chapter has been devoted to the description, modeling and control of the two most common families of variable reluctance machines, namely Synchrel machines and SRMs. In both cases, the structures have been tackled similarly by starting with a brief description of the generic topology and a description of the principle and operating conditions. The analytical models that have been developed, often under not insignificant simplifying hypotheses, are those that are most often used to study operation and establish control laws. The control procedures presented are those that are the most common or the most frequently studied. Finally, where possible, care has been given to the introduction of examples of machines used in industry.

This chapter in no way aims to provide an in-depth study of these two families of machines, but rather a "simplified" presentation of these structures and their associated controls in order to give an idea of their controlled operation.

In fact, these machines are still a topic of research today, from a topological point of view and with regards to their supply and control. The content of this chapter, as a consequence, is far from being exhaustive, and different aspects (saturation effects, induced currents, vibrations, noises, etc) have been omitted.

8.5. Bibliography

[AXE] http://www.csiro.au/solutions/aXessaustralia.html

[BET 93] BETZ R.E., LAGERQUIST R., JOVANOVIC M., MILLER T.J.E., MIDDLETON R.H. "Control of synchronous reluctance machine", *IEEE Transactions on Industry Applications*, vol. 29, no. 6, pp. 1110-1122, 1993.

[BOL 96] BOLDEA I., *Reluctance Synchronous Machines and Drives*, Clarendon Press, Oxford, 1996.

[FAH 04] FAHIMI B., EMADI A., SEPE R.B. Jr., "A switched reluctance machine-based starter/alternator for more electric cars", *IEEE Transactions on Energy Conversion*, vol. 19, no. 1, pp. 116-124, 2004.

[FAI 05] FAIZ J., MOAYED-ZADEH K., "Design of switched reluctance machine for starter/generator of hybrid electric vehicle", *Electric Power Systems Research*, vol. 75, no. 2-3, pp. 153-160, 2005.

[GAM 08] GAMEIRO N.S., CARDOSO A.J.M., "Fault tolerant control strategy of SRM drives", *International Symposium on Power Electronics, Electrical Drives, Automation and Motion, SPEEDAM'08*, pp. 301-306, June 2008.

[HAM 09] HAMITI M. O., Réduction des ondulations de couple d'une machine synchrone à réluctance variable. Approches par la structure et la commande, PhD thesis, Nancy I University, France, June 2009.

[HAN 08] HANNOUN H., Etude et mise en œuvre de lois de commande de la machine à réluctance variable à double saillance, PhD thesis, Paris-sud XI University, Gif-sur-Yvette, France, October 22, 2008.

[HAN 10] HANNOUN H., HILAIRET M., MARCHAND C., "Design of a SRM speed control strategy for a wide range of operating speeds", *IEEE Transactions on Industrial Electronics*, vol. 57, no. 9, pp. 2911-2921, 2010.

[HEN 06] HENNEN M.D., BAUER S.E., DE DONKER R.W., "Influence of continuous conduction mode on converters in SRM drives", 22nd *International Electric Vehicle Symposium, (EVS22)*, pp. 1848-1857, Yokahoma, Japan, October 2006.

[HIL 09] HILAIRET M., HANNOUN H., MARCHAND C., "Design of an optimized SRM control architecture based on a hardware/software partitioning", 35st *Annual Conference of IEEE Industrial Electronics Society (IECON'09)*, November 2009.

[HOL] http://www.holden.com.au.

[IND 02] INDERKA R.B., MENNE M., DE DONCKER R.W.A.A., "Control of switched reluctance drives for electric vehicle applications", *IEEE Transactions on Industrial Electronics*, vol. 49, no. 1, pp. 48-53, 2002.

[IND 05] INDERKA R.B., KEPPELER S., "Extended power by boosting with switched reluctance propulsion", 21th *International Electric Vehicle Symposium, (EVS21)*, Monaco, April 2005.

[ISG 11] http://www.isgev.com.

[KAL 02] KALAN B.A., LOVATT H.C., PROUT G., "Voltage control of switched reluctance machines for hybrid electric vehicles", *IEEE Power Electronics Specialists Conference PESC'02*, vol. 4, pp. 1656-1660, June 2002.

[KES 02] KESTELYN X., FRANÇOIS B., HAUTIER J.P., "Torque estimator for a switched reluctance motor using an orthogonal neural network", *Electrical Engineering Research Report*, no. 14, pp. 8-14, 2002.

[KJA 97] KJAER P.C., GRIBBLE J.J., MILLER T.J.E., "High-grade control of switched reluctance machines", *IEEE Transactions on Industry Applications*, vol. 33, no. 6, pp. 1585-1593, 1997.

[KRI 06] KRISHNAMURTHY M., EDRINGTON C.S., EMADI A., ASADI P., EHSANI M., FAHIMI B., "Making the case for applications of switched reluctance motor technology in automotive products", *IEEE Transactions on Power Electronics*, vol. 21, no. 3, pp. 659-675, 2006.

[KRI 01] KRISHNAN R., *Switched Reluctance Motor Drives: Modeling, Simulation, Analysis, Design and Applications*, CRC Press, 2001.

[LES 81] LESENNE J., NOTELET F., SEGUIER G., *Introduction à l'Électrotechnique Approfondie*, Technique et Documentation, Paris, France, 1981.

[LOU 06] LOUDOT S., "Method for controlling a heat engine vehicle driving assembly", *International Patent, W0 2006/05028 A2*, 2006.

[LOU 04] LOUIS J.-P., FELD G., MOREAU S., "Modélisation physique des machines à courant alternatif", in: LOUIS J.-P. (ed.), *Modélisation des Machines Électriques en vue de Leur Commande – Concepts Généraux*, Traité EGEM (Série Génie Electrique), Hermes-Lavoisier, Paris, France, 2004.

[LOU 11] LOUIS J.-P., FLIELLER D., NGUYEN N. K., STURTZER G., "Synchronous Motor Controls, Problems and Modelling", in LOUIS J.-P. (ed), *Control of Synchronous Motors*, ISTE-Wiley, 2011

[LUB 03] LUBIN T., Modélisation et commande de la machine synchrone à réluctance variable, prise en compte de la saturation, PhD thesis, Nancy I University, France, April 18, 2003.

[LUC 04] LUCAS-NÜLLE SOCIETY, *Catalogue of the LUCAS-NÜLLE Society, EEM Machines Électriques*, 2004.

[MAR 00] MARGOT J.P., YECHOUROUN C., MARMET R., GALSTER P., "Entraînement avec moteur à réluctance variable", *Electronique de Puissance du Futur*, Lille, France, November 29 – December 1, 2000.

[MAT 04] MATAGNE E., DA SILVA GARRIDO M., "Conversion d'énergie : du phénomène physique à la modélisation dynamique", in: LOUIS J.-P. (ed.), *Modélisation des Machines Électriques en vue de Leur Commande – Concepts Généraux*, Traité EGEM (Série Génie Electrique), Hermès-Lavoisier, Paris, France, 2004.

[MEI 86] MEIBODY-TABAR F., Etude d'une machine à réluctance variable pour des applications à grande vitesse, PhD thesis, INPL, Nancy, France, 1986.

[MIL 01] MILLER T.J.E., *Electronic Control of Switched Reluctance Machines*, Newnes Power Engineering Series, 2001.

[MIN 08] MININGER X., LEFEUVRE E., GABSI M., RICHARD C., GUYOMAR D., "Semiactive and active piezoelectric vibration controls for switched reluctance machine", *IEEE Transactions on Energy Conversion*, vol. 23, no. 1, pp. 78-85, 2008.

[MUL 94] MULTON B., Conception et alimentation électronique des machines à réluctance variable à double saillance, thesis, Ecole Normale Supérieure de Cachan, 1994.

[MUL 99] MULTON B., BONAL J., "Les entraînements électromécaniques directs: diversité, contraintes et solutions: la conversion électromécanique directe", *Revue de l'Électricité et de l'Électronique*, no. 10, pp. 30-31 and pp. 67-80, 1999.

[NAA 05] NAAYAGI R.T., KAMARAJ V., "Shape optimization of switched reluctance machine for aerospace applications", 31st *Annual Conference of IEEE Industrial Electronics Society IECON'05*, pp. 168-173, November 2005.

[OJE 09a] OJEDA X., HANNOUN H., MININGER X., HILAIRET M, GABSI M., MARCHAND C., LÉCRIVAIN M., "Switched reluctance machine vibration reduction using a vectorial piezoelectric actuator control", *EPJ Applied Physics Journal*, vol. 47, pp. 31103, 2009.

[OJE 09b] OJEDA X., MININGER X., BEN AHMED H., GABSI M., LECRIVAIN, M., "Piezoelectric actuator design and placement for switched reluctance motors active damping", *IEEE Transactions on Energy Conversion*, vol. 24, no. 2, pp. 305-313, 2009.

[RAD 92] RADUN A.V., "High-power density switched reluctance motor drive for aerospace applications", *IEEE Transaction on Industry Applications*, vol. 28, no. 1, pp. 113-119, 1992.

[RAH 00] RAHMAN K.M., FAHIMI B., SURESH G., RAJARATHNAM A.V., EHSANI M., "Advantages of switched reluctance motor applications to EV and HEV: design and control issues", *IEEE Transactions on Industry Applications*, vol. 36, no. 1, pp. 111-121, 2000.

[RAH 02] RAHMAN K.M., SCHULZ S.E., "Design of high-efficiency and high-torque-density switched reluctance motor for vehicle propulsion", *IEEE Transactions on Industrial Applications*, vol. 38, no. 6, pp. 1500-1507, 2002.

[RIC 96a] RICHARDSON K.M., POLLOCK C., FOWLER J.O., "Design and performance of a rotor position sensing system for a switched reluctance marine propulsion unit", *IEEE Industrial Application Society Annual Meting*, pp. 168-173, October 1996.

[RIC 96b] RICHTER E., FERREIRA C., RADUN A.V., "Testing and Performances Analysis of a High Speed, 250 kW Switched Reluctance Starter Generator System", *IEEE International Conference on Electrical Machines ICEM'96*, pp. 364-369, September 1996.

[REK 07] REKIK M., Commande et dimensionnement de machines à réluctance variable à double saillance fonctionnant en régime de conduction continue, PhD thesis, Paris-sud XI University, Gif-sur-Yvette, France, May 11, 2007.

[REK 08] REKIK M., BESBES M., MARCHAND C., MULTON B., LOUDOT S., LHOTELLIER D., "High-speed-range enhancement of switched reluctance motor with continuous mode for automotive applications", *European Transactions on Electrical Power*, pp. 674-693, November 2008.

[SAR 81] SARGOS F.M., Etude théorique des performances des machines à réluctance variable, PhD thesis, INPL, Nancy, France, 1981.

[SAV 99] SAVOIE TRANSMISSIONS SOCIETE, Notice Commerciale de la Société "Savoie Transmissions, Filiale du Groupe ABB, 1999.

[SCH 09] SCHOFIELD N., LONG S.A., HOWE D., MCCLELLAND M., "Design of a switched reluctance machine for extended speed operation", *IEEE Transactions on Industry Applications*, vol. 45, no. 1, pp. 116-122, 2009.

[SRD] http://www.srdrives.com.

[STA 93] STATON D.A., MILLER T.J.E., WOOD S.E., "Maximising the saliency of synchronous reluctance motor", *IEE Part. B*, vol. 140, no. 4, pp. 249-259, July 1993.

[TOU 93] TOUNZI A. Contribution à la modélisation et à la commande de machines à réluctance variable. Prise en compte de l'amortissement et de la saturation, thesis, INPL, Nancy, France, 1993.

[VAG 00] VAGATI A., CANOVA A., CHIAMPI M., PASTORELLI M., REPETTO M., "Designrefinement of synchronous reluctance motors through finite-element analysis", *IEEE Transactions on Industrial Applications*, vol. 36, no. 4, pp. 1094-1102, 2000.

[VIS 04] VIŞA C., Commande non linéaire et observateurs: applications à la MRV en grande vitesse, thesis, Metz University, 2004.

[VRI 01] DE VRIES A., BONNASSIEUX Y., GABSI M., D'OLIVEIRA F., PLASSE C., "A switched reluctance machine for a car starter-alternator system", *IEEE International Electric Machines and Drives Conference, IEMDC 2001*, pp. 323-328, 2001.

[WU 02] WU W., LOVATT H.C., DUNLOP J.B., "Optimization of switched reluctance motors for hybridelectic vehicles", *International Conference on Power Electronics, Machines and Drives PEMD'02*, pp. 177-182, June 2002.

[ZHU 07] ZHU Z. Q., HOWE D., "Electrical machines and drives for electric, hybrid, and fuel cell vehicles", *Proceedings of the IEEE*, vol. 95, pp. 746-765, 2007.

Chapter 9

Control of the Stepping Motor

9.1. Introduction

This chapter deals with control of the stepping motor, or rather controls of stepping motors. There are, in fact, different stepping motor technologies possessing particular properties that need adapted supply converters. These different types of motors are represented by specific models. The technology used in these motors and their converters as well as the development of their models are not covered in the framework of this chapter which is devoted to control. This is why, after having briefly indicated the particularities of the different types of stepping motors, the principles of control will be presented and illustrated on the basis of the most common models: those of the permanent magnet motor and the hybrid motor.

9.2. Modeling

9.2.1. *Main technologies*

We usually distinguish three families of stepping motors:

– The motor with variable reluctance exploits the principle of natural evolution towards the minimum magnetic potential energy or maximum flux rule. The torque being a function of the square of the supply current intensity, a unipolar converter that is simple and economical is sufficient. On the other hand, to increase the number of stable positions, we need to increase the number of teeth, which requires precise and costly machining. For this reason, it is not often used.

Chapter written by Bruno ROBERT and Moez FEKI.

330   Control of Non-conventional Synchronous Motors

– The permanent magnet motor implements the interaction force between the magnetic field generated by the stator windings and the magnetic field of a magnet inserted at the rotor. According to the stator technology, simple or double windings, may or may not require a bipolar supply converter. The technology of the rotor influences the inertia moment and thus the dynamic performance. The claw-pole motor, with its cylindrical magnet with radial polarization, is economical but less precise and slower than the motor with a radially magnetized disk, which costs more to produce.

– Finally, the hybrid motor carries an axial magnet on its rotor axis. The magnet polarizes two toothed disks, the north and south disks being shifted by π electrical radians. This configuration allows us to reach higher torques than previously and the angular combinations between the stator teeth and those of the disks increase precision, thanks to the higher number of steps per revolution. These advantages, which are particular to the hybrid structure, mean that this is the type of motor that is most often used [ABI 91].

9.2.2. *The modeling hypotheses*

Stepping motors are linked to synchronous machines whose main structures and models are known [SAR 04]. In this chapter we consider a generic two-phase stepping motor, whose phases in quadrature are indexed α and β. Moreover, the following hypotheses allow us to simplify the model without reducing the understanding of the control principles.

The two main phenomena concerning conductors are:

– the resistance to the electrical current; and

– the skin effect.

At motor functioning frequencies (of a few dozens of Hertz), the skin effect can be disregarded, because the skin is about 1 cm thick. When the motor has a voltage supply, we have to take into account the resistance of the conductors. It is this, in fact, which limits the current in the windings when the motor is in static position on a step.

There are three main defects of the ferromagnetic material in the magnetic circuit. These are:

– its non-zero conductivity, which allows the circulation of eddy currents;

− the existence of a saturation phenomenon and a hysteresis cycle in the magnetic characteristic; and

− a relative magnetic permeability that is not infinite.

The first two defects lead to losses in variable field regime. The third requires a non-zero magnetization current to generate the flux. The effects of the eddy currents and of the hysteresis can be disregarded if the model is not being used for a precise energy study, but rather aims to define the controls.

These hypotheses about the magnetic circuit confer linear properties that allow us to model the use of notions of self-inductance and mutual inductance. To retain their meaning, i.e. to not tend towards infinity, it is however necessary to retain the third defect mentioned, about permeability.

When the magnet exists, it is usual to assume that its functioning point, which is situated near the point with maximum specific energy, evolves under the joint action of current and reluctance variations and that it moves on a backward line. If the slope is sufficiently small, we can assume that the flux is constant. This decreases the harmonic content of the motor torque momentum.

Finally, the mechanical construction allows us to justify other simplifying hypotheses. The hypothesis of sinusoidal spatial distribution of the permeability, in particular, allows the restriction of the harmonic development of inductances. The construction symmetry, as far as it is concerned, is assumed to be sufficient for the mutual inductance fluxes in the two-phase motor to have at saliency of polar pieces of sole origin. Furthermore, as the leakage fluxes are not being involved in electromechanical conversion, the leakage inductances can be disregarded in the model.

To account for the mechanical dynamics, it is necessary to take into account the inertia moment and viscous friction of the motor. The marginal effect of dry friction can often be disregarded. It is true that dry friction shows a nonlinearity of sign type at the change in rotational direction. It is not, however, about an essential nonlinearity, i.e. one of those that naturally confer the stepping function on the motor. Moreover, at usual supply frequencies, inversion in direction does not occur and the effect of this nonlinearity, therefore, has no influence on behavior in this case[1]. It can hence possibly be disregarded in the control model.

---

1 We need to signal that at the highest frequencies the functioning becomes vibratory and the nonlinearity of dry friction type can become perceptible.

9.2.3. *The model*

The dynamics of the motor are described in a four-dimensional state space by the evolution of the state vector $X^t = [i_\alpha, i_\beta, \theta, \Omega]$. i_α and i_β are the currents in the two stator phases. θ and Ω refer to the angular position and angular speed respectively. We can write the electrical angle, θ_e, as: $\theta_e = p\theta$, where p is the number of steps per rotation of the motor[2]. The resistance of a coil is written R.

The inductance matrix is in the form $L = [L_0] + [L_2]$ with:

$$[L_0] = \begin{bmatrix} L_0 & M_0 \\ M_0 & L_0 \end{bmatrix} \text{ and } [L_2] = L_2 \begin{bmatrix} \cos(2\theta_e) & \sin(2\theta_e) \\ \sin(2\theta_e) & -\cos(2\theta_e) \end{bmatrix}. \qquad [9.1]$$

given the hypotheses: $M_0 = 0$.

By making ϕ_{aM} the maximum total flux of the magnet in a coil and $K = p \cdot \phi_{aM}$ the electromechanical constant, the flux vector of the magnet in the phases is:

$$\phi_a = \phi_{aM} \begin{bmatrix} \cos(\theta_e) \\ \sin(\theta_e) \end{bmatrix}. \qquad [9.2]$$

We then obtain the *first two* differential equations of the model inferred from Ohm's and Faraday's laws:

$$\begin{bmatrix} u_\alpha \\ u_\beta \end{bmatrix} = \left[R\mathbb{I}_2 + \frac{d\mathbf{L}}{dt} \right] \begin{bmatrix} i_\alpha \\ i_\beta \end{bmatrix} + \mathbf{L} \begin{bmatrix} \frac{di_\alpha}{dt} \\ \frac{di_\beta}{dt} \end{bmatrix} + K\Omega \begin{bmatrix} -\sin(\theta_e) \\ \cos(\theta_e) \end{bmatrix}. \qquad [9.3]$$

By writing the current vector as $I^t = [i_\alpha \; i_\beta]$ as well as the magnetic coenergy with all the currents switched off as $w_c(\theta)$ [MAT 04], given the linearity hypothesis of the magnetic circuit, the momentum of the torque developed by the motor is expressed by superposition in the following way:

---

[2] According to the structure of the motor, the expression of the electrical angle, which depends on the number of rotor teeth, the number of stator teeth and the number of magnet pole pairs can vary.

$$T = \mathbf{I}^t \frac{\partial \phi_a}{\partial \theta} + \frac{\partial w_c(\theta)}{\partial \theta} + \frac{1}{2} \mathbf{I}^t \frac{\partial \mathbf{L}(\theta)}{\partial \theta} \mathbf{I}.$$ [9.4]

Under the hypotheses given in the previous section, by assuming the flux in the magnet is constant and by restricting the development of inductances and coenergy to the first harmonic, $w_c(\theta) = w_{c0} + w_{cm} \cos(m\theta_e)$, we get:

$$T(\theta, i_\alpha, i_\beta) = \underbrace{-K\left[i_\alpha \sin\theta_e - i_\beta \cos\theta_e\right]}_{\text{Electrodynamic torque}} \underbrace{-K_d \sin(m\theta_e)}_{\text{Cogging torque}}$$

$$\underbrace{+L_2\left[i_\alpha^2 - i_\beta^2\right]\sin(2\theta_e) + 2L_2 i_\alpha i_\beta \cos(2\theta_e)}_{\text{Reluctance torque}}$$ [9.5]

with m being the number of contacts of the stator and K_d the coefficient of the cogging torque.

The torque in the motor has three origins:

– The electrodynamic torque is generated by the interaction between the stator field generated by the currents and the rotor field of the magnet.

– The detent torque comes from the external permeability variation of the magnet according to the position that introduces a flux variation in the absence of current.

– Finally, the reluctance torque originates from the variations of self and mutual inductances during the rotation.

Equation [9.5] develops the most complete expression of the torque in a permanent magnet motor. However, in this type of motor, it is the electrodynamic torque that is dominant, the reluctance torque being insignificant. The reluctance torque is the main torque of the motor with variable reluctance. Finally, the cogging torque is taken advantage of along with the electrodynamic torque, in the hybrid motor. The model of the hybrid motor is therefore very similar to that of a permanent magnet motor.

The reluctance motors, on principle insensitive to the polarity of magnetizing currents, implement more complex electromechanical structures so that it remains possible to control their rotational direction. In multiple stator-stack motors, for instance, we have at least three magnetically independent stators within which a single rotor is inserted. Another solution, that is more economical but does not produce such a high performance in terms of dynamics, is the single stator-stack toothed motor. This has a single rotor but three or four phases. The models of these motors require specific developments that are not directly derived from expression [9.5].

Table 9.1 summarizes the expressions of torque of a few motors and their essential features.

Type of motor	Torque	Advantage
		Drawback
With magnet (with claws), two-phase	$-K\left[i_\alpha \sin\theta_e - i_\beta \cos\theta_e\right]$	Very economical
		High inertia
Hybrid, two-phase	$-K\left[i_\alpha \sin\theta_e - i_\beta \cos\theta_e\right] - K_d \sin(m\theta_e)$	Torque/cost ratio
		Limited speed
With variable reluctance (with multiplestator-stack), three-phase	$-K_r \sum_{k=1}^{3} I_k^2 \sin\left(\theta_e - \frac{2\pi}{3}k\right)$	Torque/inertia ratio
		Cost of mechanical implementation

Table 9.1. *Features of the main types of motors*

To ensure the coherence and clarity of the presentation the principles of control in open loop (section 9.3) and closed loop (section 9.4) will be presented in relation to permanent magnetic motors, namely by putting $L_2 = 0$ and $K_d = 0$ into expression [9.5].

By applying the fundamental principle of dynamics and, by taking into account dry friction, we can determine the *third equation* of the model:

$$J\frac{d\Omega}{dt} + f\Omega + D\,\text{sgn}(\Omega) = -K\left(i_\alpha \sin\theta_e - i_\beta \cos\theta_e\right) - T_c, \qquad [9.6]$$

where J is the inertia moment, f the fluid friction coefficient, D the dry friction and T_c the load torque.

The *fourth equation* simply links speed and position:

$$\Omega = \dot{\theta}.$$ [9.7]

Finally, we obtain the dynamic model of the permanent magnet stepping motor in the form of a nonlinear non-autonomous four-dimensional differential system:

$$\begin{cases} \dot{i}_\alpha = -\frac{R}{L} i_\alpha + \frac{K}{L} \Omega \sin(\theta_e) + \frac{u_\alpha(t)}{L} \\ \dot{i}_\beta = -\frac{R}{L} i_\beta - \frac{K}{L} \Omega \cos(\theta_e) + \frac{u_\beta(t)}{L} \\ \dot{\Omega} = -\frac{K}{J}[i_\alpha \sin(\theta_e) - i_\beta \cos(\theta_e)] - \frac{f\Omega + D\,\text{sgn}(\Omega)}{J} - \frac{T_c}{J} \\ \dot{\theta} = \Omega \end{cases}$$ [9.8]

It is quite common that the dry friction torque associated with the sign function is disregarded in the model. This approximation is all the more justified when the motor is speed controlled without pausing at the equilibrium positions. The dry friction can then possibly be accumulated with the resistant torque. In the neighborhood of the stop at the equilibrium positions or in a vibratory functioning mode at an average speed of 0, the approximation must be discussed according to the relative orders of magnitude of the different torques present.

9.3. Control in open loop

9.3.1. *The types of supply*

As for every electromechanical converter, the safest way to control the dynamics consists of imposing flux via current sources. The cost of producing such a motor, which is linked to the sophistication of the supply converter, depends on their dynamic and energy performances. In this case, with currents being imposed, the dynamic model [9.8] becomes two-dimensional by being restricted to the last two equations.

However, working either at fixed or free chopping frequency, the current supplies always present a limitation in terms of step frequency. This limitation is due to the settling time of the current, which inevitably degrades their functioning in a voltage source functioning for frequencies that are too high. In fact, if the derivatives of currents and rotational speed are too high, the first two equations in [9.8] imply that we have sources with non-realistic voltage at our disposal.

336 Control of Non-conventional Synchronous Motors

The simplest way to control the stepping motors is to use the voltage sources. According to the type and technology of the motor, they can be unipolar or bipolar. The structure of the supply converter also influences the dynamic performance, by ensuring a relatively efficient demagnetization of phases and a relatively quick rise in current. We can infer from [9.8] that the final values of the currents depend only on supply voltages and the coil resistance. If the electrical dynamics is much faster than the mechanical dynamics, the decoupling of the two gives us the option of ignoring the first and considering that the currents are established almost instantaneously, which is then equivalent to controlling the current. In the opposite case, i.e. at high speed, the whole model rules the evolution of the variables, both electrical and mechanical.

The current control is the most effective control. It is well adapted to slow movements but we need to be cautious about the reality as soon as the speed of the motor increases. The voltage control, which is a less effective control, is more economical to produce. This is not a secondary aspect of the problem because stepping motors are specially appreciated for their ability to control position at the lowest cost.

9.3.2. *The supply modes*

Solving singular points of the differential system [9.8], in which we have disregarded dry friction, allows us to express the angular position of the rotor as a function of currents established in the stator coils. The speed being 0, it is hence about equilibrium positions that also depend on the load torque.

$$\begin{cases} I_\alpha = \dfrac{U_\alpha}{R} \\ I_\beta = \dfrac{U_\beta}{R} \\ \Omega = 0 \\ \theta_e = \arctan\dfrac{I_\beta}{I_\alpha} + (-1)^k \arcsin\dfrac{T_c}{K\sqrt{I_\alpha^2 + I_\beta^2}} + (k+1)\pi \end{cases} \quad [9.9]$$

By retaining only the stable equilibrium points, the last equation becomes:

$$\theta_e = \arctan\dfrac{I_\beta}{I_\alpha} - \arcsin\dfrac{T_c}{K\sqrt{I_\alpha^2 + I_\beta^2}} + 2k\pi \quad [9.10]$$

The first term marks the position defined by the control; whereas the second term translates the shifting introduced by the resistant torque with respect to this reference. The intensities of supply currents usually take their values in $\left\{-\sqrt{2}I_n, -I_n, 0, I_n, \sqrt{2}I_n\right\}$, with I_n and $\sqrt{2}I_n$ being fixed by a dual voltage supply in the absence of a current supply.

The supply modes are defined by the sequence of values given to I_α and I_β. Table 9.2 summarizes the equilibrium positions inferred from equation [9.10] by disregarding the load torque. The maximum load torque – also called the holding torque – and the corresponding position are inferred from equation [9.6] at equilibrium.

N°	I_α	I_β	No load equilibrium position θ_{e0}	Holding torque	Position at maximum torque θ_{eM}
1	I_n	0	0	KI_n	$\dfrac{3\pi}{2}$
2	I_n	I_n	$\dfrac{\pi}{4}$	$\sqrt{2}KI_n$	$\dfrac{7\pi}{4}$
3	0	I_n	$\dfrac{\pi}{2}$	KI_n	0
4	$-I_n$	I_n	$\dfrac{3\pi}{4}$	$\sqrt{2}KI_n$	$\dfrac{\pi}{4}$
5	$-I_n$	0	π	KI_n	$\dfrac{\pi}{2}$
6	$-I_n$	$-I_n$	$\dfrac{5\pi}{4}$	$\sqrt{2}KI_n$	$\dfrac{3\pi}{4}$
7	0	$-I_n$	$\dfrac{3\pi}{2}$	KI_n	π
8	I_n	$-I_n$	$\dfrac{7\pi}{4}$	$\sqrt{2}KI_n$	$\dfrac{5\pi}{4}$

Table 9.2. *Sequencing of the stepping motor*

The logic sequencer drives the motor supplies by describing the lines of this table according to an order that defines the four main supply modes:

– Mode 1: only one of two phases is supplied at a time by alternating the polarity. The sequencer describes the odd lines of the table, which leads to the incremental displacement by electrical quarter turn.

– Mode 2: the two phases are always supplied simultaneously, the sequencer only describing the even lines. The incremental displacement remains the same as previously, but the equilibrium positions are intermediate to the previous ones. The obvious advantage of this mode lies in the increase in torque by a factor of $\sqrt{2}$. For this reason, it is the most frequent supply mode.

– Mode 3: in this mode, the half-step advance is done by sequentially describing all the lines of the table. The positioning is more precise but this mode has the drawback of having different static (and dynamic) behaviors from one half-step to the next, the holding torque alternating between two values.

– Mode 4: this mode attempts to remedy the drawback of mode 3, by increasing the intensity of supply current on the odd lines by a factor of $\sqrt{2}$. We thus ally the advantage of mode 2 in terms of torque and that of mode 3 in terms of precision. However, this is only possible at the expense of a bivoltage supply or a current supply with variable reference.

Generalizing the approach introduced by mode 4, it is possible to define a fifth supply mode, referred to as a microstep. In fact, equation [9.10] allows us to establish that for no load the equilibrium position is entirely defined by the intensity ratio of the two supply currents. It is enough to have a supply able to do:

$$I_\alpha = \sqrt{2} I_n \cos\theta_e \quad \text{and} \quad I_\beta = \sqrt{2} I_n \sin\theta_e \qquad [9.11]$$

on a set of discrete values of θ_e, thus theoretically defining as many microsteps as desired with the same performance as in mode 2. However, this approach has several limits that quickly put a strain on economic interest in the stepping motor:

– the precision and stability of the intensities of constant currents, which can lead to the need to resort to linear supplies;

– the high quality of mechanical construction necessary to approach a really sinusoidal spatial evolution of the motor torque;

– the precise knowledge of the load torque, which must not fluctuate otherwise incremental resolution desired loses all sense.

9.3.3. *Case of slow movement*

The previous section has dealt with control by only considering the equilibrium states, their successions and their static properties. In this section, the control of displacements is analyzed within the framework of the hypothesis of quasi-stationarity. As detailed in section 9.3.1, it consists of disregarding the influence of the mechanical dynamics on currents. We consider that the currents have quickly reached their final values, in the case of a voltage supply, or their reference values in the case of a current supply, with the reservations mentioned. The last two equations of model [9.8] then suffice to describe the mechanical dynamics by assuming that the currents are piecewise constant intervals and that the dry friction can be disregarded.

9.3.3.1. *Start*

Relationship [9.6] allows us to establish the maximum load torque that the motor can support at equilibrium. This is the holding torque, which is denoted T_m. Its value depends on the supply mode, as can be seen in Table 1.1. Here, we establish the maximum load torque that will allow the motor to advance by one step when the supplies switch. This is the start torque, which is denoted T_d.

From relationship [9.10], we can infer the equilibrium position before the advance impulse. For instance, on line 1 of Table 1.1:

$$\theta_i = -\arcsin\frac{T_c}{T_m} \;;\; T_m = KI_n \qquad [9.12]$$

Immediately after switching, the motor still being stopped, the start torque is inferred from relationship [9.6]:

$$T_d = -K\left(I_\alpha \sin\theta_i - I_\beta \cos\theta_i\right) - T_c = \sqrt{T_m^2 - T_c^2} - T_c \qquad [9.13]$$

The condition to fulfill being $T_d > 0$, the advance control on one step is only possible if:

$$T_c < T_m \cos\frac{\pi}{4} \;;\; T_d \approx 0.707 T_m \qquad [9.14]$$

340   Control of Non-conventional Synchronous Motors

Similarly, we infer that the advance control on a half-step supports a higher load torque:

$$T_c < T_m \cos\frac{\pi}{8} \; ; \; T_d \approx 0.923 T_m \qquad [9.15]$$

9.3.3.2. *Oscillating response*

To generalize the sentiments, the dynamic equation can be reformulated by using the following dimensionless variables:

– electrical angle: $\theta_e = p \cdot \theta$;

– the load torque scaled with respect to the holding torque: $\Gamma = \dfrac{T_c}{T_m}$;

– the time scaled with respect to the natural period: $\tau = \dfrac{\omega_n t}{2\pi}$.

Moreover, by defining the following parameters:

– the natural pulsation $\omega_n = 2\pi \sqrt{p\dfrac{T_m}{J}}$; and

– the damping coefficient $\xi = \dfrac{f}{2\sqrt{pJT_m}}$;

we obtain:

$$\frac{d^2\theta_e}{d\tau^2} + 2\xi \frac{d\theta_e}{d\tau} + u_\alpha \sin\theta_e - u_\beta \cos\theta_e = -\Gamma \qquad [9.16]$$

Variables u_α and u_β here represent the control states and take their values in $\{-1, 0, 1\}$, in accordance with Table 1.1. Without loss of generality, we consider the advance in the particular position defined by the first line of this table:

$$\frac{d^2\theta_e}{d\tau^2} + 2\xi \frac{d\theta_e}{d\tau} + \sin\theta_e = -\Gamma \qquad [9.17]$$

The damped oscillating response, which can be inferred is a major handicap for quick and precise positioning, and is greater when the motor is not heavily loaded. The expression of the damping coefficient shows the negative influence of inertia on response, which could require the addition of a reducer in order to be attenuated. We can also consider increasing the viscous friction on purpose to increase the damping or we can obtain a certain electrical damping by adding free wheel diodes at the terminals of the coils. However, these solutions considerably reduce the performances of the motor at high speed. The best approach consists of dealing with the damping of oscillations using the control.

9.3.3.3. *Microstep control*

The first approach consists of decreasing the oscillation amplitude by decreasing the steps of the motor torque generated by the current. If we have a current supply at our disposal, we can opt for control in mode 5. Let us assume that this control imposes currents of amplitude I_n with a microstep index n defining $n-1$ intermediate positions, denoted θ_i, between the natural equilibria of the no load rotor. From equations [9.11] and [9.16], we can infer:

$$\frac{d^2\theta_e}{d\tau^2} + 2\xi\frac{d\theta_e}{d\tau} + \sin(\theta_e - \theta_i) = -\Gamma \qquad [9.18]$$

If index n is sufficient, it is possible to linearize the sinus. Furthermore, by translation we can define the angular gap with respect to the equilibrium around the resistant torque: $\delta = \theta_e + \Gamma$. This leads us to a linear equation with constant coefficient:

$$\frac{d^2\delta}{d\tau^2} + 2\xi\frac{d\delta}{d\tau} + \delta = \theta_i \;,\; \delta \in \left[\theta_i - \frac{\pi}{2n}, \theta_i\right], \qquad [9.19]$$

from which we know that the step response shows an overshoot proportional to the gap between the initial and final positions. This gap being inversely proportional to n, we can conclude that microstep control allows us more effectively to dampen the oscillation with a larger number of microsteps.

This approach holds, however, provided that the equilibrium is reached between each microstep, which leads to a relatively slow control.

9.3.3.4. *Bang-Bang control on one step*

This section is about optimizing control to enable the advance of one step, without overshooting and in a minimum period of time. The principle consists of breaking down the advance of one step into two intervals. In the first interval, taking a duration of τ_1, we develop the accelerating torque by applying the control corresponding to the next step. In the second interval, of a duration of τ_2, we perform electrical braking by reapplying the previous control. If the two intervals are properly distributed, the speed cancels out at exactly the time the next equilibrium position is reached. It is then sufficient to reapply the control that corresponds to the position to keep it in equilibrium.

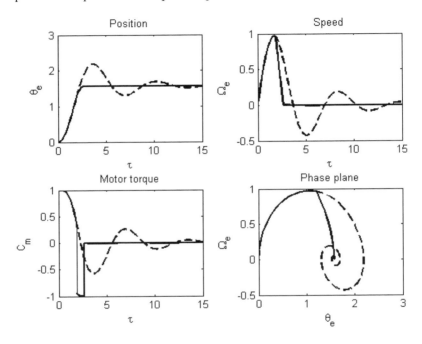

Figure 9.1. *Comparison between the advance by simple switching (dotted line) and Bang-Bang control (unbroken line)*

The dotted line in Figure 9.1 illustrates the no load response on a step by applying mode 1 control on the step going from 0 to $\dfrac{\pi}{2}$. The damping coefficient, ξ, is 0.25. With the unbroken line, we notice the application of a negative torque impulse during the braking phase. The speed very quickly decreases and the phase portrait shows that the motor quickly returns to its equilibrium position. The response time is considerably decreased, as can be seen with the position curves.

The optimal time to reach equilibrium without overshoot does not depend much on the load torque. It is generally in the neighborhood of 3.0. At no load, the ratio of the first and second interval, $\dfrac{\tau_1}{\tau_2}$ is about 2.0. It increases with the load torque.

In practice, the duration of the two intervals can be evaluated in the following way.

By applying an impulse of step advance of variable duration, τ_1, we look for the tangency point of the position with its final value. The graphs in the top line of Figure 9.2 illustrate the results obtained for different durations of advance impulse. When optimal duration τ_1 is determined, we can calibrate the duration of braking impulse to obtain that the motor stops exactly at the equilibrium position, exactly at the end of braking pulse. The graphs on the bottom line in Figure 9.2 illustrates the consequences of breaking that takes too long or is too fast, and the result when duration τ_2 is optimal.

Using the same approach, similar controls have been designed to instigate control in a minimum time in two or three steps [GOE 84].

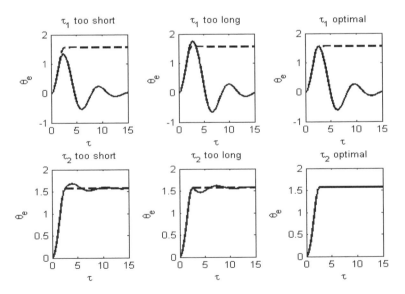

Figure 9.2. *Settings of impulse durations of Bang-Bang control (optimal settings shown in dotted lines)*

9.3.4. *Case of quick movement*

9.3.4.1. *Start–stop zone*

A displacement at high speed begins with an acceleration phase and ends with a deceleration phase. Yet, the most elementary control of a stepping motor involves applying step impulses at a fixed frequency to the sequencer and stopping it by switching off the impulses. In this case, the control frequency is temporally discontinuous whereas, for energy reasons, the speed of the motor is continuous.

The first steps being the slowest, the control goes ahead of the motor. The loss of synchronism can only be avoided when the torques developed in the following steps are sufficient to allow the rotor to catch up with its delay. This phenomenon therefore imposes a maximum start frequency, which depends on total inertia and load torque. It is usually specified in the form of a boundary curve in the torque–frequency plane, given for the inertia of the motor alone.

The motor being totally reversible, the start zone is also a stopping zone as shown in Figure 9.3. For a load torque neighbor of the starting torque, the start frequency is close to zero; whereas the highest start frequency is possible when the motor has no load.

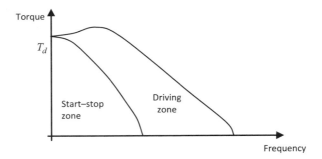

Figure 9.3. *Start–stop zone and driving zone of the motor (given for the inertia of the motor alone)*

9.3.4.2. *Driving zone*

Once the motor is in rotation, it is possible to use the inertia torque to pass more quickly from one step to the other, provided we do not stop at any equilibrium positions. The functioning analysis can then be led by using the notion of mean torque, as we do for any other rotating machine in an established regime.

The instantaneous electrodynamic torque is defined by relationship [9.5] and takes different expressions according to the combinations of currents i_α and i_β. The maximum mean torque is obtained by switching the currents in order to select the torque function of highest value at each time, exactly as we do with voltages in a diode rectifier. Any shift in switching angles with respect to these particular values leads to a decrease in mean torque, and then to the canceling out or even inversion of the torque. Figure 9.4 illustrates the sentiment in mode 1.

We now spot the switching angle that is used to supply a phase with respect to the specific angle that would cancel out the mean torque. We write this as δ. It is similar to the internal angle of a synchronous machine. Mean motor torque $\bar{T}$ is inferred by the following relationship:

$$\bar{T} = -\frac{2\sqrt{2}}{\pi} T_m \sin \delta \qquad [9.20]$$

The maximum mean torque is inferred from this. It is contained between the starting torque and the holding torque:

$$\frac{\sqrt{2}}{2} T_m < \bar{T}_{\max} = \frac{2\sqrt{2}}{\pi} T_m < T_m \qquad [9.21]$$

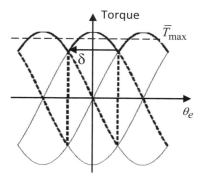

Figure 9.4. *Variation in the mean torque according to the switching angle; with 0 mean torque shown by the dotted line and maximum torque by the bold unbroken line in the upper part of the graph*

On one hand, this analysis of mean torque is valid provided that it really benefits the dynamic smoothing effect brought by inertia, i.e. quite a long way from the start, in the neighborhood of the maximum of the curve presented in Figure 9.3. On the other hand, for higher frequencies, the currents tend to be degraded because the decoupling hypothesis between the mechanical and electrical dynamics is no longer justified. The rise and fall times of currents no longer being insignificant compared to the duration of the step, the mean torque is thus degraded.

In the driving zone, different transients of speed can be put in place to lead the motor from the initial frequency f_i to the desired speed, i.e. to control frequency, f_c. The corresponding frequency transients can be programmed by taking into account the available accelerating (or decelerating) torque. We usually take a safety margin below the boundary curve of the driving zone, as shown in Figure 9.5.

Inertia being the most important phenomenon in driving, the dynamics in this zone can be summarized with:

$$J\frac{d^2\theta}{dt^2} = T - T_c \qquad [9.22]$$

The switching angles can be calculated and programmed so that at each time, during the variations of speed and hence frequency, the motor torque remains below the safety margin. If the programmed movement requires an accelerating torque greater than that available at a given frequency, the motor loses steps.

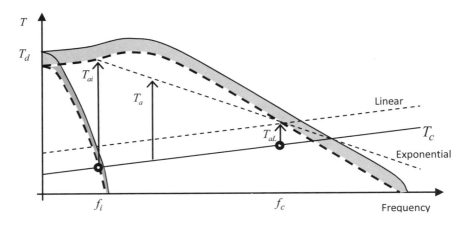

Figure 9.5. *Speed transients in the driving zone*

9.3.4.3. *Linear acceleration*

To obtain a uniformly varied movement, the motor must be controlled so that the accelerating torque is constant. Let T_{aL} be the accelerating torque, θ_i and Ω_i the initial position and speed, and θ_p the angular step. By integration of equation [9.22], we obtain:

$$\theta = \frac{T_{aL}}{2J} t^2 + \Omega_i t + \theta_i \qquad [9.23]$$

On the one hand, the uniformly varied movement is such that: $\Omega = \frac{T_{aL}}{J} t + \Omega_i$. On the other hand, the synchronism warrants: $\Omega = f\theta_p$. The elimination of time in equation [9.23] gives:

$$f^2 - f_i^2 = 2\frac{T_{aL}}{J} \frac{\theta - \theta_i}{\theta_p^2} \qquad [9.24]$$

The accelerating torque that we want to use is the torque available at frequency f_c. We infer the number of steps to program in acceleration phase as well as its duration by using:

$$N_a = \frac{J}{T_{aL}} \frac{\theta_p}{2} \left(f_c^2 - f_i^2 \right) \text{ and } t_a = J\theta_p \frac{f_c - f_i}{T_{aL}} \qquad [9.25]$$

The control of the acceleration can be easily implemented but has the drawback of not exploiting all the accelerating torque available at intermediate frequencies between f_i and f_c.

9.3.4.4. *Exponential acceleration*

For a faster increase in speed, the control must remain close to the maximum torque. The torque line:

$$T_a = T_{ai} - a(\Omega - \Omega_i) \quad ; \quad a = \frac{T_{ai} - T_{aL}}{\theta_p (f_c - f_i)} \qquad [9.26]$$

generally remains close to the maximum torque, as shown in Figure 9.5.

By integration of equation [9.22], we obtain:

348   Control of Non-conventional Synchronous Motors

$$\Omega = \frac{T_{ai}}{a}\left(1 - e^{\frac{-a}{J}t}\right) + \Omega_i \qquad [9.27]$$

$$\theta = \left(\Omega_i + \frac{T_{ai}}{a}\right)t + J\frac{T_{ai}}{a^2}e^{\frac{-a}{J}t} + \theta_i \qquad [9.28]$$

The time constant of the exponential is negative.

The duration of the acceleration phase as well as the number of steps to be programmed are inferred in the following way:

$$t_a = \frac{J\theta_p}{T_{ai} - T_{aL}}(f_c - f_i)\ln\frac{T_{ai}}{T_{aL}} \qquad [9.29]$$

$$N_a = \frac{J}{a\,\theta_p}\left[\left(\Omega_i + \frac{T_{ai}}{a}\right)\ln\frac{T_{ai}}{T_{aL}} + \frac{T_{aL}}{a}\right] \qquad [9.30]$$

9.3.4.5. *Programming of a speed profile*

The previous developments about accelerations can be applied to deal with the linear and exponential decelerations in the same way. In the case of linear deceleration, the motor torque and resistant torque combine their effects and generate braking faster than acceleration. Let us also note that, in the case of exponential deceleration, the time constant of the exponential is positive, which causes very efficient braking.

To carry out a quick displacement between two positions, we separately program a phase of N_a acceleration steps, followed by N_v steps at constant speed to finish with a deceleration phase of N_d steps.

It appears obvious in Figure 9.6 that the exponential controls provide better performance than the linear controls, both in acceleration and braking. The implantation of the exponential control being noticeably more complicated than that of the linear control, however, it is frequently approximated for a control giving linear response by intervals by using a series of constant accelerating torques distributed at an interval of $f_c - f_i$.

Calculation of the series of switching times, or more precisely step duration, can be programmed recurrently.

Let t_i be the step duration: $t_i = \dfrac{1}{f_i}$.

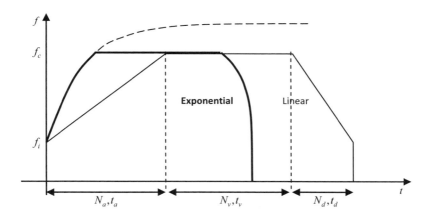

Figure 9.6. *Linear and exponential speed profiles*

Applied to two consecutive steps, formula [9.24] gives:

$$f_{i+1}^2 - f_i^2 = 2\frac{T_{aL}}{J\theta_p} \qquad [9.31]$$

$$t_{i+1} = \frac{t_i}{\sqrt{1 + 2\dfrac{T_{aL}}{J\theta_p} t_i^2}} \qquad [9.32]$$

If the displacement is predetermined, the switching times calculated are tabulated in memory to be used by a low-cost processor of the dsPIC type. If the control must be calculated in real time, the previous expression can be simplified by a development limited to the first order to limit the calculation time:

$$t_{i+1} = t_i\left(1 - \frac{T_{aL}}{J\theta_p} t_i^2\right) \qquad [9.33]$$

T_{aL}, J and θ_p being constants.

We proceed similarly to control the N_d deceleration steps with a braking torque that is negative. During the constant speed phase, however, the control is reduced to: $t_{i+1} = t_i$.

9.4. Controls in closed loop

Control in open loop is the favorite way to control the step motor because it has a very economical implantation that is consistent with the low cost of many stepping motors. As we have seen, however, it is necessary to implement a torque margin to secure functioning by avoiding loss of step. This leads to overdimensioning of the motor. To avoid overdimensioning but also to improve reliability, we resort to control in closed loop [KUO 79].

Controls in closed loop all require the use of an incremental coder, except in the case of voltage supplied motors. In this case, the observation of back electromotive force (emf) is sufficient to design the self-control, provided that the rotational speed is fast enough. Generally, the performances of these controls are largely dependent on those of the coder. Given its price and that of the calculation unit, the total cost of the installation can be doubled. Such coders must therefore be kept for demanding applications and associated with motors that are themselves high performance.

Certain motors integrate an economic coder. In this case, it is not used to design the control, but simply to control the amplitude of movement *a posteriori*.

9.4.1. *Linear models*

The application of a continuous analog control to a process that is essentially discrete remains a topic of discussion.

Is the stepping motor not, by nature, an incremental electromechanical converter? The answer is yes, when we take the time to stop at each equilibrium position. This is less obvious when the motor continuously evolves (i.e. without stopping) from its initial position to its final position. The position and the rotor speed are, in any case, continuous variables.

What about control? It can be synthesized analogically via continuous correctors. It is finally control impulses, however, that are generated in discrete form and transmitted to the sequencer. In fact, these impulses are specified by an angle or by their frequency, so it is clear that their features only mean something in the interval defined by the polar step.

Under this subheading, we present analog controls whose validity is restricted to an intermediate speed range. At very low speed, incremental dynamics are dominant and the notion of continuous control is questionable. At high speeds, electrical dynamics are not insignificant compared to the mechanical dynamics and the simplifications introduced in the control models are no longer valid.

9.4.1.1. *Modeling of the torque*

The two controls that follow are designed from linear dynamic models with the usual restrictions about validity around the functioning points where the parameters are evaluated. The hypothesis common to these models assumes that the currents are very quickly established in the phases, which allows us to abandon the two electrical equations of the initial model [9.8].

The model being established around a mean speed, the instantaneous electrodynamic torque formulated in equation [9.5] can be abandoned for the benefit of mean torque, $\overline{T}$, which is expressed in equation [9.20]. However, it appears that angle δ being orientated from a rotor position not yet reached, it cannot be exploited from the control point of view.

We therefore define a new switching angle:

$$\psi = \pi - \gamma$$

where angle γ, orientated from the switching angle to the detent position of the supplied phase, is referred to as control angle. Figure 9.7 completes Figure 9.4 by including these two new angles.

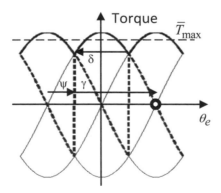

Figure 9.7. *Control angle and switching angle*

The third equation of model [9.8] then becomes:

$$J\dot{\Omega} + f\Omega + D\,\mathrm{sgn}(\Omega) + T_c = \overline{T}(\gamma) \qquad [9.34]$$

In an established regime, this relationship allows us to express the relationship between the speed and switching angle, according to the load torque:

$$f\Omega_0 + D\,\mathrm{sgn}(\Omega_0) + T_{c0} = \overline{T}(\gamma_0) \qquad [9.35]$$

This relationship establishes that with angle control imposing the mean torque, it is the load torque that sets the rotational speed. Reciprocally, when a frequency control sets the speed, the switching angle is able to evolve in order to balance the torque.

To obtain the dynamic model, we consider the speed and angle variations around the functioning point by noting:

$$\Omega = \Omega_0 + \tilde{\Omega} \;;\; \gamma = \gamma_0 + \tilde{\gamma} \;;\; T_c = T_{c0} + \tilde{T}_c \;;\; \overline{T} = \overline{T}_0 + \tilde{\overline{T}}$$

By noticing that for all $\Omega_0 \neq 0$, $\mathrm{sgn}\,\Omega = \mathrm{sgn}\,\Omega_0$, we get[3]:

$$J\dot{\tilde{\Omega}} + f\tilde{\Omega} + \tilde{T}_c = \tilde{\overline{T}}(\tilde{\gamma}) \qquad [9.36]$$

9.4.1.2. *Model for angle control*

Within the framework of restriction at low speeds (hypothesis of piecework constant currents), since the mean torque is dependent on the switching angle alone, its expression can be linearized by differentiation: $\tilde{\overline{T}} = A \cdot \tilde{\gamma}$. The linearization of [9.36] and the Laplace transform lead to the dynamic model:

$$\tilde{\Omega}(s) = \frac{A \cdot \tilde{\gamma}(s)}{Js + f} - \frac{\tilde{T}_c(s)}{Js + f} \qquad [9.37]$$

Consistent with the hypothesis, the dynamics of the speed in response to variations in the switching angle are of a first-order system reduced to the sole mechanical time constant.

---

3 This remark excludes the analysis of functioning in the neighborhood of stable equilibrium, which is out of context for the model in mean torque, but also for vibratory type functioning, i.e. at a mean speed of 0.

At high speed, the settling time of currents being of the same order of magnitude as the mechanical time constant, the expression of mean torque is noticeably more complex. It requires us to calculate the instantaneous fluxes and infer the instantaneous torque before expressing its mean. The method is rigorous and takes into account the electrical transients [GOE 84]. Another possibility consists of restricting the dynamics of currents to the first harmonic. This allows us to analytically integrate the differential system [9.8] with greater ease [HAM 92]. In any case, the mean torque is then dependent on the rotational speed, also since it influences the current transients.

In this case the linearization of equation [9.36] becomes:

$$\tilde{\bar{T}}(\tilde{\gamma},\tilde{\Omega}) = A \cdot \tilde{\gamma} + B \cdot \tilde{\Omega}$$

Model [9.37] is modified in the following way:

$$\tilde{\Omega}(p) = \frac{A \cdot \tilde{\gamma}(s)}{Js + f - B} - \frac{\tilde{T}_c(s)}{Js + f - B} \qquad [9.38]$$

The responses to the perturbations of control angle or load torque remain of the first order; the mechanical time constant depending on the mean rotational speed being:

$$\tau_m = \frac{J}{f - B(\Omega_0)} \qquad [9.39]$$

Since the mean torque decreases with speed, coefficient B is negative. As a consequence, the mechanical time constant tends to decrease at high speed.

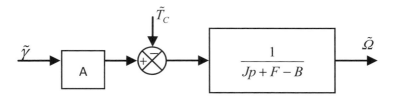

Figure 9.8. *Angle control structure*

Let us remind ourselves that the performance of this model is compromised by:

− the approach of the torque in mean value;

− linearization; and

− processing of the control angle in the same way as a continuous variable, although it only makes sense discretely on a step of the motor.

9.4.1.3. *Model for frequency control*

Even if control of the angle appears to be the most natural, the stepping motor is the most often controlled in frequency in order to set its rotational speed using the frequency of its impulses. A step of the control angle corresponds to a Dirac impulse on frequency.

The control angle is normally defined discretely in an electrical step. To harmonize the nature of this angle with a continuous approach to control, however, it is preferable to redefine the angle by the difference between the electrical angle (continuous variable linked to the mechanical position of the rotor[4]) and the switching angle ψ.

$$\gamma = \theta_e - \psi = p\theta - \psi \qquad [9.40]$$

The mechanical speed is derived from the electrical angle and the switching angle follows the variations in impulse frequency:

$$\tilde{\Omega} = s\frac{\tilde{\theta}_e}{p} \qquad [9.41]$$

$$\tilde{\psi} = \frac{2\pi}{s}\tilde{f}_c \qquad [9.42]$$

From relationships [9.40] to [9.42], we can infer the dynamics of the control angle:

$$\tilde{\gamma} = \tilde{\theta}_e - \tilde{\psi} = \frac{p}{s}\tilde{\Omega} - \frac{2\pi}{s}\tilde{f}_c \qquad [9.43]$$

Figure 9.9 shows how we can naturally complete the structure of frequency control from the control angle by using equation [9.43]. After reorganization, Figure 9.10 presents the block diagram of the frequency control.

---

4 See footnote 2 on page 332.

By noticing that $A < 0$, we get:

$$K_1 = \frac{2\pi}{p} \quad ; \quad K_2 = \frac{-1}{A \cdot p}$$

$$\omega_m = \sqrt{\frac{-A \cdot p}{J}} \quad ; \quad \xi_m = \frac{F - B}{2}\sqrt{\frac{-1}{J \cdot A \cdot p}}$$

[9.44]

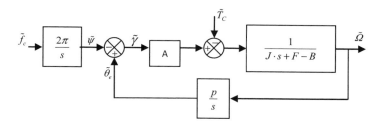

Figure 9.9. *Structure of frequency control*

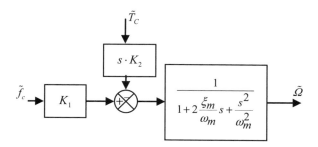

Figure 9.10. *Block diagram for frequency control*

Figure 9.11a shows the damped oscillating response of speed in response to successive steps of the frequency which controls the motor. Figure 9.11b shows that a perturbation of load torque causes a temporary oscillation in instantaneous speed. The control angle evolves naturally and cancels out the speed error.

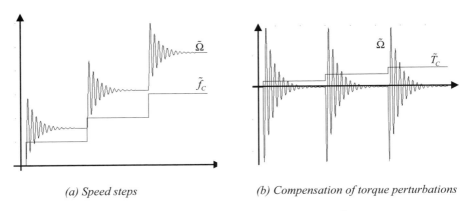

(a) Speed steps   *(b) Compensation of torque perturbations*

Figure 9.11. *Responses of frequency control*

9.4.1.4. *Choice criteria of control*

One of the advantages of the stepping motor being its low cost, the economic aspect is an important criterion in the choice of control. By taking into account the demands of the application, the easiest control is obviously preferred.

For that reason, the choice frequency control appears to be more natural because it is the easiest. This is done economically, in open loop. The second-order model allows us to predetermine the response of the motor to the frequency variations and torque perturbations, around a functioning point. It is therefore well suited to the load displacements whose parameters (mass, inertia, friction, etc.) are known and are constant.

The angle control is more difficult and costly to implement. It requires the addition of an incremental coder whose quality and resolution influence directly the performance of control. Moreover, this control, operated in closed loop, requires precise adjustment of the coder with respect to the detent positions of the motor. We must also foresee a procedure for starting in open loop, since the control impulses are generated from those of the coder. We then understand that angle control is kept for the most demanding applications, which require us to take full advantage of the capabilities of the motor.

Functioning safety is another factor to take into account in the choice of control. As we have seen, frequency control does not show any static error in the speed. When some load parameters vary, however, particularly but not only the torque, the motor can suddenly stall and stop. This can happen due to an overload or due to a simple torque cough whose amplitude prevents the motor from regaining normal functioning. In the case of an angle control that controls the torque, the speed is free to decrease during the cough in order to adapt itself to the characteristic of the motor, then to return to its initial value after the disappearance of the perturbation. This automatic adaptation of speed works until the speed is 0 and to guarantee that the motor does not lose a step. The angle control is hence the favored control from the point of view of running safety.

9.4.2. *Servo-control of speed*

Here we consider the stepping motor whose step frequency is controlled by servo-control of its speed. A servo-control of the rotational frequency is shown in Figure 9.12. It requires a speed sensor or, at least, something to determine the instantaneous rotational frequency that is required prior to the motor being started. Input V_0 sets the reference speed (frequency) of the motor.

Let the frequency transfer function of the motor be:

$$M(s) = \frac{1}{1 + \frac{2\xi_m}{\omega_m}s + \frac{s^2}{\omega_m^2}} \qquad [9.45]$$

The synthesis of the corrector can be carried out by algebraic or other methods for analog or digital implementation. The choices are made taking into account specifications, the cost of control and use of the technology available.

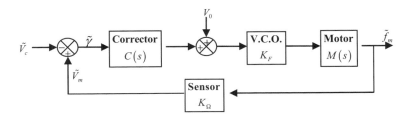

Figure 9.12. *Servo-control of speed*

9.4.2.1. *Development in Taylor series*

A simple approach proposed in [HAM 92] determines the parameters of transfer function $C(s)$ of the corrector, knowing model $M(s)$ of the motor, so that the transfer function of the system in closed loop, $T_S(s)$, is close to a theoretical second-order system, $T_T(s)$. We assume that the voltage controled oscillator (VCO) is modeled by a gain K_F (Hz/V), similarly to the speed sensor (frequency), K_Ω (V/Hz).

The choices of damping coefficient ξ_0 and natural pulsation ω_0 completely define the second order of reference:

$$T_T(s) = \frac{1}{1 + \frac{2\xi_0}{\omega_0}s + \frac{s^2}{\omega_0^2}} \qquad [9.46]$$

A development limited to third order allows us to synthesize a three-parameter corrector such as a PID (propotional integral derivate). It is obtained by polynomial division:

$$T_T(s) = 1 - \frac{2\xi_0}{\omega_0}s + \frac{4\xi_0^2 - 1}{\omega_0^2}s^2 - \frac{4\xi_0}{\omega_0^3}\left(2\xi_0^2 - 1\right)s^3 + \mathcal{O}(s^4) \qquad [9.47]$$

The transfer function of the system in closed loop is given by:

$$T_S(s) = \frac{K_F C(s) M(s)}{1 + K_F K_\Omega C(s) M(s)} = \frac{1 + b_1 s + b_2 s^2 + b_3 s^3 + \mathcal{O}(s^4)}{1 + a_1 s + a_2 s^2 + a_3 s^3 + \mathcal{O}(s^4)} \qquad [9.48]$$

Its limited development is obtained similarly. Namely:

$$T_S(s) = 1 + \alpha_1 s + \alpha_2 s^2 + \alpha_3 s^3 + \mathcal{O}(s^4)$$
$$\begin{cases} \alpha_1 = b_1 - a_1 \\ \alpha_2 = b_2 - a_2 - a_1 \alpha_1 \\ \alpha_3 = b_3 - a_3 - a_2 \alpha_1 - a_1 \alpha_2 \end{cases} \qquad [9.49]$$

The identification from [9.49] to [9.46] gives the parameters of the corrector.

The application of this method to setting classical regulators such as PID is not of any interest to us. It provides good results for developments in series at higher orders and then concerns more sophisticated correctors with more parameters that need to be set.

9.4.2.2. *Naslin polynomial*

Another approach consists of using the Naslin polynomial [HAU 97, NAS 58]. The transfer function of the system in open loop is:

$$H_S(s) = K_F K_\Omega C(s) M(s) = \frac{N(s)}{D(s)} \qquad [9.50]$$

Let the Naslin polynomial be:

$$N(s) = N(s) + D(s) = a_0 + a_1 s + a_2 s^2 + \cdots + a_k s^k \qquad [9.51]$$

Naslin defines the following characteristic ratios:

$$\alpha_1 = \frac{a_1^2}{a_0 a_2} \; ; \; \alpha_2 = \frac{a_2^2}{a_1 a_3} \; ; \; \alpha_k = \frac{a_k^2}{a_{k-1} a_{k+1}} \qquad [9.52]$$

The Naslin polynomial is referred to as *normal* when all the ratios are equal. Each of the ratios depends on the parameters of the corrector. The Naslin criterion sets the equality of as many ratios as the number of parameters of the corrector allow. It is about solving an algebraic system of n equations with n unknowns, if the corrector possesses n setting parameters.

By setting the first equal ratios to 2, we obtain a behavior close to that of a second-order system with a damping coefficient close to 0.7. It remains to check that the ratios of higher ranks, which are not adjustable, are greater than 2. If this is not the case, we can increase the first ratios a bit without going beyond 2.2.

A drawback of this method is the great sensitivity of the response to the value of the first characteristic ratios. For instance, going beyond 2.3 leads to a closed loop system that is over-damped; and below a ratio of 1.7, the damping of oscillations are insufficient.

9.4.2.3. *Correction by phase advance*

A corrector with phase advance is sometimes enough to efficiently damp the oscillations of the speed of the motor.

The transfer function of this corrector is:

$$C(s) = K_{AV} \frac{1 + T_{AV} \cdot s}{1 + \alpha T_{AV} \cdot s} \qquad [9.53]$$

Setting the phase margin in open loop between 50° and 60° is generally satisfactory. Coefficient α is calculated to obtain phase advance φ necessary for this result. Time constant T_{AV} is then calculated to center the action of the corrector on the cutoff pulsation ω_c of the transfer function in open loop before correction. We set the gain K_{AV} so that the value of the cutoff frequency of the closed loop transfer function after correction is the same as the value of the cutoff frequency of the closed loop transfer function before correction. This leads to the following relationships:

$$\alpha = \frac{1-\sin\varphi}{1+\sin\varphi} \quad ; \quad T_{AV} = \frac{1}{\omega_c \sqrt{\alpha}} \quad ; \quad K_{AV} = \sqrt{\alpha} \qquad [9.54]$$

9.4.2.4. *PID correction*

A PID industrial corrector can also be used. The integral term added by this corrector allows us to cancel out the static error that the corrector with phase advance has the tendency to induce. The transfer function of this corrector is:

$$C(s) = K_P \frac{1+T_i \cdot s}{T_i \cdot s} \frac{1+T_d \cdot s}{1+\alpha T_d \cdot s} \qquad [9.55]$$

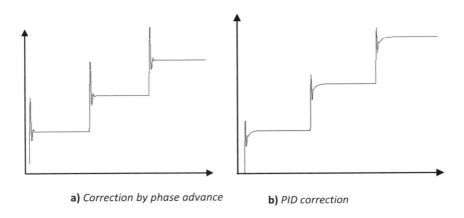

a) *Correction by phase advance* **b)** *PID correction*

Figure 9.13. *Speed servo-controls*

This corrector can be set empirically. The time constant of the derived term T_d can be set in the same way as previously described. The integral time constant is set so that it does not excessively degrade the response time and maintains it in the same order of magnitude as previously. The proportional gain is then adjusted to act on the overshoot. From this initial setting, it is possible to more finely adjust these three parameters to refine the result.

9.5. Advanced control: the control of chaos

9.5.1. *Chaotic behavior*

As has been indicated above, the most natural control of the stepping motor is control in open loop. It has been shown in [ROB 00] and [PER 00], however, that performances at high speed are considerably degraded. It has been established in [REI 03] that functioning modes that are not only quasi-periodic but also chaotic appear when the supply frequency enters the instability zone.

The control of chaos consists of stabilizing one of the unstable periodic orbits that are embedded in infinite number in the chaotic attractor. The trajectory describing the dynamics in state space must be kept in this orbit when a system parameter is varied. In the case of the stepping motor, controling the chaos means forcing of the electrical angle to produce periodic behavior, i.e. to impose a rotational speed that remains synchronous with the supply frequency.

The control of chaos has aroused a lot of interest since the seminal works presented in [OTT 90]. Different control strategies have been proposed since [BAS 97, BLE 96, PYR 92, PYR 95]. The synthesis of the controller can be implemented by a linear [BAS 98] or a nonlinear approach [FEK 03, YAN 02].

In the next section, we present a method allowing us to control the chaotic behaviors of the stepping motor in the instability zone at high speed. By using the theory of absolute stability [KHA 92], it is possible to synthesize a linear controller capable of stabilizing one of the periodic orbits initially embedded in the chaotic attractor over a large supply frequency range. The motor is supplied by voltage, in mode two, and the control variables are the supply voltages. We therefore have control in closed loop requiring the control of the output voltage of the supply choppers. This control remains relatively economical because it does not require any position/speed sensor but only current sensors to measure the instantaneous intensities of the supply currents.

9.5.2. *The model*

The great instability of simulations in chaotic mode, owing to the sensitivity to initial conditions of the dynamic system, requires us, among other things, to conform with model [9.8] in order to solve certain numerical problems [ALI 02]. Basically, it is about using dimensionless variables as well as a dimensionless temporal variable with respect to the control period. Moreover, to improve the generality with respect to model [9.8], here we use a model taking into account the cogging torque, $K_d \neq 0$.

Let the generic model of the hybrid model be:

$$\begin{aligned}
\frac{dI'_\alpha}{dt'} &= \mu I'_\alpha + \rho\Omega' \sin(\theta_e) + U'_\alpha(t') \\
\frac{dI'_\beta}{dt'} &= \mu I'_\beta - \rho\Omega' \cos(\theta_e) + U'_\beta(t') \\
\frac{d\Omega'}{dt'} &= -\gamma I'_\alpha \sin(\theta_e) + \gamma I'_\beta \cos(\theta_e) + \lambda \sin(4\theta_e) + \varphi\Omega' \\
\frac{d\theta_e}{dt'} &= \Omega'
\end{aligned} \qquad [9.56]$$

The rectangular supply voltages are defined as follows:

$$U_\alpha = E \operatorname{sgn}\left[\cos(2\pi f_c t)\right] \;;\; U_\beta = E \operatorname{sgn}\left[\sin(2\pi f_c t)\right]$$
$$U'_\alpha = \operatorname{sgn}\left[\cos(\theta_e)\right] \;;\; U'_\beta = \operatorname{sgn}\left[\sin(\theta_e)\right]$$

The dimensionless state variables are:

$$t' = f_c \cdot t \;;\; I'_\alpha = \frac{Lf_c}{E} I_\alpha \;;\; I'_\beta = \frac{Lf_c}{E} I_\beta \;;\; \theta_e = p\theta \;;\; \Omega' = \frac{\Omega}{f_c}$$

The parameters are expressed in the following way:

$$\mu = -\frac{R}{Lf_c} \;;\; \rho = \frac{Kf_c}{pE} \;;\; \gamma = \frac{pKE}{JLf_c^3} \;;\; \lambda = -\frac{pK_d}{Jf_c^2} \;;\; \varphi = -\frac{F}{Jf_c}$$

By writing the state vector as $\xi = (I'_\alpha, I'_\beta, \Omega', \theta_e)$, the input vector as $U = (U'_\alpha, U'_\beta)$ and the output as Y, system [9.55] can be put in the form:

$$\begin{aligned}
\dot{\xi} &= f(\xi) + B_1 U \\
Y &= C_1 \xi
\end{aligned} \qquad [9.57]$$

in which B_1 is the injection matrix and C_1 is the output matrix:

$$f(\xi) = \begin{pmatrix} \mu I'_\alpha + \rho \Omega' \sin(\theta_e) \\ \mu I'_\beta - \rho \Omega' \cos(\theta_e) \\ -\gamma I'_\alpha \sin(\theta_e) + \gamma I'_\beta \cos(\theta_e) + \lambda \sin(4\theta_e) + \varphi \Omega' \\ \Omega' \end{pmatrix}$$

$$B_1 = \begin{bmatrix} 1 & 0 \\ 0 & 1 \\ 0 & 0 \\ 0 & 0 \end{bmatrix}, \quad C_1 = \begin{bmatrix} 1 & 0 & 0 & 0 \\ 0 & 1 & 0 & 0 \end{bmatrix}.$$

9.5.3. *Orbit stabilization*

The goal here is not to stabilize the motor at an equilibrium point but to stabilize it along an orbit of period $T = 1$, in scaled time. We can extend the notion of equilibrium to the periodic dynamics in the following way. Let us first notice that, if the natural dynamics of the motor at control frequency are really chaotic, there is always at least one unstable orbit, $\xi_p(t)$, of period-one embedded in the attractor. The stability of this orbit can be analyzed in the Lyapunov sense.

We define a new state vector $x(t) = \xi(t) - \xi_p(t)$, so that the origin $x = 0$ is an equilibrium point of the following system:

$$\dot{x} = f\left(x + \xi_p(t)\right) - f\left(\xi_p(t)\right) \quad [9.58]$$

The behavior of the solutions to equation [9.57] close to $\xi_p(t)$ is equivalent to that of solutions to equation [9.58] close to $x = 0$. As a consequence, the stability of $\xi_p(t)$ can be similarly characterized from the stability of equilibrium, $x = 0$. In particular, the periodic solution $\xi_p(t)$ of equation [9.57] is uniformly asymptotically stable if $x = 0$ is an equilibrium point of equation [9.58] that is uniformly asymptotically stable.

The stability issue can be dealt with by linearizing equation [9.57] around $x = 0$ and adding an excitation input to stabilize the periodical orbit.

$$\dot{x} = \frac{\partial f\left(x+\xi_p(t)\right)}{\partial x}\bigg|_{x=0} \cdot x + B_1 u \qquad [9.59]$$

The partial derivative with respect to the state can be broken down into two parts: a linear invariant part (linear time invariant (LTI)) called A, expressed in equation [9.60] and a linear variable part (linear time variable (LTV)), called A_v.

$$A = \begin{bmatrix} \mu & 0 & 0 & 0 \\ 0 & \mu & 0 & 0 \\ 0 & \gamma & \varphi & 4\lambda \\ 0 & 0 & 1 & 0 \end{bmatrix} \quad ; \quad A_v = \frac{\partial f\left(x+\xi_p(t)\right)}{\partial x}\bigg|_{x=0} - A \qquad [9.60]$$

The reformulation is done by defining a matrix $\Delta_p(t)$ which is a diagonal of rank 7 whose elements are the non-zero elements of A_v:

$$A_v = B\Delta_p(t)C \qquad [9.61]$$

$$B = \begin{bmatrix} 1 & 1 & 0 & 0 & 0 & 0 & 0 \\ 0 & 0 & 1 & 1 & 0 & 0 & 0 \\ 0 & 0 & 0 & 0 & 1 & 1 & 1 \\ 0 & 0 & 0 & 0 & 0 & 0 & 0 \end{bmatrix} \quad ; \quad C = \begin{bmatrix} 0 & 0 & 1 & 0 \\ 0 & 0 & 0 & 1 \\ 0 & 0 & 1 & 0 \\ 0 & 0 & 0 & 1 \\ 1 & 0 & 0 & 0 \\ 0 & 1 & 0 & 0 \\ 0 & 0 & 0 & 1 \end{bmatrix}$$

System [9.59] is now written in the following form:

$$\begin{aligned} \dot{x} &= Ax + B_1 u + B\psi(t,z) \\ y &= C_1 x \\ z &= Cx \\ \psi(t,z) &= \Delta_p(t)z \end{aligned} \qquad [9.62]$$

Figure 9.14 represents the linearized system whose transfer function, $G(s)$, is expressed in the following way:

$$G(s) = \begin{bmatrix} C \\ C_1 \end{bmatrix}(sI - A)^{-1}[B \mid B_1] = \begin{bmatrix} G_{zv}(S) & G_{zu}(S) \\ G_{yv}(S) & G_{yu}(S) \end{bmatrix} \quad [9.63]$$

The goal is to synthesize a linear controller admitting -y as input and u as output, as shown in Figure 9.14.

The transfer function between z and v in closed loop is given by:

$$H_{zv}(s) = G_{zv}(s) + G_{zu}(s)\left(I + K(s)G_{yu}(s)\right)^{-1} K(s) G_{yv}(s) \quad [9.64]$$

We need to notice that $\psi(t,z)$ is a class of nonlinearities dependent on time and without memory effect. It depends on the periodical orbit, and hence on the supply frequency. The analysis of the stability of a system interconnecting a linear part with a nonlinear function of this type can be led by means of the theory of absolute stability.

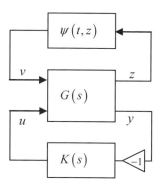

Figure 9.14. *Block diagram of linear control dissociated from the nonlinearities*

9.5.4. *Absolute stability*

Here we remind ourselves briefly of the principal elements of the theory of absolute stability.

DEFINITION 9.1.– (sector condition)

A memoryless nonlinearity $\psi : [0,\infty) \times \mathbb{R}^p \to \mathbb{R}^p$ is said to satisfy a sector condition if:

$$\forall t \geq 0, \forall z \in \Gamma \subset \mathbb{R}^p, \left[\psi(t,z) - K_{\min} z\right]^T \left[\psi(t,z) - K_{\max} z\right] \leq 0 \qquad [9.65]$$

For some real matrices, $K_{\min}$ and $K_{\max}$, so that $K_{\sec} = K_{\max} - K_{\min}$ is a symmetric positive definite matrix and the interior of Γ is connected and contains the origin. If $\Gamma = \mathbb{R}^p$, then $\psi(\cdot,\cdot)$ globally fulfills the sector condition, in which case it is said that $\psi(\cdot,\cdot)$ belongs to a sector $[K_{\min}, K_{\max}]$.

DEFINITION 9.2.– (absolute stability)

Consider the nonlinear system where $\psi(\cdot,\cdot)$ satisfies the sector condition. The system is absolutely stable if the origin is globally, uniformly, asymptotically stable for every nonlinearity belonging to the given sector.

THEOREM 9.1.– consider the nonlinear system described above, where triplet $\{A,B,C\}$ is controllable and observable and $\psi(\cdot,\cdot)$ globally fulfills the sector condition. The system is absolutely stable if:

$$\tilde{H}_{sv}(s) = \left[I - H_{sv}(s) K_{\min}\right]^{-1} H_{sv}(s) \qquad [9.66]$$

is Hurwitz and:

$$Z(s) = I - K_{\sec} \tilde{H}_{sv}(s) = \left[I - K_{\max} \tilde{H}_{sv}(s)\right]\left[I - K_{\min} \tilde{H}_{sv}(s)\right]^{-1} \qquad [9.67]$$

is strictly positive-real.

REMARK 9.1.– if the system is absolutely stable, then it suffices to analyze the linear system $H_{zv}(s)$ to comprehend the behavior of the nonlinear system. This is interesting because the unknown periodic orbits are not needed for the controller design.

9.5.5. *Synthesis of the controller*

By applying sector condition [9.65] to nonlinearities of the motor, we get:

$$\begin{aligned}\left[\psi(t,z) - K_{\min} z\right]^T \left[\psi(t,z) - K_{\max} z\right] &\leq 0 \\ z^T \left[\Delta_p(t,z) - K_{\min}\right]^T \left[\Delta_p(t,z) - K_{\max}\right] z &\leq 0\end{aligned} \qquad [9.68]$$

We can choose: $K_{\min} < \inf_{t>0} \Delta_p(t,z)$; $K_{\max} > \sup_{t>0} \Delta_p(t,z)$

Among other chaos properties, we need to remind ourselves that a strange attractor is bounded. As a consequence, whether the motor dynamics are periodical or chaotic, the excursion of variable is limited in the state space. We infer that $K_{\min}$ and $K_{\max}$ exist and are bounded. The sector condition is hence globally fulfilled in the state space of the motor.

Without loss of generality [FEK 05], we put $K_{min} = 0$. From equations [9.66] and [9.67], we obtain:

$$\tilde{H}_{zv}(s) = H_{zv}(s) \text{ and } Z(s) = I - K_{\max} H_{zv}(s) \qquad [9.69]$$

We notice, from equation [9.64], that the synthesis of $K(s)$ from a general constraint on $H_{zv}(s)$ will be very difficult. To simplify things, we can use a parameterized form of $K(s)$. From equation [9.64], by putting:

$$Q(s) = \left(I + K(s)G_{yu}(s)\right)^{-1} K(s), \qquad [9.70]$$

we obtain:

$$H_{zv}(s) = G_{zv}(s) + G_{zu}(s)Q(s)G_{yv}(s). \qquad [9.71]$$

All calculations done, we get:

$$K(s) = \left[I - G_{yu}(s)Q(s)\right]^{-1} Q(s) \qquad [9.72]$$

Transfer function, $Q(s)$, is a matrix of the following form:

$$Q(s) = \begin{bmatrix} Q_{11}(s) & Q_{12}(s) \\ Q_{21}(s) & Q_{22}(s) \end{bmatrix} \qquad [9.73]$$

The Pyraguas control principle consists of not creating new orbits or modifying the pre-existing orbits, but in stabilizing one of them. As a consequence, transfer matrix $K(s)$ must be:

$$K(jn\omega_c) = 0 \quad ; \quad \omega_c = 2\pi f_c \qquad [9.74]$$

With this goal in mind, elements $Q_{ii}(s)$ are in the form $Q_{ii}(s) = Q_0(s)\hat{Q}(s)$, where:

$$Q_0(s) = \frac{s}{1+\tau s} \prod_{n=1}^{N_0} \frac{1+s^2\left(\dfrac{n}{\omega_c}\right)^2}{(1+\tau s)^2} \qquad [9.75]$$

and τ is a positive constant usually chosen to be equal to $1/\omega_c$ and $\hat{Q}(s)$ is determined to satisfy conditions [9.65] and [9.66] imposed by the theorem. In the case of the stepping motor to control, the development of $Q_0(s)$ can be limited to $N_0 = 1$ without compromising the performances of the controller. Parameters q_0, q_1 and q_2 are dependent on the control frequency.

$$\hat{Q}(s) = q_0(\omega_c) + \frac{q_1(\omega_c)}{1+\tau s} + \frac{q_2(\omega_c)}{(1+\tau s)^2} \qquad [9.76]$$

9.5.6. *Examples*

The first example presents the stabilization of a periodical orbit in a zone where the functioning mode is quasiperiodic in the absence of control. It is not about chaos control but, with clarity in mind, it is preferable to begin by illustrating the action of the controller on dynamics that are easier to represent than chaotic dynamics.

The trajectory of the state vector should be represented in a four-dimensional space, which is obviously not possible. To ease the interpretation of the observed curve, we represent the projection of this trajectory in the plane of supply currents (i_α, i_β). Figure 9.15 shows, in fine line, the projection of the trajectory in the absence of control. The shape of the (open) curve wrapping at the surface of a hypertorus is characteristic of the quasiperiodic dynamics [ROB 00]. By synthesizing the controller as indicated in section 9.5.5, without knowing it *a priori*, we stabilize the naturally unstable periodical trajectory represented in a bold line.

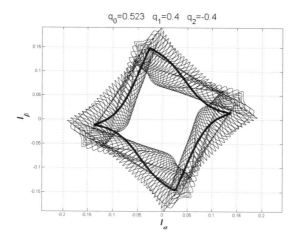

Figure 9.15. *Periodic stabilization of a quasiperiodic regime*

The second example presents the case of chaotic behavior. The trajectory of a chaotic dynamics converges towards a geometrical object of fractal nature, referred to as a *strange attractor*. The extreme (geometrical) complexity of these attractors does not allow us to directly represent the trajectory or even what the project is. In this case we resort to a *Poincaré section* and, more particularly, in the case of an excited driven system with fixed period (supply frequency), to a *stroboscopic* Poincaré section. With this mode of representation, it is obvious that a periodical functioning at the driven frequency will be reduced to a point in the plane (and a finite number of points in the case of sub-harmonic regimes).

The graph in Figure 9.16 is achieved by sampling the currents once per supply period. The cloud of 20,000 points sketches the strange attractor, giving us a glimpse of the self-similarity and the folding that give it the laminated aspect characteristic of this type of attractor. Figure 9.16b is a close-up of the portion of Figure 9.16a that is surrounded by a bex. These visual considerations must be confirmed by the estimation of a certain number of geometrical and dynamic invariants that allow us to more formally characterize the chaotic dynamics [CAS 10a, CAS 10b]. The target placed on the figure spots the location of the unique point corresponding to the section of the stabilized periodical trajectory.

370   Control of Non-conventional Synchronous Motors

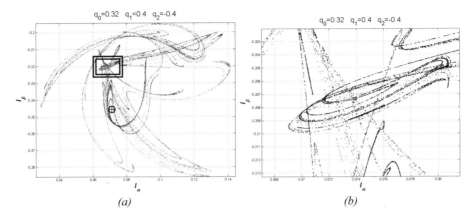

Figure 9.16. *(a) Periodical stabilization of a chaotic regime; and (b) close up of part of the chaotic regime*

Another way to visualize the action of the control consists of representing the waveform of one of the two currents before and after the application of control. This is shown in Figure 9.17a.

For each value of the supply frequency, the three parameters of the corrector (q_0, q_1, q_2) are to be determined. Parameter q_1 is constant throughout the whole domain studied. Figure 9.17b represents the evolutions of the two other parameters.

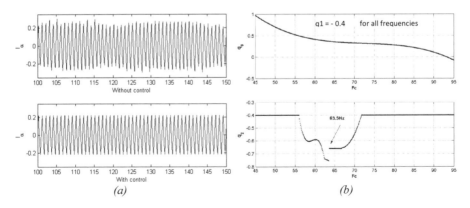

Figure 9.17. *(a) Waveforms of a current with and without control; and (b) corrector parameters*

To finish, Figure 9.18 shows the superposition of the Feigenbaum (improper bifurcation) diagrams before the application of controls and after. For each value of supply frequency (as abscissa), the sampled values of one of the currents are reported to the vertical of the abscissa. All aperiodical modes functioning takes on the appearance of a line segment densely populated with points. Only the Poincaré sections and the analysis of invariants allow us to distinguish chaotic regimes from quasiperiodic regimes. The bold line shows us the superimposition of the Feigenbaum diagram with control application. For each abscissa (except for one) a single point is depicted on the vertical. This means that to within an exception, the controller allows us to stabilize a periodical orbit across the whole functioning domain. The shift between the two diagrams, in the naturally periodical zones, is due to the development of $Q_0(s)$, which has been limited to $N_0 = 1$.

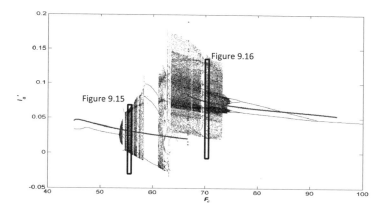

Figure 9.18. *Periodical stabilization of a mode: (a) quasiperiodic; and (b) chaotic*

9.6. Bibliography

[ABI 91] ABIGNOLI M., GOELDEL C., "Moteurs pas à pas", *Techniques de l'Ingénieur*, Génie Electrique, D3 III, D3690-1 – D3690-21, Techniques de l'Ingénieur, Paris, 1991.

[ALI 02] ALIN F., ROBERT B., GOELDEL C., "On the limits of chaotic simulations by classic software – Application to the step motor", *Proceedings of IEEE International Conference on Industrial Technology*, CD-ROM, Bangkok, Thailand, December 11-14, 2002.

[BAR 94] BARBOT J.P., BIC J-C., GOLLREITER R., GRIGAT M., KADEL G., LEVY A.J., LORENZ R., MOHR W., STRASSER G., WALBERER A., "Channel modelling for advanced TDMA mobile access", *Proceedings RACE MPLA Workshop*, Amsterdam, pp. 683-687, May 1994.

[BAS 97] BASSO M., GENESIO R., TESI A., "Stabilizing priodic orbits of forced systems via generalized pyragas controllers", *IEEE Trans. Circuits Systems I*, vol. 44, no. 10, pp. 1023–1027, 1997.

[BAS 98] BASSO M., GENESIO R., GIOVANARDI, L. TESI, A. TORRINI G., "On optimal stabilization of periodic orbits via time delayed feedback control", *Int. J. Bifurcation Chaos*, vol. 8, no. 8, pp. 1699–1706, 1998.

[BER 95] BERG J.-E., "A recursive method for street microcell path loss calculations", *PIMRC'95*, Toronto, Canada, pp. 140-143, September 1995.

[BLE 96] BLEICH M.E., SOCOLAR J.E.S, "Stability of periodic orbits controlled by time-delay feedback", *Physics Letters A*, vol. 210, pp. 87–94, 1996.

[CAS 10a] DE CASTRO M.R., ROBERT B.G.M., GOELDEL C., "Experimental chaos and fractals in a linear switched reluctance motor", *EPE-PEM Conference*, Ohrid, Republic of Macedonia, September 6-8, 2010.

[CAS 10b] DE CASTRO M.R., ROBERT B.G.M., GOELDEL C., "Analysis of aperiodic and chaotic motions in a switched reluctance linear motor", *Proceeding of IEEE International Symposium on Circuits and Systems, ISCAS Conference*, Paris, France, May 30–June 2, 2010.

[FEK 03] FEKI M., "An adaptive feedback control of linearizable chaotic systems", *Chaos, Solitons& Fractals*, vol. 15, pp. 883–890, 2003.

[FEK 05] FEKI M., ROBERT B., ALIN F., GOELDEL C., "Chaotic ehavior of the stepper motor", *Proceedings of the Electrimacs Conference*, Hammamet, Tunisia, April 17-20, 2005.

[GOE 84] GOELDEL C., Contribution à la modélisation, à l'alimentation et la commande de moteurs pas à pas, PhD thesis, Nancy, p. 173-179, 1984.

[HAU 97] J.-P. HAUTIER, J.-P. CARON, *Systèmes Automatiques, Tome 2: Commande des Processus*, Ellipses, 1997.

[HAM 92] HAMZAOUI A., Modèles dynamiques et commandes en boucle fermée d'un moteur pas à pas, PhD thesis, Reims, 1992.

[KHA 92] KHALIL H.-K., *Nonlinear Systems*, Macmillan, New York, 1992.

[KUO 79] KUO B.C., *Step Motors and Control System*, SRL Publishing, 1979.

[LOU 04a] LOUIS J-P. (ed.), *Modélisation des Machines Électriques en vue de Leur Commande*, Hermes-Lavoisier, 2004.

[LOU 04b] LOUIS J-P., *Modèles pour la Commande des Actionneurs Électriques*, Hermes-Lavoisier, 2004.

[MAT 04] MATAGNE E., DA SILVA GARRIDO M., "Conversion électromécanique d'énergie : du phénomène physique à la modélisation dynamique",in: LOUIS J-.P. (ed.), *Modélisation des Machines Électriques en vue de Leur Commande*, Hermes-Lavoisier, 2004.

[NAS 58] NASLIN P., *Technologie et Calcul Pratique des Systèmes Asservis*, Dunod, 1958.

[OTT 90] OTT E., GREBOGI C., YORKE J.A., "Controlling chaos", *Physical Review Letters*, vol. 64, pp. 1196–1199, 1990.

[PER 00] PERA M-C., ROBERT B., GOELDEL C., "Nonlinear dynamics in electromechanical systems-application to a hybrid stepping motor", *Electromotion*, vol. 7, pp. 31–42, 2000.

[PYR 92] PYRAGAS K., "Continuous control of chaos by self-controlling feedback", *Physics Letters A*, vol. 170, pp. 421–428, 1992.

[PYR 95] PYRAGAS K., "Control of chaos via extended delay feedback", *Physics Letters A*, vol. 206, pp. 323–330, 1995.

[REI 03] REISS J., ROBERT B., ALIN F., SANDLER M., "Flip phenomena and co-existing attractors in an incremental actuator", *Proceedings of IEEE International Conference on Industrial Technology*, CD-ROM Maribor, Slovenia, December 9-12, 2003.

[ROB 00] ROBERT B., PERA M.-C., GOELDEL C., "Dynamiques apériodiques et chaotiques du moteur pas à pas", *Revue Internationale de Génie Electrique*, vol. 3, pp. 375–410, 2000.

[SAR 04] SARGOS F-M., MEIBODY-TABAR F., "Modèles dynamiques des machines synchrones", in: LOUIS J-P. (ed.), *Modélisation des Machines Électriques en vue de Leur Commande*, Hermes-Lavoisier, 2004.

[YAN 02] YANG S-K., CHEN C-L., YAU H-T., "Control of chaos in Lorenz system", *Chaos, Solitons & Fractals*, vol. 13, pp. 767–780, 2002.

Chapter 10

Control of Piezoelectric Actuators

10.1. Introduction

10.1.1. *Traveling wave ultrasonic motors: technology and usage*

Traveling wave ultrasonic motors make use of the mechanical vibration of a stator to drive – by friction – a rotor, strongly, which is held against the stator [SAS 93]. The vibration is obtained by the excitation of small piezoelectric elements which deform under the action of a transverse electric field. Figure 10.1 shows a diagram of the piezoelectric motor on which the deformations of the stator are exaggerated.

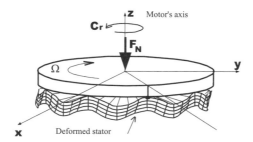

Figure 10.1. *Principle of functioning of a traveling wave ultrasonic motor*

Traveling wave piezoelectric motors are considered to be 2-phase motors which are supplied with two alternating sinusoidal voltages and are referred to as v_α and v_β.

Chapter written by Frédéric GIRAUD and Betty LEMAIRE-SEMAIL.

Under the inverse piezoelectric effect, the electric field generates in the exciters' deformations; for each voltage considered independently, a standing deformation wave is established which is combined with the other to propagate a traveling wave. It is this traveling wave which seems to rotate around the axis of the motor that allows the generation of torque.

Thus, contrary to electromagnetic motors, the stresses which are the cause of the electromechanical conversion are not generated remotely, via an air-gap, but at the heart of the matter. Besides, even if the stator vibrates at a frequency close to its resonating frequency, the amplitudes of the deformations generated based on the piezoelectric effect remain small, typically in the micrometer range.

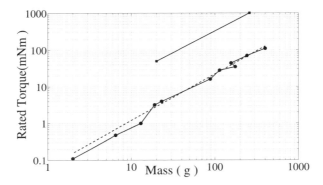

Figure 10.2. *Comparison of nominal torques of motors according to their mass, for brushed DC motors of small power (Faulhaber catalog), and traveling wave piezoelectric motors (Shinsei catalog)*

In order to obtain commercial motors, another conversion is added to this electromechanical conversion – mechanomechanical this time – to transform these small vibrating displacements either in rotation or in linear movement of large amplitude [NOG 96]. The piezoelectric motors generally possess advantages which make them attractive for the motorization of kinematic chains which is more or less complex, like robotic actuators for instance. In fact, these motors have naturally strong torque/low speed. Thus, it is often pointless to associate them with a speed reducer which allows us to save space or mass. We have shown an illustration in Figure 10.2 regarding the evolution of the nominal torque supplied with a range of small DC motors, compared to the range of traveling wave ultrasonic motors. This figure shows that a factor of 10 on the torque is obtained for two motors of identical mass and different technologies, or that at equivalent mass, the traveling wave piezoelectric motors develop 10 times more torque than their DC counterparts. Even if this comparison does not take into account the electrical supply of these motors, it does however lets us make out the potential savings in the design of actuated mechanisms.

However, if the use of traveling wave ultrasonic motors is attractive, despite the fact that it seems to be possible to restrict the bulk or the mass of the actuator in a kinematic chain, the user can be faced with functioning features which are much less easy than that of traditional electromagnetic motors, which are described in the following section.

10.1.2. *Functioning features*

The piezoelectric motor is supplied with two alternating and sinusoidal voltages. These voltages are generally dephased at an angle of $\pm \frac{\pi}{2}$ according to the selected rotation direction. Let us note that these motors possess a sensor which allows us to know the deformation of the stator at a precise point by generating a voltage which is proportional to it. Figure 10.3 defines the effective voltages of the piezoelectric motor.

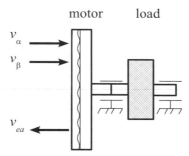

Figure 10.3. *Supplied voltages v_α and v_β of the traveling wave ultrasonic motor and v_{ea}, the measured voltage of the vibrating amplitude*

From a mechanical point of view, the output features of the traveling wave ultrasonic motor show the drawback of a strong dependence of the motor torque on the rotational speed, all other things being equal. According to the speed, these torque features, as shown in Figure 10.4, come from the contact conditions between stator and rotor, and are nonlinear. In addition, they can degrade in time, due to wear, or vary from one motor to another.

In addition, providing the resonating feature of the stator assembly, the vibration amplitude as shown in Figure 10.1 strongly depends on the frequency of the supplied voltages. As a consequence, the no load speed (for zero torque) which depends on the vibrating amplitude, has a strong dependence on the frequency, which is shown in Figure 10.4b. In order to control the frequency of the supplied voltages, setting the speed is, moreover, a classical way tuning of these motors but which is not optimal. In fact, the traveling wave ultrasonic motor is faced with the pull-out phenomenon: starting from a high frequency which decreases when the motor gets closer to its resonating point and accelerates. Beyond a maximum, the motor suddenly stalls. To restart it, we

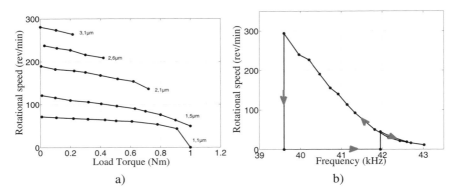

Figure 10.4. *Mechanical features of a motor USR60 traveling wave ultrasonic motor.
a) torque-speed features as a function of the amplitude of stator vibrations,
b) rotational speed as a function of the frequency of the supplied voltages*

need to increase the frequency again. The problem lies in the fact that the position of the maximum is not well known: it depends on the temperature of the motor, load, and amplitude of the supplied voltages. That is why sometimes we are bound to strongly restrict the functioning range of the motor in order to avoid it getting closer to the maximum point, which amounts to decreasing its rate of output power.

10.1.3. *Models*

10.1.3.1. *Equivalent electrical diagram*

It is sometimes interesting to model the motor only from an electronic point of view, to dimension its supply for instance. The diagram used is a capacitor, referred to as *intrinsic* in parallel with a *motional* branch which translates the resonating feature of the motor. This model, as shown in Figure 10.5, is valid around the principal resonating frequency of the motor.

In this equivalent diagram, motional current i_m is proportional to the vibrating speed, that is the derivative of the amplitude of the deformation at a point of the stator. But the voltage source V_Γ takes into account the load torque on the motor. This diagram therefore allows us to translate the electromechanical conversion of the piezoelectric transducer. The mechanomechanical conversion, described in section 10.2, is more tricky to approach. [PIÉ 95] proposes a diagram including a global representation of contact phenomena. This diagram is single phase, the supply of the other path being assumed to be in quadrature. [SAS 93] translates the contact nonlinearities by adding nonlinear electrical elements, like diodes.

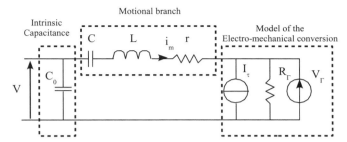

Figure 10.5. *Electrical circuit equivalent to a traveling wave ultrasonic motor. Adapted from [PIÉ 95]*

These diagrams allow us to approach the real features of the motor in steady state, and are simple to use. On the other hand, it is necessary to adjust the value of parameters in a nonlinear fashion in order to widen their range of validity. Finally, they are essentially meant to model established steady-state regimes, and their exploitation during a transient regime is not assured.

10.1.3.2. *"Hybrid" model*

Hybrid models are made up of four modules, shown in Figure 10.6, and the variables exchanged between each module are also shown.

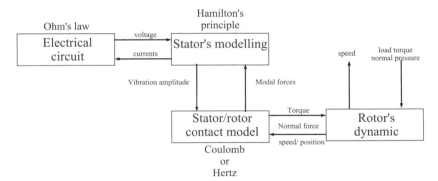

Figure 10.6. *Modeling by hybrid model*

First of all, a model of the stator is worked out. The main objective is to determine the amplitude of standing waves as a function of supplied voltages. The effect of the external efforts is taken into account through *modal reaction forces*. This calculation is done initially by determining the form of stator deformations. We resort to Hamilton's principle and Raileygh-Ritz discretization which allow us to give an approximate analytical form of the deformed equations. At this level, a finite element method can be used to calculate the stator parameters.

380   Control of Non-conventional Synchronous Motors

Then, a model of the contact interface is worked out. This model supplies the effort applied to the load and its reaction on the stator. It receives the normal and tangential speeds and the load position, as well as the vibrating speeds and the amplitudes of the standing waves as input. Several approaches seem to be possible, according to the precision of the model desired. For instance, we can use Hertz theory [BUD 03] or Coulomb's law [MAA 95, HAG 95, GHO 00].

Finally, the dynamics of the load, as well as the energy source and its likely imperfections are taken into account.

These models are complete. They allow us to characterize the torque and speed behavior of the motor, to determine its performances in terms of maximum torque or output. That is why they are described as "design" models because we can imagine clearly the possible optimizations of the geometry owing to them. They also allow us to consider the transient regimes and therefore they are usable for the control of these motors. However, in practice, it quickly appears that a worked out interface model induces a control law that is difficult to implant in real time.

The following section, therefore, deals with an approximate model, sufficiently developed to take into account the phenomena to be controlled, and simple enough to supply, by inversion, a control architecture.

10.2. Causal model in the supplied voltage referential

10.2.1. *Hypotheses and notations*

The stator is assimilated to a disk embedded on a cylindrical shaft, and on which are glued piezoelectric ceramics. These ceramics are distributed in two phases which are referred to as α and β, respectively, supplied with sinusoidal voltages v_α and v_β of peak values V_α and V_β. The polarization of ceramics allows the propagation of a wave which we also assume to be purely sinusoidal (hypothesis H1). In addition, it is optimized to propagate a flexion mode of order k, that is we can find k wavelengths of flexion; for instance, for commercial Shinsei motors, k amounts to 9 or 11 and the resonating frequency lies around 40 and 50 kHz. The rotor of mass m_R and rotational inertia J, put against the stator by an effort F_τ, drives the mechanical load into rotation. We represent C_r as the resistant torque of the load on the rotor.

Figure 10.7 shows a transverse view of a piezoelectric motor on which we write:
– a and b are the inner and outer radii of the stator, respectively;
– $2h$ is the height of the stator;
– $F\tau$ is the axial pre-stress of the rotor on the stator.

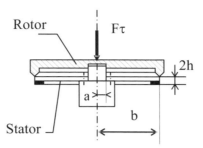

Figure 10.7. *Transverse section of the studied motor*

The traveling wave which deforms the stator structure then transmits its movement to the rotor by contact. Initially, to free oneself from the problems of mechanical contact study between the stator and the rotor, we have been led to develop the concept of the "ideal rotor". This element, storing no kinetic energy (its mass is zero-valued), is intercalated between the stator and real rotor; it is in contact with the stator according to the idealized conditions:

– punctual contacts between the stator and ideal rotor (H2);
– rolling without sliding condition (H3).

and with the real rotor, but imperfectly.

Obviously, this "ideal rotor" is virtual with the sole aim to simplify the modeling on the one hand, and later to characterize the contact phenomena. Finally, we assume that the stator materials are linear (H4).

We define a fixed referential $\mathfrak{R}_1 = (O, \vec{x}, \vec{y}, \vec{z})$ and a rotating referential $\mathfrak{R}_\theta = (M, \vec{u}_r, \vec{u}_\theta, \vec{z})$ for this motor. We prefer to express the coordinates of a point M in polar coordinates for referential $\mathfrak{R}_1$, whereas we retain Cartesian coordinates (u, v, w) for $\mathfrak{R}_\theta$. Figure 10.8 shows the organization of the two frames.

10.2.2. Kinematics of the ideal rotor

10.2.2.1. Deformation of the stator

When only phase α is supplied with a sinusoidal AC voltage, the standing wave relative to α is established. Each point M at the surface of the stator of polar coordinates $\overrightarrow{OM} = (r, \theta)$ moves toward M' under the action of constraints generated by the piezoelectric elements. We write $\overrightarrow{MM'} = (u, v, w)$ relative to $\mathfrak{R}_\theta$ (Figure 10.8). We can then show that the deformation according to $\vec{z}$ is written as:

$$w(\theta, t) = w_\alpha(t)\cos(k\theta) \qquad [10.1]$$

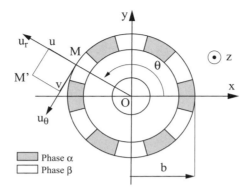

Figure 10.8. *Definition of the frames used*

Similarly, when only phase β is supplied, a standing wave shifted in space propagates; the deformation according to $\vec{z}$ is then written as:

$$w(\theta, t) = w_\beta(t)\sin(k\theta) \qquad [10.2]$$

which provides the localization of piezoelectric elements of phase β.

Under the hypothesis of a linear behavior of the mechanical resonator (hypothesis H_4), the total deformation observed at each point of the stator is written as:

$$w(\theta, t) = w_\alpha(t)\cos(k\theta) + w_\beta(t)\sin(k\theta) \qquad [10.3]$$

The hypothesis of thin plates [HAG 95] can be applied in the case of the stator. A relationship is then inferred which allows us to calculate the deformation according to $\vec{u}_\theta$, written as v:

$$v = -\frac{h}{b}\frac{\partial w}{\partial \theta} = -k\frac{h}{b}(-w_\alpha\sin(k\theta) + w_\beta\cos(k\theta)) \qquad [10.4]$$

From these equations [10.4] and [10.3], we can show in Figure 10.9 the deformation of the stator according to $\vec{u}_\theta$ and $\vec{u}_z$, as a function of θ for a given time.

10.2.2.2. *Definition of the contact point*

The rotor, pushed against the stator, contacts the latter at several points distributed on its periphery. Each point of the stator can be a contact point, it is enough that the coordinate according to $\vec{u}_z$ is maximum. We can write as θ_c the position of the contact point at time t, this condition leads to $\frac{dw}{d\theta}\big|_{\theta=\theta_c} = 0$, or also:

$$-kw_\alpha\sin(k\theta_c) + kw_\beta\cos(k\theta_c) = 0 \qquad [10.5]$$

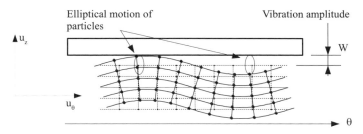

Figure 10.9. *Stator's traveling wave*

We can then infer the angular position of the contact point at time t as a function of the amplitudes of the two standing waves w_α and w_β:

$$k\theta_c = atan\frac{w_\beta(t)}{w_\alpha(t)} \quad [10.6]$$

The amplitude of the deformation at the contact point is then given by:

$$W = w(k\theta_c, t) = w_\alpha cos(k\theta_c) + w_\beta sin(k\theta_c)$$
$$= \sqrt{(w_\alpha^2 + w_\beta^2)} \quad [10.7]$$

10.2.2.3. *Speed of the ideal rotor*

In perfect contact conditions, that is, one contact point and rolling without sliding condition, the stator communicates its speed to the rotor on the normal and tangential axes. This speed is derived from the elliptical trajectory of the points of the stator; we can note that $\vec{v}_{Nid}$, the normal speed of the contact point, is written as:

$$\vec{v}_{Nid} = V_{Nid}\vec{z} = \frac{dw}{dt}\bigg|_{\theta=\theta_c}\vec{z} = (\dot{w}_\alpha cos(k\theta_c) + \dot{w}_\beta sin(k\theta_c))\vec{z} \quad [10.8]$$

According to the tangential axis $\vec{u}_\theta$, the speed of displacement of the contact point between the stator and the ideal rotor is written as:

$$\vec{v}_{Tid} = V_{Tid}\vec{u}_\theta = \frac{dv}{dt}\bigg|_{\theta=\theta_c}\vec{u}_\theta = -k\frac{h}{b}(-\dot{w}_\alpha sin(k\theta_c) + \dot{w}_\beta cos(k\theta_c))\vec{u}_\theta \quad [10.9]$$

Let us put $V'_{Tid} = -\dot{w}_\alpha sin(k\theta_c) + \dot{w}_\beta cos(k\theta_c)$. This then leads to relationship [10.10]:

$$V_{Tid} = k\frac{h}{b}V'_{Tid} \quad [10.10]$$

384   Control of Non-conventional Synchronous Motors

We can introduce here the rotation matrix of angle θ, $\Re(\theta)$, defined in the following way:

$$\Re(\theta) = \begin{pmatrix} \cos(\theta) & -\sin(\theta) \\ \sin(\theta) & \cos(\theta) \end{pmatrix} \qquad [10.11]$$

to write:

$$R_W: \begin{pmatrix} V_{Nid} \\ V'_{Tid} \end{pmatrix} = R(-k\theta_c) \begin{pmatrix} \dot{w}_\alpha \\ \dot{w}_\beta \end{pmatrix} \qquad [10.12]$$

Finally, if Ω_{id} refers to the rotational speed of the ideal rotor, we can write owing to [10.10] that:

$$R_{Rm1}: \Omega_{id} = \frac{1}{b} V_{Tid} = \frac{kh}{b^2} V'_{Tid} \qquad [10.13]$$

Thus, we have just described how the existence of two standing waves enables us to generate normal and tangential speeds of a rotor in ideal contact with the stator. Initially, we express these speeds in a simple case of supply. Then, we show how a torque can be generated from these deformations.

10.2.2.4. *Speeds in steady state in a particular case of excitation*

In the general case, the motor supplied with two alternating sinusoidal voltages is dephased by $\frac{\pi}{2}$ and of the same amplitude, thereby leading to the propagation of two standing waves in quadrature and also of the same amplitude. In these conditions, we can put that:

$$\begin{cases} w_\alpha = W \cos(\omega t) \\ w_\beta = W \sin(\omega t) \end{cases} \qquad [10.14]$$

with ω being the pulsation of supplied voltages. Then, equation [10.6] allows us to write that:

$$k\theta_c = \operatorname{atan}\left(\frac{W \sin(\omega t)}{W \cos(\omega t)} \right) = \omega t \qquad [10.15]$$

In steady state, that is, when W is a constant, equations [10.8] and [10.9] allow us to calculate the normal and tangential speeds of the motor:

$$V_{Nid} = (-\omega W \sin(\omega t) \cos(\omega t) + \omega W \cos(\omega t) \sin(\omega t)$$
$$= 0 \qquad [10.16]$$

$$V_{Tid} = k\frac{h}{b^2} (\omega W \sin(\omega t) \sin(\omega t) + \omega W \cos(\omega t) \cos(\omega t)$$
$$= k\frac{h}{b^2} \omega W \qquad [10.17]$$

On the one hand, this result means that the altitude of the ideal rotor remains constant in a steady state. In fact, as there is always an anti-node of the vibration in contact with the rotor, it keeps at constant altitude corresponding to the gap necessary for the propagation of the traveling wave. In addition, we show that the tangential speed of the ideal rotor is directly proportional to the vibration amplitude. This is an important feature of traveling wave piezoelectric motors.

The following section describes a model allowing us to calculate approximately the efforts generated by the motor to drive the real rotor into movement.

10.2.3. *Generation of the motor torque*

The real rotor of a traveling wave piezoelectric motor possesses two degrees of freedom. The first allows its rotation around its axis and gives rise to rotational speed Ω. The second is a displacement according to the axis $\vec{z}$ of the flexion wave of amplitude W at the contact point. In fact, the real rotor possesses some flexibility which allows it to rise to a necessary quantity, whereas the stator is assumed to be much more rigid. We can therefore admit the existence of a normal speed of translation, which can be written as V_N. The dynamics linked to these speeds are given by the fundamental principle of dynamics applied both in translation and rotation:

– in translation:

$$R_8 : m_R \frac{dV_N}{dt} = F_N - F_\tau \qquad [10.18]$$

with F_N being the force developed by the stator to raise the rotor;

– in rotation:

$$R_9 : J \frac{d\Omega}{dt} = C - C_r \qquad [10.19]$$

with C, the motor torque of the stator on the rotor.

It is a question of determining the way motor torque C and force F_N are generated as a function of vibration amplitudes of the stator. Several modelings try to describe the tribological phenomena at the stator-rotor interface. In a certain category of models, we consider the stator as perfectly rigid at the surface, and that it penetrates inside a friction band whose rigidity in compression is much weaker than that of the rotor. We can then consider not a point but a contact zone. Thus, for each point of the contact zone, we calculate the friction force by using Coulomb friction law, which then gives rise to the motor torque. As the contact zone varies according to the vibration amplitude, we can calculate the torque-speed characteristics of the actuator [MAA 95, GHO 00]. This model can also be completed to take into account other phenomena like the tangential rigidity of the friction band [WAL 98, LU 01]. It can be put in the form of an approximate analytical solution or by resorting to finite element software [TOU 99]. However, in general, these models are well adapted in a design approach, but result in a solution too complex to lead to a simple model which will be able to be inverted to end up with a control structure.

For another category of models, the perspective is different. It concerns modeling globally the behavior of the stator-rotor contact to end up with an approximate form of the expressions of F_N and C. Thus, the chosen model proposes to take into account the elastic potential energy stored in the normal stiffness of the friction layer, which leads to:

$$R_{CN}: F_N = K \int (V_{Nid} - V_N) dt \qquad [10.20]$$

with K being the elasticity of the friction layer. Obviously, value K must vary according to the functioning point.

In addition, on the tangential axis, we show a modeling which allows us to approach the torque-speed characteristics (like those of Figure 10.4) by lines whose slope is fixed and equal to the mean slope observed over the whole domain of functioning:

$$R_{CT}: C = f_0(\Omega_{id} - \Omega) - C_f \qquad [10.21]$$

with f_0, the mean slope of torque-speed characteristics with constant Ω_{id} and a friction torque C_f.

In equation [10.21], torque C_f allows us to take into account the gluing effect at low speed. In fact, when the wave amplitude is weak, the stator is glued on the rotor, and the amplitude necessary for the degluing is greater than that which maintains the rotation of the rotor. This mechanism can be assimilated to a dry friction requiring particular care in the modeling [DAI 09]. However, for simplification purposes, we will put in this treatise $C_f = 0$.

10.2.4. *Stator's resonance*

The stators of the traveling wave ultrasonic motors are mechanical resonators. Owing to the dimensions which characterize them and the materials used in their fabrication, an infinite number of flexion modes can propagate. Around a resonating frequency, between 40 and 50 kHz, a particular mode propagates that for which the stator has been optimized. Under the action of piezoelectric ceramics, the vibration amplitude of a standing wave varies according to time and excitation frequency. The relationship linking the supplied voltages and amplitude w_α and w_β of standing waves, is a second order equation forced by voltages v_α and v_β. If we consider that there is no coupling between the two phases, we will put [GIR 98]:

$$m \begin{pmatrix} \ddot{w}_\alpha \\ \ddot{w}_\beta \end{pmatrix} + d_s \begin{pmatrix} \dot{w}_\alpha \\ \dot{w}_\beta \end{pmatrix} + c \begin{pmatrix} w_\alpha \\ w_\beta \end{pmatrix} = N \begin{pmatrix} v_\alpha \\ v_\beta \end{pmatrix} \quad [10.22]$$

This equation makes several parameters appear, whose analytical expression can be determined in [GIR 98]:

– m: vibrating mass for the mode considered;
– c: modal stiffness of the mode considered;
– N: force factor. This translates the inverse piezoelectric effect which operates in piezoelectric ceramics.

This free-stator model does not take into account the effect of the rotor on the stator. For that purpose, [HAG 95] and [MAA 00] propose to add efforts, variable in time, written as $f_{r\alpha}$ and $f_{r\beta}$. These *modal reaction forces* translate the effect of external effort on the wave. In fact, in order to complete the electromechanical conversion, the electrical power supplied to the stator allows us on the one hand to put it in vibration, but must also allow the rotation of the rotor. Thus, equation [10.22] of putting in resonance of the stator must be completed to take into account the energy transfer:

$$m \begin{pmatrix} \ddot{w}_\alpha \\ \ddot{w}_\beta \end{pmatrix} + d_s \begin{pmatrix} \dot{w}_\alpha \\ \dot{w}_\beta \end{pmatrix} + c \begin{pmatrix} w_\alpha \\ w_\beta \end{pmatrix} = N \begin{pmatrix} v_\alpha \\ v_\beta \end{pmatrix} - \begin{pmatrix} f_{r\alpha} \\ f_{r\beta} \end{pmatrix} \quad [10.23]$$

The following section shows how to obtain the expression of these *modal reaction forces*.

10.2.5. *Calculation of modal reaction forces*

The stator and the ideal rotor exchange power at the level of k contact points distributed along the stator ring. The power exchanged at these points must be equal to the product of modal reaction forces by the vibrating speeds. We, therefore, write:

$$p_{ext} = (\dot{w}_\alpha, \dot{w}_\beta) \begin{pmatrix} f_{r\alpha} \\ f_{r\beta} \end{pmatrix} \quad [10.24]$$

In addition, this power is equal to the power received by the rotor, both on the vertical and horizontal axes. We can then write:

$$p_{ext} = (V_{Nid}, \Omega_{id}) \begin{pmatrix} F_N \\ C \end{pmatrix} \quad [10.25]$$

Then, from equations [10.12], [10.10], and [10.13], we can write that:

$$(V_{Nid}, \Omega_{id}) \begin{pmatrix} F_N \\ C \end{pmatrix} = (V_{Nid}, V'_{Tid}) \begin{pmatrix} F_N \\ k\frac{h}{b^2}C \end{pmatrix}$$

$$= \left[R(-k\theta_c) \begin{pmatrix} \dot{w}_{r\alpha} \\ \dot{w}_{r\beta} \end{pmatrix} \right]^t \begin{pmatrix} F_N \\ k\frac{h}{b^2}C \end{pmatrix} \qquad [10.26]$$

$$= (\dot{w}_\alpha, \dot{w}_\beta) R^t(-k\theta_c) \begin{pmatrix} F_N \\ k\frac{h}{b^2}C \end{pmatrix}$$

Thus, by using [10.26] and [10.24] we can express by identifying the expression of modal reaction forces:

$$R_F: \begin{pmatrix} f_{r\alpha} \\ f_{r\beta} \end{pmatrix} = R^t(-k\theta_c) \begin{pmatrix} F_N \\ F'_T \end{pmatrix} \qquad [10.27]$$

Having put initially:

$$R_{m2}: F'_T = k\frac{h}{b^2}C \qquad [10.28]$$

10.2.6. *Complete model*

We have just established the equations of functioning of the traveling wave ultrasonic motors, and a model is proposed here. The approach relies on the principle of system energies; therefore, we have identified the zones where they accumulate, while determining their nature, potential, or kinetics.

Thus, the system can be represented by a causal ordering graph (COG) [HAU 98, GIR 01]. This graph is constituted by processes of different natures, connected with one another by action or reaction links. A causal processor stores energy (potential or kinetic); it is epitomized by a simple arrow directed from the input to the output of the processor. The rigid processors are epitomized by a double arrow; they are referred to as energy dissipators. Finally, the modulators (two-coupled processors) ensure the power transfer with neither loss nor accumulation. This model is shown in Figure 10.10. Relationship R_{xx} appearing in each processor has been defined previously.

Due to the analysis this model is divided into four domains:
 – the domain of electrical variables ($v_{\alpha\beta}$, $i_{m\alpha\beta}$);
 – the stator domain ($w_{\alpha\beta}$);
 – the domain of the ideal rotor (F_N, V_{Nid}, C, Ω_{id});
 – the domain of the real rotor (F_N, V_N, C, Ω).

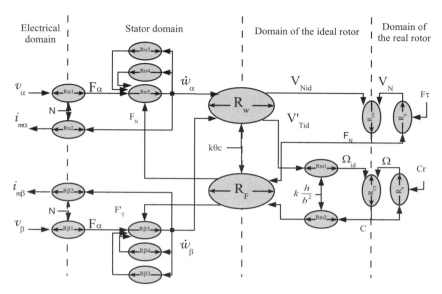

Figure 10.10. *COG representation of the piezoelectric motor*

The graph emphasizes the energy transfer on the two axes, normal and tangential. The virtual efforts and their associated power are found at the boundary between each domain. In addition, we can notice that efforts f_α and f_β are proportional to voltage, via couplers $R_{\alpha 1}$ and $R_{\beta 1}$. This property is dual to that of electromagnetic machines, in which the efforts are proportional to current. Conversely, currents $i_{m\alpha}$ and $i_{m\beta}$ read on this graph are proportional to the vibrating speeds, and property is also dual of electromagnetic machines where the speed generates a back electromotive force analogous to a voltage.

This modeling is valid during both the transient regimes and in steady state. Let us notice here that the two rotation matrices appearing both in the kinematics of the ideal rotor and on the calculation of modal reaction forces (relationships R_F: and R_W:) remind us of models of classical electromagnetic machines, for which a representation in a rotating referential is proposed. That is why we propose a causal model in the rotating referential of the traveling wave which is adapted to the traveling wave ultrasonic motor.

10.3. Causal model in the referential of the traveling wave

10.3.1. *Park transform applied to the traveling wave motor*

In electromagnetic machines, the referential chosen is linked to the rotating field: an axis of the referential is then placed in the direction where the field is maximum.

By analogy, we are going to link the axis of the referential to the vector of the traveling wave. We therefore write:

$$\begin{pmatrix} v_\alpha \\ v_\beta \end{pmatrix} = R(k\theta_c) \begin{pmatrix} V_d \\ V_q \end{pmatrix} \quad [10.29]$$

and:

$$\begin{pmatrix} i_{m\alpha} \\ i_{m\beta} \end{pmatrix} = R(k\theta_c) \begin{pmatrix} I_d \\ I_q \end{pmatrix} \quad [10.30]$$

This transformation can be described in Figure 10.11. In this figure, we start by drawing a referential (O, α, β) orthonormal and fixed. Then, on each axis, we report the measure at time t of vibration amplitudes w_α and w_β. We can then infer the position of the phasor $\underline{w}$ whose angle with respect to (O, α) amounts to $k\theta_c$: this phasor is hence a vector rotating in the fixed referential.

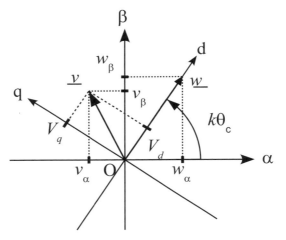

Figure 10.11. *Representation of voltage and wave phasors in the fixed referential (O, α, β) and rotating referential (O, d, q)*

This phasor $\underline{w}$ serves as reference axis to another referential – thus rotating – which is also orthonormal (O, d, q). Then, from the measure of $v_\alpha(t)$ and $v_\beta(t)$, we can infer the position of phasor $\underline{v}$ which gives, by projection, coordinates V_d and V_q in rotating referential (O, d, q).

We will observe thereafter that the change of referential is judicious because it allows us to obtain constant variables in steady state for two standing waves of the same amplitude and in quadrature. The modeling has to be carried on to get to the equations ruling the evolutions of speed of the ideal rotor as a function of variables expressed in the referential of the wave.

10.3.2. *Transformed model*

From equation [10.12], we can initially write the derivatives of the amplitudes of the two standing waves w_α and w_β in the form of equation [10.31]:

$$\begin{pmatrix} \dot{w}_\alpha \\ \dot{w}_\beta \end{pmatrix} = R(k\theta_c) \begin{pmatrix} V_{Nid} \\ V'_{Tid} \end{pmatrix} \quad [10.31]$$

On the one hand, we can express the motional currents as a function of the speeds of the ideal rotor:

$$\begin{pmatrix} i_{m\alpha} \\ i_{m\beta} \end{pmatrix} = R(k\theta_c) \begin{pmatrix} I_d \\ I_q \end{pmatrix} = N \begin{pmatrix} \dot{w}_\alpha \\ \dot{w}_\beta \end{pmatrix} = NR(k\theta_c) \begin{pmatrix} V_{Nid} \\ V'_{Tid} \end{pmatrix} \quad [10.32]$$

hence:

$$R_{d2}: I_d = NV_{Nid} \quad [10.33]$$

$$R_{q2}: I_q = NV'_{Tid} \quad [10.34]$$

On the other hand, we can express the second derivative terms as a function of the speeds of the contact point:

$$\begin{pmatrix} \ddot{w}_\alpha \\ \ddot{w}_\beta \end{pmatrix} = \frac{d}{dt}\left[R(k\theta_c)\begin{pmatrix} V_{Nid} \\ V'_{Tid}\end{pmatrix}\right]$$

$$= R(k\theta_c)\begin{pmatrix} \dot{V}_{Nid} \\ \dot{V}'_{Tid}\end{pmatrix} + k\dot{\theta}_c R\left(k\theta_c + \frac{\pi}{2}\right)\begin{pmatrix} V_{Nid} \\ V'_{Tid}\end{pmatrix} \quad [10.35]$$

Finally, by using equations [10.8] and [10.5] we can write:

$$\int V'_{Nid} dt = \int \left(\dot{w}_\alpha(t)\cos(k\theta_c) + \dot{w}_\beta(t)\sin(k\theta_c)\right) dt$$

$$= \left[w_\alpha \cos(k\theta_c) + w_\beta \sin(k\theta_c)\right]$$

$$- \underbrace{\int \left(-w_\alpha \sin(k\theta_c) + w_\beta \cos(k\theta_c)\right) dt}_{0} \quad [10.36]$$

Then:

$$R(-k\theta_c)\begin{pmatrix} w_\alpha \\ w_\beta \end{pmatrix} = \begin{pmatrix} \cos(k\theta_c)w_\alpha + \sin(k\theta_c)w_\beta = \int V'_{Nid} dt \\ -\sin(k\theta_c)w_\alpha + \cos(k\theta_c)w_\beta = 0 \end{pmatrix} \quad [10.37]$$

We obtain a new expression of amplitudes of the standing waves:

$$\begin{pmatrix} w_\alpha \\ w_\beta \end{pmatrix} = R(k\theta_c)\begin{pmatrix} \int V'_{Nid} dt \\ 0 \end{pmatrix} \quad [10.38]$$

392 Control of Non-conventional Synchronous Motors

We can then rewrite equation [10.23] by making these changes of variables reappear:

$$mR(k\theta_c)\left[\begin{pmatrix}\dot{V}_{Nid}\\ \dot{V}'_{Tid}\end{pmatrix} + k\dot{\theta}_c R\left(\frac{\pi}{2}\right)\begin{pmatrix}V_{Nid}\\ V'_{Tid}\end{pmatrix}\right]$$
$$+ dsR(k\theta_c)\begin{pmatrix}V_{Nid}\\ V'_{Tid}\end{pmatrix} + cR(k\theta_c)\begin{pmatrix}\int V_{Nid}dt\\ 0\end{pmatrix}$$
$$= NR(k\theta_c)\begin{pmatrix}V_d\\ V_q\end{pmatrix} - R(k\theta_c)\begin{pmatrix}F_N\\ F'_T\end{pmatrix}$$

Matrix $R(k\theta_c)$ is an invertible matrix; we can thus simplify on the left and right by R, and end up with the equations ruling the evolution of normal and tangential speeds of the ideal rotor:

$$m\dot{V}_{Nid} + dsV_{Nid} + c\int V_{Nid}dt = NV_d - F_N + mk\dot{\theta}_c V_{Tid} \qquad [10.39]$$

$$m\dot{V}'_{Tid} + dsV'_{Tid} = NV_q - F'_T - mk\dot{\theta}_c V_{Nid} \qquad [10.40]$$

These equations are important insofar as we have been able to decouple the actions according to the axis where they are applied. Axis d, as it makes the normal effort appear, can be referred to as normal axis. Axis q, will be referred to as tangential axis, because it is the tangential effort which intervenes. These equations need, however, to be rearranged to make the coupling terms on the speeds totally disappear. In fact, [GIR 02] shows that we can write, by means of a few hypotheses, that:

$$V'_{Tid} = k\dot{\theta}_c \int V_{Nid}dt \qquad [10.41]$$

$$k\dot{\theta}_c V'_{Nid} \approx \dot{V}'_{Tid} \qquad [10.42]$$

Then, relationship [10.39] can be rewritten, by replacing the coupling term $mk\dot{\theta}_c V'_{Tid}$ by $m(k\dot{\theta}_c)^2 \int V'_{Nid}$:

$$m\dot{V}'_{Nid} + dsV'_{Nid} + (c - m(k\dot{\theta}_c)^2)\int V'_{Nid}dt = NV_d - F'_N \qquad [10.43]$$

Concerning axis q, the derivation of equation [10.41] leads to:

$$2m\dot{V}'_{Tid} + dsV'_{Tid} = NV_q - F'_T \qquad [10.44]$$

Given that the rest of the equations remain unchanged, we can show in Figure 10.12 a new COG representation of this model in the new referential. We put for this purpose:

$R_{d1}: F_d = NV_d$

$R_{q1}: F_q = NV_q$

$R_{d3}: V_{Nid} = \dfrac{1}{m} \int \left(F_d - F'_N - d_s V'_{Nid} - c \int V'_{Nid} \right) dt$

$R_{q3}: V'_{Tid} = \dfrac{1}{2m} \int \left(F_q - F'_T - ds V'_{Tid} \right) dt$

$R_{\theta 1}: k\dot{\theta}_c = \dfrac{V'_{Tid}}{\int V'_{Nid} dt}$

$R_{\theta 2}: k\theta_c = \int k\dot{\theta}_c dt$

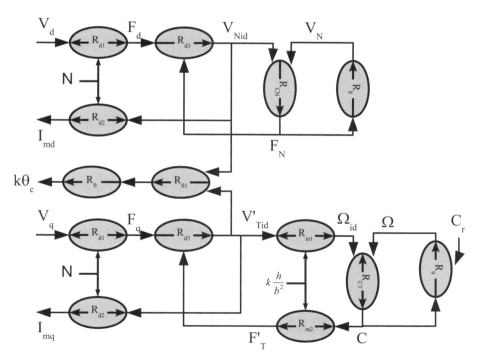

Figure 10.12. *Causal ordering graph of the piezoelectric motor in the rotating reference frame attached to the wave*

10.3.3. *Study of the motor stall*

In this section, we describe the phenomenon of motor stall such as explained in section 10.1.2, and we attempt to give an explanation and a means to protect us against it. We put ourselves in the case of two standing waves in quadrature and of the same amplitude, so that equations [10.14, 10.15, 10.16], and [10.17] still apply. In these conditions, equation [10.41] allows us to write that $\int V_{Nid} dt = W$. Then, equation [10.43] becomes, in steady state, and for two waves in quadrature:

$$(c - m\omega^2)W + F_N = NV_d \qquad [10.45]$$

and equation [10.44] becomes:

$$d_s \omega W + k\frac{h}{b^2}C = NV_q \qquad [10.46]$$

Equations [10.45] and [10.46] allow us to draw the affix of the voltage vector in the rotating referential in Figure 10.13, if we provide ourselves with vibration amplitude W, external efforts C and F_N and knowing voltage pulsation ω. Generally, the amplitude of the supplied voltages is constant, so that the affix of the voltage vector moves on a circle centered at the origin of the rotating referential. We can notice the position of voltage vector by angle Ψ which it makes with the horizontal axis of this referential.

Then, we study different cases of supply conditions, according to the voltage pulsation. We define resonating pulsation ω_r as being that for which the amplitude of the traveling wave is maximum ($\omega_0 = \sqrt{\frac{c}{m}}$ is the resonating pulsation of the stator alone). We distinguish in Figure 10.13 three cases according to whether the supplied voltages of the motor are of a frequency smaller or greater than the resonating frequency.

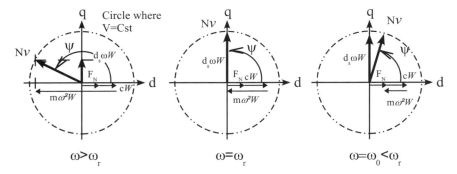

Figure 10.13. *Affix of the voltage vector in the rotating reference frame for different supplied pulsations*

Given the weak variation of the supplied voltage pulsation, and by analyzing Figure 10.13, the amplitude of the traveling wave is maximum when Ψ amounts to $\frac{\pi}{2}$. In this case, relationship [10.45] imposes that ω_r is greater than ω_0. Let us note here that the role played by F_N tends to shift the resonance of the motor from that of the stator alone. Furthermore, in some models, taking into account the normal effort made by modifying the stiffness of the stator: parameter c is changed to $c + \tilde{c}$ with $\tilde{c} = \frac{F_N}{W}$ [MAA 97b].

In the case of frequencies greater than the resonance, term $m\omega^2 W$ of equation [10.45] is greater than the sum $cW + F_N$. Then V_d becomes negative, and $\underline{v}$ lies to the left of axis q. We read the amplitude of the traveling wave by projecting on axis q the voltage vector. As the pulsation decreases, the term $m\omega^2 W$ decreases, forcing the voltage vector to get closer to axis q, but the amplitude of the wave increases.

If we decrease ω further, the voltage vector goes to the right of axis q. The tendency is inverted and the amplitude of the traveling wave decreases when the pulsation decreases. Generally, the motor stalls when we go beyond this resonating point, which can be explained with the model which we have presented [GIR 03] or a close model [MAA 97a], thereby generating the readings of Figure 10.4.

But the stall can also intervene for vibration amplitudes much weaker than that obtained from the resonance of the motor. To show this, we now study the influence of a load torque for a motor at the resonance, that is when $V_d = 0$, and when the amplitude of the supplied voltages is fixed. According to equation [10.46], voltage V_q is made of a term proportional to the amplitude of the traveling wave, and of a term proportional to the transmitted torque; when the height of the wave is large, the part remaining to the torque is decreased (Figure 10.14). If the load torque becomes too large, the motor cannot supply the effort necessary, and it stalls.

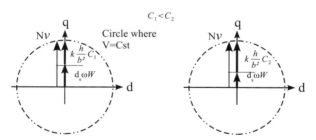

Figure 10.14. *Functioning at the resonance: the maximum torque available decreases when the height of the wave increases*

Equation [10.46] gives in the boundary case of resonance $d_s \omega W = NV - k\frac{h}{b^2}C$, or also:

$$W = \frac{NV}{d_s \omega} - k \frac{h}{b^2} \frac{C}{d_s \omega} \qquad [10.47]$$

Thus, in plane $W(C)$, the vibration amplitude is limited by a line beyond which the motor stalls. Or also, at fixed W, there is a boundary torque beyond which the motor stops abruptly. This limit depends linearly not only on W but also on the supplied voltage. This remark is to be compared to Figure 10.4 where we can observe clearly that beyond a certain load torque the motor stops (break of the curve $N(C)$).

10.3.4. *Validation of the model*

This model can be validated experimentally, by drawing measurements at steady state in a rotating referential. In fact, the motor having had an auxiliary electrode put on the stator, we can measure at a point the propagating traveling wave. The voltage delivered by this sensor is proportional to the measured deformation, the gain being equivalent to $36V/\mu m$. Then, by measuring the dephasings between the supplied voltages and the voltage coming from this electrode, we can redraw the diagrams of Figure 10.13. Let us note that the auxiliary electrode is not aligned on the path α, but is shifted in space by a constant electrical angle equivalent to $-180° + 29°$. The information is hence shifted by the same amount on the diagram. We have thus shown in Figure 10.15 the chronograms of the supplied voltages and of the deformation at a point of the stator for a traveling wave of amplitude 2 μm. From the dephasings readings, we can then draw the Fresnel diagram showing the voltage coming from the auxiliary electrode and the supplied voltages.

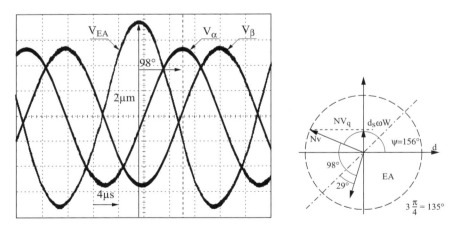

Figure 10.15. *Oscillographic reading and corresponding diagram for a vibration amplitude of 2 μm and two supplied voltages in quadrature of peak amplitude of 130V*

Thus, these measures allow us to identify the angle Ψ which amounts to 156° in this case. Then, we perform the same experiment, this time by changing the frequency of supplied voltages in order to set the vibration amplitude to 3 μm. Again, we read

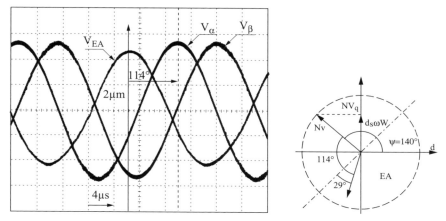

Figure 10.16. *Oscillographic reading and corresponding diagram for a vibration amplitude of 3 µm and two supplied voltages in quadrature of peak amplitude of 130 V*

the chronograms and the dephasings associated with this functioning point in Figure 10.16. We then note angle Ψ which we identify is decreased, indicating that we are getting closer to the resonance.

Finally, when the torque on the motor shaft increases, we can compare the readings of Figure 10.15 with those of Figure 10.17 which have also been generated at an amplitude of 2 µm, but by imposing a load torque of $C_r = 0.5N$ on the motor shaft. Again, angle Ψ decreases, in order to allow both the propagation of the wave and the electromechanical conversion generating the torque.

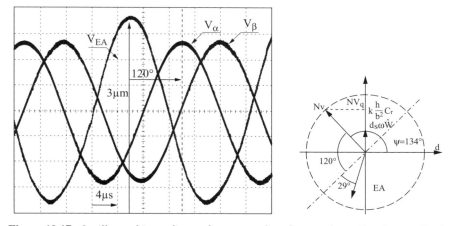

Figure 10.17. *Oscillographic reading and corresponding diagram for a vibration amplitude of 2 µm and two supplied voltages in quadrature of peak amplitude of 130 V. $C_r = 0.5$ nm*

These readings therefore confirm the functioning of the motor, and illustrate the model in the rotating reference frame associated with the traveling wave. As an application example of this model, we propose a torque estimator in the following section.

10.3.5. *Torque estimator*

The modeling of traveling wave piezoelectric motors shows that the chain of energy conversion linking the torque generated by the motor to the electrical supply is neither simple nor direct. It is, therefore, not conceivable to control an electrical variable to control the torque, like the classical current controls proposed for electromagnetic motors.

However, it is possible to supply a torque estimator, which allows us to know in real time the torque generated by the motor, in order to servo-control or limit it. We describe in this section how to obtain the structure of a torque estimator by inversion of the causal ordering graph in Figure 10.12, which we take again in Figure 10.18 to show only path q as this is the one which carries the torque.

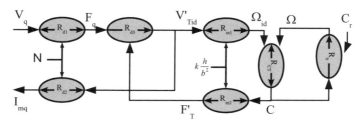

Figure 10.18. *Causal ordering graph in axis q*

The principle of a torque estimator can be described by the causal ordering graph of Figure 10.19 where we obtain, by inversion of pathways which lead to motor torque C, three different strategies:

– from relationship [10.19], which requires us to know inertia J of the load and its resistant torque C_r;

– by inversion of relationship [10.21] and of the measure of Ω and Ω_{id} which has been proposed by [MAA 97b]; this estimator requires us to know all the torque speed characteristics of the motor;

– from relationship [10.44] and the measure of V_q and V'_{Tid}.

Solutions 1 and 2 require us to know the speed of the motor. In order to avoid the speed sensor, we will use the solution based on the inversion of relationship [10.44]. However, the inversion of this equation is not direct, because of the differentiated term $2m\dot{V}'_{Tid}$ which has to be evaluated. But if we restrict ourselves to the cases of steady

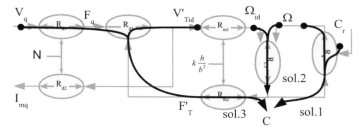

Figure 10.19. *The three families of torque estimators*

state or even for functionings in which V'_{Tid} varies slowly, this term can be disregarded: equation [10.44] then becomes:

$$dsV'_{Tid} = NV_q - k\frac{h}{b^2} \qquad [10.48]$$

Equation [10.48] then allows us to supply based on the torque estimator:

$$\tilde{C} = \frac{b^2}{kh}\left[NV_q - d_s V'_{Tid}\right] \qquad [10.49]$$

This principle has been used on an experimental bench, detailed in Figure 10.20. This bench uses a Shinsei USR30 traveling wave ultrasonic motor 30 mm in diameter, to develop a nominal torque of 0.05 Nm coupled to a DC motor which allows us to impose a resistant torque brake or motor. A torque sensor is also inserted between the load and motor in order to compare the estimation with the torque really developed.

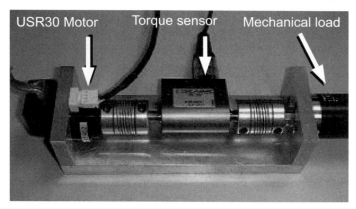

Figure 10.20. *Experimental bench for the estimation of the motor torque*

The results of the estimator are shown in Figure 10.21. In this figure, we show the evolution of estimated torque $\tilde{C}$ as a function of time when vibration amplitude W is maintained at constant and the load torque C_r is slowly varying in time.

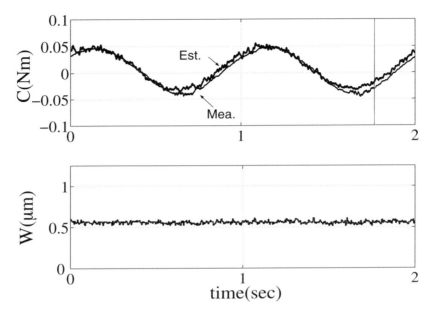

Figure 10.21. *Comparison estimation-measure of the motor torque when the vibration amplitude is constant*

This figure shows that the estimator supplies a reliable value for the motor torque, both in the case of running as a motor ($C > 0$ when $\omega > 0$) and as a generator ($C < 0$ when $\omega > 0$). This functionality is important because it shows that even if the torque is not a simple function of the electrical supply variables – as is often the case for electrical motors – its estimation is possible without additional sensors to that already present on the motor. It allows us to consider its use in applications with a limited or even controlled torque.

10.4. Control based on a behavioral model

The control of piezoelectric actuators follows two trends: the first, and the most used, consists of considering the behavior of the motor according to the "black box" approach and to compensate for the poorly known and strongly nonlinear effects by online identification methods, and adaptive and/or predictive controllers. The second relies on a physical analysis of the functioning of the actuator and a precise model, although usable in control. This section is linked to the models of the first approach. A summary of the variables for controlling the rotational speed of a traveling wave ultrasonic motor can be found in [NOG 96]. Clearly, the supply frequency, around the resonance, the amplitude of the supply 2-phase voltages and the dephasing between these two voltages are the variables for setting the mechanical variables of the motor.

The dephasing between the two supply voltages allows the rotation in one direction or the other, this setting variable therefore is particularly used for the positioning. The relationships between the rotational speed and these variables are however not linear, in particular, the speed-dephasing relationship shows a dead zone around the origin which depends on the load torque [SEN 02]. The speed-frequency relationship also depends on this load torque [PET 00]. Most control strategies rely on one (or even two) of these variables to set the rotational speed of the motor, from the initial online identification of speed-frequency, speed-phase or speed-voltage relationships for given functioning conditions of the motor [CHE 08]. The general approach then consists of approaching the behavior linking the rotational speed (or the position) and the setting variable by a simple function. For instance, in [SEN 02], the authors link the position of the motor to the dephasing of supply voltages by a linear function of second order. Owing to an adaptive control by reference model, they manage to control the position of the rotor; however, the dead zone around the origin not being taken into account for the model control, they call on a fuzzy corrector in order to compensate for it. Fuzzy logic is also used in [BAL 04] to make up for the lack of knowledge on the position-instantaneous frequency model of the motor. A similar approach is developed in [BIG 05], this time using a predictive corrector. Taking into account the nonlinearity brought by this dead zone can finally be done by neuronal identification, as proposed in [CHE 09], or by the exploitation of neuronal and fuzzy methods like in [CHA 03]. Other more thorough research proposes an explicit control of the motor torque [CHU 08]; in this case, the variation range of the torque can even be separated into two parts, one where the motor is supplied with two voltages in phase, the other where the motor is supplied with two voltages in quadrature. In the first part, the motor undergoing a standing wave, it can only offer a resistant torque tuneable according to the supplied voltage, in the second part, it supplies a motor torque. The setting of the torque is ensured by the variation of supplied voltage, according to an initial identification in steady state of the torque characteristics. By performing these methods it may seem that the black box approach knows its limits in terms of practical and industrial implantation, of reproducibility and understanding. They can be helped by a complementary approach which consists of taking advantage of the physical knowledge of the system and relying on knowledge models.

10.5. Controls based on a knowledge model

Less wide spread than the previous ones, these controls generally rely on a coupled analytical model, taking into account the double electromechanical and mechanomechanical conversion which is done in the motor [MAA 00]. The mechanomechanical conversion making tribological phenomena intervene at the stator-rotor contact is often approached for the needs of the control by models which are linearized [GIR 03] or tabulated [MAA 00]. Once these models are put in place, the control consists of inverting the chain of action in order to infer, from the desired references on mechanical variables, the action to be taken on the setting variables.

10.5.1. *Inversion principle*

This general inversion principle is put into application here via the CIG formalism which illustrates graphically the systematic obtaining of the control structure. The control is based on the inversion principle which consists of defining the input setting according to the desired trajectory of its output in virtue of the principle: find the good cause to generate the good effect. This inversion principle is applied to each elementary sub-system which appears to be a complete model in each chain of action. The control then consists of determining the inverse physical functionality of the process considered [BAR 06]. The inversion method depends on the elementary process to be inverted: if it concerns a rigid processor (time independent), the inversion is direct; for instance:

$$s(t) = ke(t) \longrightarrow e_{reg} = \frac{1}{k} s_{ref} \qquad [10.50]$$

If the process is causal (integrally dependent on its input), the direct inversion is replaced with a servo-control of the variable in order to make the real output s_{mes} converge to the desired output s_{ref}; for instance with a proportional corrector:

$$s(t) = k \int e(t) dt \longrightarrow e_{reg} = k_p \left(s_{ref} - s_{mes} \right) \qquad [10.51]$$

10.5.2. *Control structure inferred from the causal model: emphasis on self-control*

The causal ordering graph of Figure 10.12 makes two axes to be servo-controlled appear in order to control the motor, normal axis, and tangential axis.

10.5.2.1. *Inversion of the tangential axis*

Figure 10.22 shows the inversion of path q of the motor in order to control the torque. This control structure is based on the speed control of the ideal rotor Ω_{id}, which is obtained by servo-controlling voltage V_q. Works have allowed us to determine the setting method of the servo-control [DAI 09], which we will not reproduce here.

10.5.2.2. *Inversion of the normal axis*

We can apply the same method on path d, by inversion of CIG, as on path q. It leads to three strategies, according to the variable controlled:

– Control of the normal speed of the ideal rotor. The inversion of the CIG on path d shows that it is possible to servo-control V'_{Nid}, the normal speed of the ideal rotor.

However, we have already noticed in equation [10.16] that in the case of an excitation in quadrature of two waves of same amplitude and for the steady state $V'_{Nid} = 0$. In fact, another non-zero constant value would make the rotor lift off from the stator, which is a case that is difficult to conceive in practice. We therefore have a limited choice of value of V'_{Nidref}, which has to be at least zero-valued in steady state.

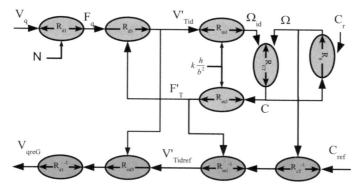

Figure 10.22. *Inversion of the tangential axis; torque regulation*

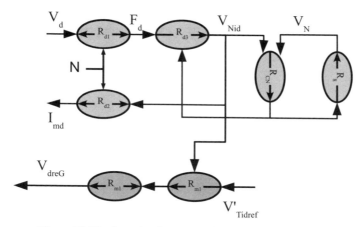

Figure 10.23. *Causal ordering graph of the control of axis d*

Furthermore, we are not interested in the transient regime on axis d. As a consequence, we will not control this variable, but the controls proposed later will ensure that $V_{Nid} = 0$ in steady state;

– Servo-control of the pulsation of the supplied voltages.

If we carry on the analysis of CIG of Figure 10.12, we notice that $k\dot{\theta}_c$ is an action variable. Its servo-control is thus possible. Yet, in steady state, $k\dot{\theta}_c = \omega$ (relationship [10.15]), the pulsation of supplied voltages. It is sometimes interesting to control this pulsation, for instance, for some resonating supplies which show an optimal functioning at a given frequency [MAA 00].

The CIG of Figure 10.24 shows a control structure of this pulsation in axis d.

This graph is made of two interwoven loops. It also requires the compensation of F_N, which can be easily done by putting $F_N = F_\tau$, which is valid in steady

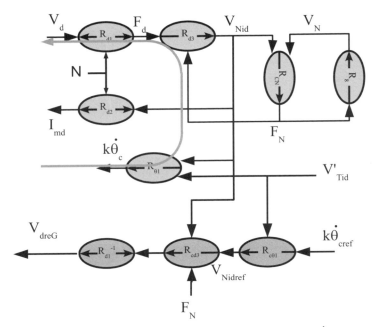

Figure 10.24. *Causal ordering graph of the control of $k\dot{\theta}_c$*

state. Furthermore, an additional simplification of this servo-control is possible, and a calculation method of the correctors used can be found in [GIR 02].

If it is conceivable to servo-control the frequency at a desired value, we have, however, to notice that it cannot be chosen randomly. Too low, it is possible that the motor will stall; too high, the effective value of voltages to apply becomes too large for the supply. Thus, this control must be supplied with an online identification of the resonating frequency, to be able to follow the possible variations due to the temperature. The influence of the resonating pulsation is seen on CIG by intervention of ω_0. The principle of this control has however been done in practice by [MAA 97b];

– Open loop on path d. We have just mentioned and presented two strategies which implement a setting of V_d by a servo-control loop. In electromagnetic machines, we encounter this configuration where an axis is devoted to the torque control, whereas the other allows us to control another variable, like a flux for instance.

However, this is not always the case: for a synchronous machine with smooth poles, for instance, we set $I_d = 0$, because this allows us to decrease the Joule losses. We can transpose this reasoning in the case of a piezoelectric motor. By noticing that in established steady state, equation [10.44] allows us to write that V_d is constant, we can choose to directly impose $Vd = cste$. This concerns an open loop control, because no servo-control on path d is undertaken. The choice of this value can be done according to several criteria. For instance, we can choose to work always at resonance, in order to limit the amplitude of the supplied voltages. Then Figure 10.13 shows that in this case, $V_d = 0$ is the value to impose. Or, some supplies being easy to make if the value of the effective voltages which they deliver remains constant, we can impose $V_d^2 + V_q^2 = V^2$ constant. The amplitude of voltages v_α and v_β will then be equal to V. Let us note, however, that for a given V and V_q, two values of V_d are possible: one is positive and the other is negative. We keep the solution negative, because it makes it possible to work beyond the resonating frequency (Figure 10.13). Let us note that the pulsation of supplied voltages is no longer imposed by the control; it is free not only to evolve according to V_d but also to external conditions. We have simulated in Figure 10.25 the responses in transient regime at a scale interval of V_q, V'_{Tid}, and frequency of supplied voltages, the effective value of voltages being maintained at constant.

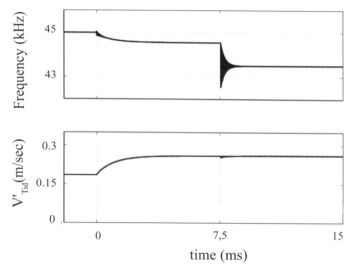

Figure 10.25. *Self-adaptation of the frequency of supplied voltages to the variations of resonating frequency*

For $t = 7.5ms$, we impose a variation of -5% of parameter c; this gap simulates a variation of the resonating frequency of the motor under its own over-heating. The fluctuation speed is exaggerated to test the robustness of this strategy with respect to the resonating frequency. We note that this variation is without effect on the value in steady state of V'_{Tid}. The reason for this is that the frequency of supplied voltages is adapted to the new value of the resonating frequency. We will come back to this property later on.

It is this feature of self-adaptation associated with the simplicity of the supply which led us to choose this control strategy by self-control.

10.5.3. *Practical carrying out of self-control*

From a practical point of view, the self-control of a traveling wave ultrasonic motor is not an easy thing since the inversion of the rotation matrix must be done at a frequency which is higher than that of the rotating field electromagnetic motors, parameter θ_c of equation [10.15] varying at the frequency of about 40 kHz. However, by noticing that $V_d^2 + V_q^2 = V^2$ is constant, controlling V_q is done if one controls Ψ, the dephasing between the traveling wave and the supplied voltages. That is why, the self-control presented in this chapter is done based on a frequency multiplier with loop and analog phase locking. The principle, represented in Figure 10.26, consists of synchronizing a clock signal S_{HF} with the measure of the voltage coming from the deformation sensor of the motor. This signal, of frequency N times greater than the stator deformation, is the base clock of a counter which counts until N. This counter possesses a load input of adjustable value N_Ψ, and its output supplies a sinus table which, associated with an external digital analog converter, allows us to generate an alternating sinusoidal voltage which can be written as $v_\alpha(t)$. Thus, the frequency of $v_\alpha(t)$ is always equal to that of the deformation measure of the motor, whereas the dephasing is a function of

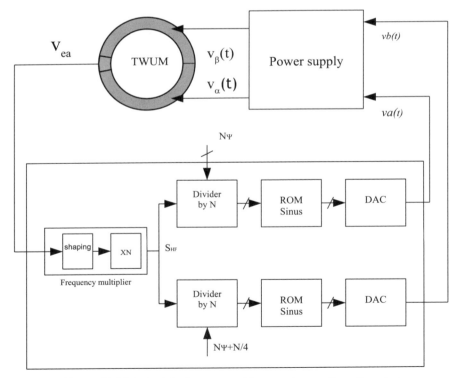

Figure 10.26. *Principle diagram used for carrying out the self-control*

the pre-load input of counter N_Ψ. The traveling wave motors being 2-phase, $v_\beta(t)$ is obtained similarly to $v_\alpha(t)$, but by modifying the pre-load input of the counter of path β by adding a quarter of the period $(N/4)$.

The performances of the self-control will be better because the response time of the locking loop will be small, and the number N is large. Better results are obtained with a response time in closed loop of the phase locking loop of 700 μsec, and with $N = 128$. Thus, the base clock frequency of counters S_{HF} were of the order of 4 Mhz; an FPGA solution was then suitable to integrate the set made of two counters, sinus roms and a part of the frequency multiplier.

10.6. Conclusion

The traveling wave ultrasonic motors, without involving electromagnetic fields, ensure a double energy conversion, an electromechanic conversion owing to the indirect piezoelectric effect, then a mechanomechanical conversion through contact. The analytical modeling of these two conversions, by means of certain simplifying hypotheses, emphasizes similarities and a duality with the models classically used for alternating electromagnetic machines. The traveling wave has the role of a rotating field and the position of its maximum can be used to self-control the motor. The use of the representation tool COG allows us to emphasize the decoupled model on two paths d and q, and to analyze the reasons for the stalling of the motor under certain load and supply conditions. As a consequence, the control structure obtained by inversion allows an automatic adaptation of the supply frequency, able to compensate for these phenomena and also to protect us against nonlinear effects because of the thermal drift.

10.7. Bibliography

[BAL 04] BAL G., BEKIROGLU E., DEMIRBAS S., COLAK I., "Fuzzy logic based DSP controlled servo position control for ultrasonic motor", *Energy Conversion and Management*, vol. 45, p. 3139-3153, 2004.

[BAR 06] BARRE P., BOUSCAYROL A., DELARUE P., DUMETZ E., GIRAUD F., HAUTIER J., KESTELYN X., LEMAIRE-SEMAIL B., SEMAIL E., "Inversion-based control of electromechanical systems using causal graphical description", *Proceedings of IEEE-IECON*, vol. 6, p. 5276-5281, 2006.

[BIG 05] BIGDELI N., HAERI M., "Simplified modeling and generalized predictive position control of an ultrasonic motor", *ISA Transactions*, vol. 44, p. 273-282, 2005.

[BUD 03] BUDINGER M., Contribution à la conception et la modélisation d'actionneurs piézoélectriques cylindriques à deux degrés de liberté de type rotation et translation, PhD thesis, INP Toulouse, May 2003.

[CHA 03] CHAU K., CHUNG S., CHAN C., "Neuro-fuzzy speed tracking control of traveling-wave ultrasonic motor drives using direct pulsewidth modulation", *IEEE Transactions on Industry Applications*, vol. 39, no. 4, July 2003.

[CHE 08] CHEN T., YU C., TSAI M., "A new driver based on dual-mode frequency and phase control for traveling-wave type ultrasonic motor", *Energy Conversion and Management*, vol. 49, p. 2767-2775, 2008.

[CHE 09] CHEN T., YU C., "Motion control with deadzone estimation and compensation using GRNN for TWUSM drive system", *Expert Systems with Applications*, vol. 36, p. 10931-10941, 2009.

[CHU 08] CHUNG S., CHAU K., "A new compliance control approach for traveling-wave ultrasonic motors", *IEEE Transactions on Industrial Electronics*, vol. 55, no. 1, January 2008.

[DAI 09] DAI Z., Actionneurs piézo électriques dans des interfaces homme-machine à retour d'effort, PhD thesis, University of Lille 1, March 2009.

[GHO 00] GHOUTY N.E., Hybrid modeling of a traveling wave piezoelectric motor, PhD thesis, Aalborg University, Department of Control Engineering, May 2000.

[GIR 98] GIRAUD-AUDINE C., Contribution à la modélisation analytique d'actionneurs piézo électriques en vue de leur conception et dimensionnement, PhD thesis, INP Toulouse, no. 1501, December 1998.

[GIR 01] GIRAUD F., LEMAIRE-SEMAIL B., HAUTIER J.-P., "Modèle dynamique d'un moteur piezo électrique à onde progressive", *RIGE*, vol. 4, no. 3, p. 411-430, April 2001.

[GIR 02] GIRAUD F., Modélisation Causale et commande d'un actionneur piézoélectrique à onde progressive, PhD thesis, University of Lille 1, no. 3147, July 2002.

[GIR 03] GIRAUD F., LEMAIRE-SEMAIL B., "Causal modeling and identification of a travelling wave ultrasonic motor", *The European Physical Journal Applied Physics*, vol. 21, no. 2, p. 151-159, February 2003.

[HAG 95] HAGOOD N.W., MCFARLAND A.J. IV, "Modeling of a piezoelectric rotary ultrasonic motor", *IEEE Transactions on Ultrasonics, Ferroelectrics and Frequency Control*, vol. 42, no. 2, March 1995.

[HAU 98] HAUTIER J., CARON J., *Les convertisseurs statiques: méthodologie causale de modélisation et de commande*, Editions Technip, Paris, 1998.

[LU 01] LU F., LEE H., LIM S., "Contact modeling of viscoelastic friction layer of traveling wave ultrasonic motors", *Smart Material Structure*, vol. 10, p. 314-320, 2001.

[MAA 95] MAAS J., IDE P., FRÖHLEKE N., GROSTOLLEN H., "Simulation model for ultrasonic motors powered by resonant converters", *IAS '95*, IEEE, vol. 1, p. 111-120, October 1995.

[MAA 97a] MAAS J., GROSTOLLEN H., "Averaged model of inverter-fed ultrasonic motors", *IEEE Power Electronics Specialists Conference (PESC)*, IEEE, vol. 1, p. 740-786, June 1997.

[MAA 97b] MAAS J., SCHULTE T., GROSTOLLEN H., "Optimized drive control for inverter-fed ultrasonic motors", *IEEE-IAS Annual Meeting*, IEEE, vol. 1, p. 690-698, October 1997.

[MAA 00] MAAS J., SCHULTE T., FRÖHLEKE N., "Model-based control for ultrasonic motors", *IEEE/ASME Transactions on Mechatronics*, vol. 5, no. 2, June 2000.

[NOG 96] NOGARÈDE B., "Moteurs piézoélectriques", *Techniques de l'ingénieur*, vol. D3765, p. 1-20, 1996.

[PET 00] PETIT L., RIZET N., BRIOT R., GONNARD P., "Frequency behaviour and speed control of piezomotors", *Sensors and Actuators*, vol. 80, p. 45-52, 2000.

[PIÉ 95] PIÉCOURT E., Caractérisation électromécanique et alimentation électronique des moteurs piézoélectriques, PhD thesis, INP Toulouse, no. 1037, July 1995.

[SAS 93] SASHIDA T., KENJO T., *An Introduction to Ultrasonic Motors*, Clarendon Press, Oxford, 1993.

[SEN 02] SENJYU T., KASHIWAGI T., UEZATO K., "Position control of ultrasonic motors using MRAC and dead-zone compensation with fuzzy inference", *IEEE Transactions on Power Electronics*, vol. 17, no. 2, March 2002.

[TOU 99] TOUHAMI H.O., DEBUS J., BUCHAILLOT L., "Contact modelling by the finite element method: application to the piezoelectric motor", *Journal de Physique IV*, vol. 9, p. 217-226, 1999.

[WAL 98] WALLASCHEK J., "Contact mechanics of piezoelectric ultrasonic motors", *Smart Material Structure*, vol. 7, p. 369-381, 1998.

List of Authors

Pierre-Jean BARRE
LSIS
Arts et Métiers ParisTech
Aix-en-Provence
France

François BAUDART
CEREM
Université Catholique de Louvain
Louvain-la-Neuve
Belgium

Mohamed Fouad BENKHORIS
IREENA
University of Nantes
Saint Nazaire
France

Damien FLIELLER
GREEN
INSA of Strasbourg
France

Moez FEKI
ICOS
École Nationale d'Ingénieurs de Sfax
Tunisia

Frédéric GIRAUD
L2EP
Lille 1 University
Villeneuve d'Ascq
France

Mickael HILAIRET
LGEP
Paris Sud University
Gif sur Yvette
France

Xavier KESTELYN
L2EP
Arts et Métiers ParisTech
Lille
France

Francis LABRIQUE
CEREM
Université Catholique de Louvain
Louvain-la-Neuve
Belgium

Betty LEMAIRE-SEMAIL
L2EP
Lille 1 University
Villeneuve d'Ascq
France

Jean-Paul LOUIS
SATIE
ENS Cachan/UniverSud-Paris
Cachan
France

Thierry LUBIN
GREEN
Henri Poincaré University
Nancy
France

Ernest MATAGNE
CEREM
Université Catholique de Louvain
Louvain-la-Neuve
Belgium

Ngac Ky NGUYEN
MIPS
Haute Alsace University
Mulhouse
France

Nicolas PATIN
LEC
Université Technologique de Compiègne
France

Ghislain REMY
LGEP
Paris Sud University
Gif sur Yvette
France

Bruno ROBERT
CReSTIC
Reims Champagne-Ardenne University
France

Hervé SCHWAB
Robert Bosch GmbH
Bühl (Baden-Württemberg)
Germany

Éric SEMAIL
L2EP
Arts et Métiers ParisTech
Lille
France

Guy STURTZER
GREEN
INSA of Strasbourg
France

Abdelmounaïm TOUNZI
L2EP
Lille 1 University
Villeneuve d'Ascq
France

Lionel VIDO
SATIE
University of Cergy-Pontoise
France

Index

A, B, C

absolute stability (theory of), 361, 365, 366
acceleration, 244, 245, 250, 254, 262, 270, 274, 276-278, 344, 347, 348
analytical model, 241, 288, 323, 401, 407
armature circuit, 18-22, 25, 27, 34, 35, 45, 50, 51, 57, 60, 63, 64, 227
bang-bang control, 262
causal ordering graph, 241, 248, 388, 393, 398, 403, 404
coenergy, 54, 64, 131, 292, 308, 309, 332, 333
cogging torque, 67, 77, 78, 80-82, 86, 88, 89, 114, 116, 117, 119, 120, 122, 241, 251, 252, 254, 265, 279, 333, 362
continuous mode, 326
control of chaos, 361, 373
coupling, 41, 115, 119, 126, 129, 130, 132, 135, 148, 149, 151, 153, 162, 173, 178, 182, 185-187, 226, 227, 296, 386, 392
current
    control, 122, 157, 174, 195-200, 204, 205, 231, 258, 259, 262, 336
    inverter, 33

D, E, F

damper, 3, 4, 22-24, 26, 27, 33-35, 41, 42, 48, 56, 58, 60, 63-65, 127, 136
diagonalization, 140, 150-155
dimensionless, 2, 340, 362
double star synchronous machine, 125, 158
driving zone, 344, 346
emf, 5-7, 13, 29, 40, 48, 51-53, 63-65, 78, 81, 85, 102, 116, 148, 165, 169, 173, 176-182, 185, 190-203, 234-235, 248-250, 268-270, 276-277, 317-318, 350
failure, 68, 69, 71-77
Feigenbaum diagram, 371
flux, 13, 40, 45, 48, 51, 55, 60, 65, 114, 115, 120, 163, 164, 166, 205, 207-216, 218-220, 222-224, 226, 230, 235-237, 239, 248, 249, 253, 257, 281, 290-292, 295, 299, 304, 306-308, 318, 319, 329, 331-333, 335, 404
    weakening, 210-212, 216, 218, 230, 235-237
folding, 369

H, I, L

holding torque, 337-340, 345

hybrid
  excitation synchronous machine (HESM), 207, 239
  stepping motor, 373
hypertorus, 368
hysteresis control, 193
induction machine, 159, 162, 164, 287, 289, 290, 291, 303, 249, 255, 256, 258, 259, 273, 402
inverter, 1, 2, 7, 68, 70, 72, 75, 77, 115, 127, 158, 161-163, 171, 176, 178, 182-191, 195, 197, 201, 205, 215, 232, 235-237, 259, 262, 291, 294
iterative learning, 253, 279
linear
  control, 262, 348, 361, 365
  motor, 241-248, 250, 253-255, 262-266, 268, 270-272, 274, 277-279, 282, 283, 372

M, N, O

maximum torque control, 297
microstep, 338, 341
Naslin polynomial, 359
neural network, 82, 119, 254, 324
open loop control, 404
optimal control, 121, 237
overlap, 10, 12, 15, 18, 33-36, 38, 46, 47, 60, 62, 64, 65, 97, 105, 109, 110, 112
overshoot, 264, 341, 343, 361

P, R

permanent magnet, 66, 68-70, 72, 75-77, 121-124, 127, 162-164, 178, 204, 205, 207-211, 213, 219, 220, 230, 237, 241, 243-247, 253, 264, 284, 287, 329, 330, 333-335
  motor, 121, 122, 329, 330, 333
  stepping motor, 335

synchronous motor, 72, 75, 76, 122, 123
perturbation torque, 80, 116
Poincaré section, 369, 371
pull out, 377
Pyragus control, 367
resonance, 254, 386, 387, 395, 397, 400, 404
resonant controller, 241, 256, 265-271, 279, 283, 284

S, T, V

saturation, 3, 34, 53, 60, 77, 86, 88, 89, 110, 111, 120, 128, 158, 161, 164, 189, 208, 215, 245, 250, 252, 262, 276, 288, 290, 291, 308, 313, 323, 325, 327, 331
self-controlled, 1, 2, 7, 22, 29, 33, 60, 65, 158, 292, 310
  synchronous machine, 2, 22, 29, 33, 60, 292
self-similarity, 369
sliding, 22, 23, 25, 29, 51, 53, 60, 63, 212, 254, 323, 381, 393
space vector, 159, 205
speed profile, 348, 349
stepping motor, 329, 330, 336, 338, 344, 350, 354, 356, 357, 361, 368
strange attractor, 367, 369
switched reluctance motors (SRM), 326, 327
synchronous
  actuator, 7, 29, 33, 65, 66, 122, 123, 148, 205, 206, 241-243, 245-247, 249, 284
  machine, 1, 2, 16, 22, 30, 34, 63-65, 67, 77, 115, 118, 122-124, 125-127, 131, 133, 134, 136, 145, 146, 156, 157, 159, 161-164, 176, 178, 204-207, 220, 221, 224, 235-239, 254, 266,

288, 290-293, 306, 330, 345, 404
motor, 7, 29, 33, 65, 66, 122, 123, 148, 205, 206, 241-243, 245-247, 249, 284
torque, 2, 14, 15, 28, 53-56, 58, 63, 67, 71, 77, 78-82, 86-92, 96, 100-103, 110, 111, 114-116, 118-126, 131, 135, 136, 145, 146, 155-158, 165, 166, 169, 173, 176, 178, 181, 182, 189-194, 196, 199-203, 207, 225-228, 230-238, 266, 288-301, 303, 306, 308-321, 324, 326, 329-356, 376-380, 384-386, 395-404

transformation matrix, 167, 140

vector control, 71, 123, 248, 253, 293, 294